Treatise on Concrete

the rhetoric of pier luigi nervi
concrete and ferrocement forms

Treatise on Concrete
Directed by Roberto Gargiani

roberto gargiani
alberto bologna

the rhetoric of pier luigi nervi concrete and ferrocement forms

Translated from the Italian by Juliet Haydock

EPFL Press

A Swiss academic publisher
distributed by Routledge

EPFL Press

The EPFL Press is the English language imprint of the Foundation of the Presses polytechniques et universitaires romandes (PPUR). The PPUR publishes mainly works of teaching and research of the Ecole polytechnique fédérale de Lausanne (EPFL), of universities and other institutions of higher education.

Presses polytechniques
et universitaires romandes
EPFL – Rolex Learning Center
Post office box 119
CH-1015 Lausanne, Switzerland
E-mail: ppur@epfl.ch
Phone: +41 (0)21 693 2130
Fax: +41 (0)21 693 4027

www.epflpress.org

ISBN 978-2-940222-95-7 (EPFL Press)
Printed in Italy

Routledge
Taylor & Francis Group

Taylor & Francis Group Ltd
2 Park Square, Milton Park
Abingdon, Oxford, OX14 4RN, UK

Routledge is an imprint of Taylor & Francis Group, An informa business.

Simultaneously published in the USA and Canada by Routledge, 711 Third Avenue, New York 10017

www.routledge.com

Library of Congress Cataloging-in-Publication Data
A catalog record for this book is available from the Library of Congress.

The graphical layout of the book and the cover design are by Paola Ranzini Pallavicini of Studio Pagina Milan, Italy. The photolithographic services were provided by Walter Bassani, Milan Italy.

This work was translated from the Italian into English by Juliet Haydock. Proofreading by Emily Lundin.

ÉCOLE POLYTECHNIQUE
FÉDÉRALE DE LAUSANNE

Series Editors for the publication program in Architecture are Professors Jacques Lucan and Luca Ortelli.

The publisher and authors express their thanks to the Ecole polytechnique fédérale de Lausanne (EPFL) for its generous support towards the translation of this book.

The authors are grateful to the Institute of Architecture of the EPFL, who provided additional support.

Treatise on Concrete

Directed by Roberto Gargiani

The volumes of the *Treatise on Concrete* intend to offer a new and documented vision of the evolution of construction techniques and of the creative potential and formal expressions of concrete, from antiquity to the present time. Each book in the series will be based on specific archival research; the analysis of scientific, technical and iconographic original sources; and the investigation into documents made available by companies, engineers, architects and artists.
The full spectrum of the use of concrete will be covered, including the building of foundations, ports, moles, bridges, vaults, domes, fortresses, bunkers, houses and monuments, as well as the manufacture of artificial stones and sculptures. In addition, the volumes will address the theoretical issues of *béton brut*, Brutalism and the true nature of matter.
All volumes focus on the techniques and materials for the manufacture of formwork, from wood to fabric; on the meaning of the imprint; on the surface processing of concrete by hand or with machine; and on the composition of the mixture.
The series of books aims to provide the scientific community with an unique and updated reference on the historical and contemporary uses of concrete, which will open a new chapter in the knowledge of this crucial material for architectural construction, as well as for masterpieces of engineering and art.

Volumes published:
Concrete, From Archeology to Invention, 1700–1769
Louis I. Kahn, Exposed Concrete and Hollow Stones, 1949-1959
Louis I. Kahn, Towards the Zero Degree of Concrete, 1960-1974
German Concrete. The Science of Cement from Trass to Portland, 1819-1877

Volumes in preparation:
Tadao Ando, Concrete Spaces
Mnemosyne in Concrete
Research on the Roman Cement, 1770-1779
Concrete Sculpture, from 1950 to Today: Visions of a Material
Contemporary Ornament in Concrete
Bared Concrete of Natural and Artificial Pozzolanas, 1780-1817
Formless: Beyond Formwork
The Concrete of Perret and Pupils

Table of Contents

chapter one
technical inventions and building site economies

Sidero-cemento, a reinforced concrete

Pier Luigi Nervi was born in Sondrio on 21 June 1891. After finishing high school in Ancona, he moved to Bologna in 1908. Despite a passion for aeronautics, mechanics and chemistry, he opted for a course in civil engineering at the Regia Scuola di Applicazione per Ingegneri of Bologna under the guidance of professor, engineer and entrepreneur Attilio Muggia. He graduated in 1913 after completing a design for a small villa.[1] That same year, Muggia hired him to work at his Bologna-based Società Anonima per Costruzioni Cementizie, a company Muggia had set up in 1908 together with the engineer Leone Poggi. The company held the license for the reinforced concrete structural system patented by François Hennebique for the regions of Emilia-Romagna, Tuscany and Marche. When World War I broke out, Nervi was enlisted in the 69th Sappers' Company of the Bologna Military Engineering Corps, and in 1916, he was assigned to the Hydrogen Gas Airship Battalion in Rome, where he was responsible for the gas supply service. His engineering skills enabled him to develop, in collaboration with the chemistry professor Carlo Andreucci, some solutions for improving the safety of the metal devices used to compress and purify gas for airships, and he filed three patents for these designs in 1917 (Nervi was to continue working on these ideas until 1919).[2]

Prompted by his experience working on the building sites of Società Anonima per Costruzioni Cementizie, which he continued to visit when he

[1] For a critical appraisal of Nervi's work, see the seminal work, Claudio Greco, *Pier Luigi Nervi. Dai primi brevetti al Palazzo delle Esposizioni di Torino. 1917-1948*, Lucerna, Quart Edizioni, 2008. See also Giulio Carlo Argan, *Pier Luigi Nervi*, Milano, Il Balcone, 1955; Ada Louise Huxtable, *Pier Luigi Nervi*, New York, George Braziller, Inc., 1960; Agnoldomenico Pica, *Pier Luigi Nervi*, Roma, Editalia, 1969; Tullia Iori, *Pier Luigi Nervi*, Milano, Motta Architettura, 2009; Carlo Olmo, Cristiana Chiorino, eds., *Pier Luigi Nervi. Architecture as Challenge*, Cinisello Balsamo, Silvana Editoriale, 2010; Gloria Bianchino, Dario Costi, eds., *Cantiere Nervi. La costruzione di un'identità*, Milano-Parma, Skira-CSAC, 2012; John F. Abel, Görün Arun, Mario Alberto Chiorino, eds., *Pier Luigi Nervi: art and technology in building*, special double issue of "Journal of the International Association for Shell and Spatial Structures," vol. LIV, September 2013, No. 2-3.

[2] See *Elenco dei brevetti*, in, Greco, *op. cit.*, (pp. 283-284), p. 283.

was on leave, Nervi developed a cementitious compound manufactured with the addition of metal fragments. In 1917, he filed a patent for this compound, to which he gave the general name of "*sidero-cemento.*"[3]

The patent description is a valuable document as it reflects Nervi's thoughts about the cementitious mixture, reinforced concrete and the skeleton in general. Nervi's line of reasoning stemmed from certain static considerations about the nature of concrete. While manuals dating from the early 20th century reveal that the contemporary trend in reinforced concrete engineering was to focus on standardising the optimum arrangement of metal bars, Nervi examined the material composition in an attempt to overcome the mechanical shortcomings of the concrete itself. In doing so, he virtually envisaged a compound that possessed all the qualities necessary for constructing large buildings, even without bars. That this was his goal was confirmed by the definition he chose as a synonym of *sidero-cemento*: "reinforced cementitious concrete mixture."[4] The "reinforcements" to be added to the concrete mixture during the preparation and the casting stages took the form of ironworking waste, "pieces of wire," or strips of hot-rolled iron known as "*mojetta.*" In the latter case, the compound assumed the characteristics of concrete interwoven with wire mesh. Nervi alluded to his own personal practice of identifying the proportion of metal "reinforcements" required for the works to be built.

Nervi's dreams of producing a concrete that was reinforced within its very substance led him to use the term "metal fibres" to describe the fragments that gave the compound an "overall fibrous structure." Both Nervi's experiences on building sites and the teaching of Silvio Canevazzi had helped make him aware that although mathematical calculation instruments could be used to monitor the static behaviour of reinforced concrete structures, they could not prevent the cracking effects that had to be avoided at all costs in the reservoirs, pipes, tanks or boats that he saw Muggia building and which, he believed, he could eliminate with his "reinforced cementitious concrete mixture." He termed these effects "secondary stresses dependent on elements that cannot be accurately evaluated when calculating the main stresses." As a natural conclusion, his explanation of the invention ended with a list of works that would be made possible by "reinforced cementitious concrete mixture": "reservoirs, dams, reinforced concrete hulls, etc."

One passage in the patent description allows us a glimpse of the development prospects for his "reinforced cementitious concrete mixture." Nervi juxtaposed two structural approaches: the Hennebique system of right-angled grid reinforcement and a future structure, to be built using his own concrete mixture and mainly subject to compression. This clearly indicated that his *sidero-cemento* alluded to a structural vision that went beyond the trends emerging in building practice at that time. The ground had already been laid for Nervi's research into reinforced concrete composition, which eventually led *sidero-cemento* to evolve into the ferro-cementate (*ferro cementato*) or ferrocement (*ferro cemento*) used to construct variously folded and waved membranes.

"In solids subject to bending," he wrote in the patent, "or in stretched membranes, the presence of the above metal fibres will not generally obviate

[3] Pier Luigi Nervi, patent IT 163733, *Conglomerato cementizio rinforzato da elementi metallici uniformemente distribuiti nell'impasto con speciale riguardo alla utilizzazione dei trucioli o altri cascami della lavorazione del ferro, e sua applicazione alle costruzioni in cemento armato*, 7 April 1917, ACS.

[4] *Ibid.*, p. 1.

the need for the presence of main reinforcements; despite this, if we can assume the concrete mixture's ability to withstand a certain level of tensile and shear stress and an increase in its resistance to compression, the entity of those reinforcements could be reduced, and they could even be replaced altogether in certain cases, if the percentages are properly gauged. This concrete mixture system is therefore applicable in conjunction with Hennebique and Hennebique-like reinforcement systems currently in use and can also be used to partly or fully replace concrete constructions of all types (particularly in compressed solids, arches, non-reinforced concrete walls, etc.). What is more, its use may be extended to applications that are not yet feasible".

Hyperstatic trusses for roofs and bridges

Nervi resumed his collaboration with the Società Anonima per Costruzioni Cementizie of Muggia when he joined Poggi to become one of the employees of the Florence branch when he was on furlough in 1919. The works built ranged from bridges to roofs for libraries, markets or cinemas and represented a crucial chapter in the spread of reinforced concrete throughout Tuscany. At the various building sites, concrete was used in accordance with the building customs of the early 20th century: cast in wooden formwork prepared in rudimentary fashion to obtain a rough structure that could be clad with compounds ranging from stucco to cement mortar, as required.

Nervi was involved in designing individual works and began to experiment with his own idea of structure. He tackled technical approaches to the building site, learned to invent and develop details and structures guided by the principle of economy and examined ways to improve the manufacture of concrete and disseminate the results.

Some works from the Florentine branch of the Società Anonima per Costruzioni Cementizie are characterised by the emergence of complex geometrical structures that give a glimpse into Nervi's originality. Nervi invented his first significant reinforced concrete structure for the vaulted roof of the building housing the *Gioco della Pelota* court in the Alhambra complex in Florence.

1. Società Anonima per Costruzioni Cementizie, *Gioco della Pelota*, Viale della Giovine Italia, Florence, 1919-21. Demolished (C. Greco, *Pier Luigi Nervi. Dai primi brevetti al Palazzo delle Esposizioni di Torino. 1917-1948*, Lucerna, Quart Edizioni, 2008, p. 43).

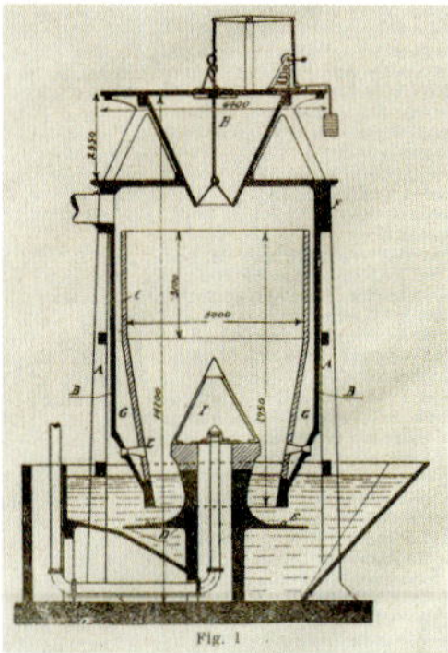

2. Società Anonima per Costruzioni Cementizie, gas holder, Castelnuovo dei Sabbioni, 1921. Cross-section ("Il Cemento. Rivista tecnica della moderna costruzione," 1923, No. 1, p. 8).

This structure, designed by Adolfo Coppedé and built between 1919 and 1921, heralded some of the technical and static aspects that are so characteristic of his masterpieces of the late 1930s. The space is covered by special arched truss girders with an elliptical profile connected to extending pillars as well as to the roof. Parallel ribs were applied to the truss girders by means of diagonal bars, all enclosed by a curved shell so that the players would see a smooth surface. The unformed lines of the structural device, including the pillar stumps, protrude from the roof like an outlandish animal skeleton, as slender as the lines of a metal truss. They bear witness to the experimental aspiration and inventive courage of their creator. Furthermore, they offer an early hint of reinforced concrete as a material able to give visible form to an intuitive perception of a flow of static forces that can be harnessed to form appropriate geometric patterns. Through the very way in which trusses, ribs and shells are juxtaposed to create a figure with its own surprising logic, this work also reflects a desire to cast doubt on the role of the engineer and explore possible geometrical juxtapositions as a means of defining a structure that is able to symbolically encapsulate the entire creative potential of reinforced concrete. What is more, this roof also offered an intuitive and rudimentary glimpse of the aspect that ultimately transformed Nervi's own concept of structure for large roofs. Evident here is the relationship between rib and shell that he was able to put together in ever more original ways, due to his capacity for invention in the field of static form and also his research into the very nature of the compound, which he had already demonstrated with the experiments summarised in his patent for *sidero-cemento*.

During his work on the roof for the *Gioco della Pelota*, Nervi became aware that temperature changes caused problems determined by the different expansion coefficients of the reinforced concrete in the pillars, which had already been poured, and the wood of the arched truss girder formwork. Until it set, the concrete deformed with the formwork and was unable to follow the movement of the pillars.[5] At the end of the 1930s, he devoted himself to the study of prefabrication systems, partly with the aim of removing this type of defect.

Another unusual structure with moulded lines provided a rational configuration for a special device that Nervi designed in 1921 for the gas holder of the Società Mineraria ed Elettrica del Valdarno, in Castelnuovo dei Sabbioni. This was one of the first attempts to apply reinforced concrete to this type of system, which was patented by Otho Robert Verity and usually made out of metal. The new approach resulted in considerable savings, demonstrated by the international repercussions of this minor work. Nervi separated the ribs of the superstructure from the thin shell of the gas bell – some 10 centimetres thick – to allow the parts to expand freely under the effect of the different temperatures to which they were subjected. In order to build the shell, he combined the usual metal reinforcements with a metal network made out of a mesh with sides of approximately 3 centimetres and a mixture of fine aggregates in order to "achieve maximum homogeneity."[6] The gas holder offered him an opportunity to rave about the unstoppable advance of reinforced

[5] Pier Luigi Nervi, *Scienza o Arte del Costruire? Caratteristiche e possibilità del cemento armato*, Roma, Edizioni della Bussola, 1945, pp. 84-85.

[6] Pier Luigi Nervi, *Gazogeno in cemento armato tipo "Verity" per distillazione di lignite*, in, "Ingegneria," I, 1 August 1922, No. 2, p. 43, and Pier Luigi Nervi, *Gazogeni in cemento armato*, in, "Il Cemento. Rivista tecnica della moderna costruzione,"

concrete over "iron constructions" in all "branches of constructive technique."[7] The competition between reinforced concrete and ironwork and sheet metal was to have decisive consequences on his view of reinforced concrete, with its geometrically structured configuration.

More truss girders, this time with a parabolic profile and stiffened by horizontal trusses, were used to form the three 25-metre spans of the bridge over the River Cecina, along the Massetana provincial road between Pomarance and Saline di Volterra. The bridge was built between 1920 and 1922, again by the Società Anonima per Costruzioni Cementizie. The details of the notching in the connections between the arched girders and those of the deck, or in the refined slenderness of the bars, now almost elastic, again revealed Nervi's pleasure in making static function expressive and effective through structural geometry, with every part pared down to an essential skeleton in order to use less material. These connections gave rise to the "nodes" that became so crucial in Nervi's structures. He himself explained the principles he followed when developing a structure: "maximum economy" and "minimum dead weight."[8] The considerable disparities that arose between the work of calculation and that of bridge construction taught Nervi about the unpredictable reactions of the parts of a reinforced concrete structure combined in a monolithic body.[9]

The piers, beams and deck were constructed out of formwork made from rudimentary wooden boards; the concrete was concealed by a layer of grey smoothed cement mortar, as was the norm for this kind of engineering work.

Nervi used I-beams hollowed out in accordance with Vierendeel geometry in order to build the bridge over the River Pescia, near Pescia, which was only 3.50 metres wide and subdivided into four spans (the longest measuring 20.60 metres), again built by the Società Anonima per Costruzioni Cementizie between January and May of 1923, and intended to carry the trucks and wagons transporting products produced by the Fratelli Marchi fertiliser factory. The beams also formed the high parapets, which were again clad with smoothed cement mortar. Studies for the bridge over the River Cecina revealed Nervi's idea of a horizontal, yet trussed, girder to be set up beneath the deck. His method of calculating the moment diagrams steered the design for the structure of the bridge over the Pescia, including its piers and deck, toward the desired "minimum dead load," low building costs and the use of minimal quantities of concrete and steel.[10] Quintessential examples of this paring back of each part to its essentials are the cantilevered connections between piers, deck and beams used to stabilise the piers with their reduced cross-section or the application of hollowed out I-beams along the neutral axis area. In some ways, the cantilevered connections between piers and beams are equivalent to the diagonals in the truss girders of the *Gioco della Pelota* roof, given their static role in the general definition of the structure. To an even greater extent than in the spans of the bridge over the River Cecina, Nervi built restraints between the reinforced concrete parts in order to define a monolithic body with all the hyperstatic characteristics that were to become the trademark of his structures. The following is almost a statement of principle, a manifesto

XX, January 1923, No. 1, (pp. 8-9), p. 8.

[7] *Ibid.*, p. 9.

[8] See Pier Luigi Nervi, *Ponte in cemento armato sul fiume Cecina tra Pomarance e Saline di Volterra*, in, "Ingegneria," I, 1 November 1922, No. 5, pp. 126-127.

[9] Pier Luigi Nervi, letter to Eugenio Miozzi, 2 April 1924, Maxxi, Epistolario.

[10] Pier Luigi Nervi, *Ponte sul fiume Pescia*, in, "Ingegneria," II, 1 August 1923, No. 8, pp. 232-233.

PONTE SUL TORRENTE CECINA

Ponte sul fiume Pescia per la fabbrica frat.lli Marchi
Pescia - 1922 -

3. Società Anonima per Costruzioni Cementizie, bridge over the River Cecina, Massetana road, Pomarance, 1920-22. Demolished (ASF, series III, No. 34).

4. Società Anonima per Costruzioni Cementizie, bridge over the River Pescia, Pescia, 1922. Demolished (ASF, series III, no. 34).

of the hyperstatic structure to come as well as a specific declaration of intent about the bridge over the River Pescia: "Due to the slenderness of the piers (designed to offer the least resistance to the free flow of water) and hence the weak resisting moment, it was not believed necessary to separate the beam from the piers, given the intrinsic deformability of the piers."[11]

Other bridges built in Tuscany by the Società Anonima per Costruzioni Cementizie bear witness to the design variations put to the test in structures built under Nervi's guidance. The bridge over the River Sterza at Salitone, for example, reveals a new and more sophisticated expression of the new role that Nervi sought for the bars in the spans built for the bridge over the River Cecina – now the arches become a powerful support structure to which the deck is anchored by vertical ties.

Slender components, hollow material, rigid connections and an increase in the number of restraints transformed the constituent parts of the *Gioco della Pelota* roof or of the trussed bridges with the nodes that rendered them hyperstatic, into a quest for the ideal essential form, beyond the rules of mathematics.

Reinforced masonry made out of cement-brick blocks

Nervi did not limit himself to designing and overseeing the works of the Società Anonima per Costruzioni Cementizie. The engineer and builder quickly revealed his intention to turn the empirical criteria adopted on building sites by his *cementisti* workers into science. One example was the way they determined the compressive strength of the concrete mixtures by measuring the depth of the impression created by a "hammer blow," reflecting the criteria developed by Johan August Brinell. This prompted Nervi to write letters to engineering journals, encouraging the setting up of an experimental office tasked with drawing up standards to improve checks of concrete quality on building sites.[12]

The early 1920s also saw Nervi's first writings in engineering journals dedicated to the discovery of reinforced concrete, which cropped up throughout the international technical literature. His commitment to the engineering debate led to his appointment as secretary of the Florentine section of the Associazione Nazionale degli Ingegneri Italiani, in 1921. He also actively encouraged the establishment of the Sindacato Nazionale Ingegneri, becoming Chairman of the Florentine section.[13]

Certain technical solutions that he learned of through industry literature surfaced in the creation of his structures. In the journal "Le Constructeur de Ciment Armé," he discovered a way of joining the iron in the concrete reinforcements by electrical welding, a method implemented in the penstock for the hydroelectric plant at Drac-Romanche.[14] He went on to use the same type of welding to assemble the prefabricated elements of his hangars. Nervi's interest in rotary kilns, described in "Revue des Matériaux de construction et des Travaux Publics," testified to his growing curiosity over the manufacture of cement, with the goal of enhancing its mechanical and technical properties.[15]

Nervi began to debate the question of the design expertise necessary for what had now become established as the basic building material of the 20th century, reinforced concrete, in the context of his commitment as an inven-

[11] *Ibid.*, p. 233.
[12] Pier Luigi Nervi, *Lettere al Direttore. E' possibile il metodo di Brinell per la prova dei cementi?*, in, "Ingegneria," I, August 1922, No. 2, p. XV.
[13] Greco, *op. cit.*, pp. 57-58.
[14] Pier Luigi Nervi, *Riviste e Giornali. Recensioni. Conduttura in cemento armato dell'officina idro-elettrica di Drac-Romanche*, in, "Ingegneria," II, 1 July 1923, No. 7, p. 107.
[15] Nervi, *Riviste e Giornali. Recensioni. Ricupero del calore nei forni rotativi, ibid.* See Pier Luigi Nervi, *I cementi ad alta resistenza*, in, "Ingegneria," II, 1 June 1923, No. 6, pp. 167-168.

tor of building systems and as an employee of Muggia's Società Anonima per Costruzioni Cementizie. A post-war change in the way contracts were awarded for public works led to the emergence of a new role: the professional responsible for designing the reinforced concrete structure other than the entrepreneur responsible for the execution. However, Nervi took a stand against this in support of his preferred role as a design engineer and entrepreneur, as personified by Muggia and his company. Only this kind of professional, whom he defined as a "designer-builder" was able to guarantee all the qualities of a reinforced concrete structure – "economy, resistance and aesthetics."[16] Nervi's inclusion of aesthetics on his list should not be overlooked, since it reveals his ambition to transform the creation of a structure into a new expressive form.

It was in the context of Nervi's work on behalf of the Società Anonima per Costruzioni Cementizie and his interest in the experimental aspects of concrete manufacturing that he began to investigate various technical approaches to improving building construction. The Florentine branch of the company invented and implemented construction systems based on cement products, such as the "tubular walls in reinforced concrete" that he used to build houses and for which he filed a patent in 1920.[17]

In February 1921, as part of international research into the various ways of using reinforced concrete and against the backdrop of post-war reconstruction, Nervi filed a patent for hollow "cement-brick" blocks to be employed as an alternative to concrete blocks and hollow floor blocks when building the reinforced walls and floors of two-storey villas or houses without having to resort to expensive carpentry for the formwork. The blocks were to be manufactured using a machine or moulds in iron or wood. They had brick sides and a concrete core and contained a cavity for thermal spaces or "tubular compartments" that could be used to insert the reinforcement and pour the concrete used to form the pillars and beams.[18]

Saw-tooth roofs and skeletons for factories

One of Nervi's last actions as an employee of the Società Anonima per Costruzioni Cementizie was to build the Società Mineraria ed Elettrica del Valdarno power station in Prato, in 1922–23. In early 1923, Nervi parted company with Muggia, and in June he set up the Società per Costruzioni Ing. Nervi & Nebbiosi Cementi Armati, with branches in Rome, Florence and Naples, together with the Roman engineer and entrepreneur Rodolfo Nebbiosi. The company had a technical department in the Rome branch, which employed other engineers specialising in reinforced concrete, including Giovanni Bartoli, Nervi's cousin, who was to replace Nebbiosi as his partner.

The first important reinforced concrete works produced by Nervi & Nebbiosi were destined for industrial factories, residential housing, bank offices, cinemas and theatres. They were built in various cities and towns of central Italy, especially in the vicinity of the three branches. Nervi & Nebbiosi did not take particular care of the way concrete was poured in the industrial

[16] Pier Luigi Nervi, *Lettere al Direttore*, in, "Ingegneria," I, 1 July 1922, No. 1, p. 24.
[17] Società Anonima per Costruzioni Cementizie, Florence branch, patent IT 189878, *Sistema di muri tubolari in conglomerato cementizio*, 4 August 1920, quoted in, Claudio Greco, *Muggia e Nervi, il fecondo e complesso rapporto tra un grande maestro e un discepolo illustre*, in, M. Beatrice Bettazzi, Paolo Lipparini, eds., *Attilio Muggia. Una storia per gli ingegneri*, Bologna, Editrice Compositori, 2010, (pp. 211-219), p. 213.
[18] Pier Luigi Nervi, patent IT 195346, *Blocchi cemento-laterizi per strutture, murature e solai*, 17 February 1921, ACS.

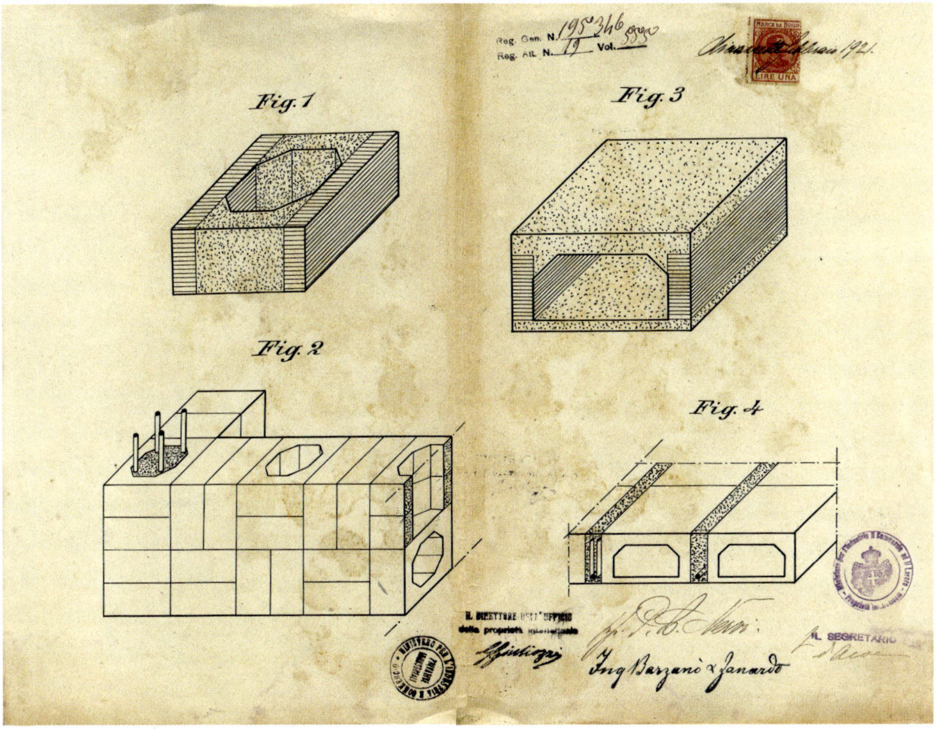

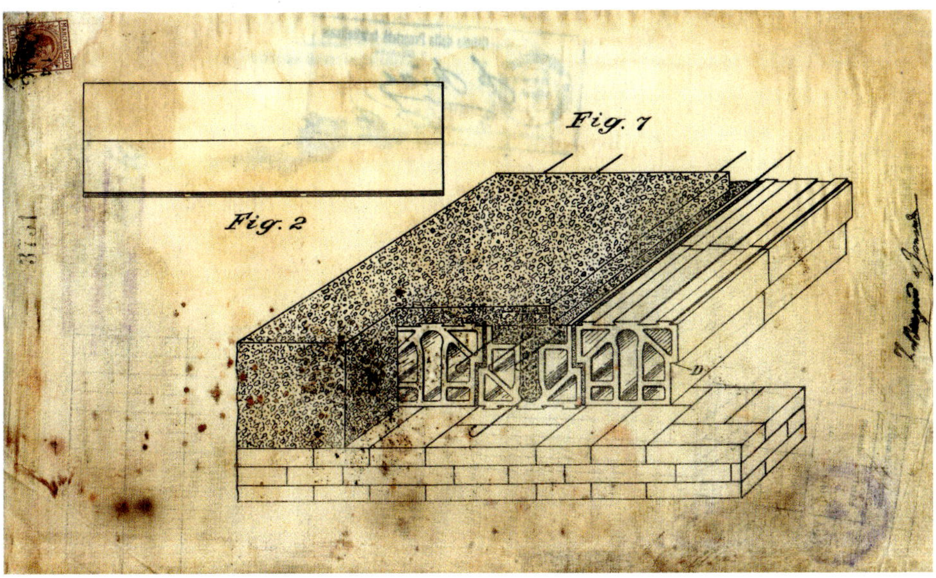

5. P. L. Nervi, patent IT 195346, *Blocchi cemento-laterizi per strutture, murature e solai*, 17 February 1921 (ACS).

6. Nervi & Nebbiosi, patent IT 258746, *Laterizio e procedimento per la formazione di travi, solai e strutture portanti in cemento armato senza armature provvisorie*, 14 May 1927 (ACS).

7. Società Nervi & Nebbiosi, G. Mazzini wool mill, [Via G. Mazzini], Prato, 1927-28. Demolished (APT).

structures. Details of the connections between the bars of the structures were often rudimentary in geometry and execution, imposed by rules of business pragmatism. The tender specifications often stated that the company should undertake to hand over the structure "in its rough state following removal of the timber formwork."[19] Painting was the only treatment adopted to improve the finish quality of the concrete.

Nervi & Nebbiosi built several sheds for the industrial processing of wool in Prato.[20] The skeletons were based on a Hennebique pillar and beam model, often made more slender to optimise material use; sometimes the beams took the form of tall lamellae, as in the Orlando Franchi, Fratelli Querci or Egisto Lenzi wool mills, while the roof truss bars, even when derived from a transposition of timber roof trusses, as in the Luigi Pecci wool mill, were made slender, in evident competition with the steelwork. In situations when he adopted a saw-tooth roof design, always arranged on a set of pillars, Nervi sometimes established, as in the case of the Giuseppe Mazzini wool mill, sequences of lines that intersected to define an overhead skeleton of braced bars, slender beams used to stiffen slabs that ran with a saw-tooth

[19] Pier Luigi Nervi, letter to Ditta Pecci, 17 July 1925, and *Ore in economia Cantiere Pecci,* attached to an invoice of the Società per Costruzioni Ing. Nervi & Nebbiosi, 14 September 1925, AP.

[20] See Giuseppe Guanci, *Costruzioni & Sperimentazione. L'attività del giovane Pier Luigi Nervi a Prato,* Campi Bisenzio, Centro Grafico Editoriale, 2008.

[21] Rodolfo Nebbiosi and Pier Luigi Nervi, patent IT 229015 *Solaio in cemento armato con tavelloni,* 26 March 1924, quoted in, Greco,

movement, heralding other future undulations in their geometry and sequencing. On other occasions, when creating wool mill roofs with skylights, Nervi used structural stiffening components located above the roof, as in the Brunetto Calamai wool mill, where he came up with his own version of a gable roof design with timber roof trusses, opening up the top for lighting and ventilation through the use of reinforced concrete – with the chains of his pared-down joists invisible from the room. It was rare for the finish quality of the concrete pouring details to display any level of sophistication, although this was the case for the Egisto Lenzi wool mill.

Nervi demonstrated his intent in the skeletons of the Prato wool mills much more clearly than he did in his works for Muggia. He was aiming for a lightness of the bars and the invention, always in a spirit of economic pragmatism and static intuition, of geometrical intersections in a quest for a complex balance over and above the visible balance that was still aulic for the Hennebique system. He now achieved this by joining welded parts in a filamentous spatial mesh, almost as though they were interwoven in a quest for a static and constructive pattern aimed at overcoming the conventional juxtaposition of pillars, beams, roof trusses or saw-tooth roofs to create a filiform three-dimensional dynamic network.

The structures built by Nervi & Nebbiosi during the 1920s included other types that were not always at a level that allowed Nervi to exercise his creative talents. A degree of experimentation, even if only in a few details, is evident in the reinforced concrete skeleton of *Gran Garage Europa*, built around 1927 at Borgognissanti 96, Florence. Nervi raised octagonal pillars and moulded a colossal pillar, with cantilever beam and faceted corners to create a large opening in the masonry facade along the road. Compared with the industrial buildings in Prato, the formwork boards used for the skeleton of the garage were installed without any geometrical control, revealing the variable quality of the work executed by Nervi & Nebbiosi.

At the end of the 1920s, as the company continued to build sheds for wool mills in Prato (always in exposed reinforced concrete, with rudimentary working), Nervi designed and built the invisible reinforced concrete skeleton of the Naples branch of Banca d'America e d'Italia around 1924. He developed a special upturned trestle to branch the pillars on the office floors and the mixed concrete and masonry skeleton of the building in Lungotevere Arnaldo da Brescia 9, in Rome, between 1928 and 1930, designed by Giuseppe Capponi and featuring a spiral staircase.

Continuing from his designs dating back to the early 1920s, Nervi went back to considering other technical solutions to make floor construction more economic. To do so, he patented beams in brick and reinforced concrete with the aim of reducing the use of temporary scaffolding,[21] as well as developing some improvements in concrete pouring (increase in concrete compactness and strength through the application of mechanical, electrical or pneumatic vibrators to the metal bars or, in the case of non-reinforced concrete, to the formwork).[22]

Pier Luigi Nervi cit., p. 283; Soc. Ing. Nervi & Nebbiosi, patent IT 258746, *Laterizio e procedimento per la formazione di travi, solai e strutture portanti in cemento armato senza armature provvisorie*, 14 May 1927, ACS.

[22] Soc. Ingg. Nervi & Nebbiosi, patent IT 278611, *Procedimento per ottenere costruzioni con conglomerato cementizio avente particolari caratteristiche di compattezza e resistenza*, 27 April 1929, ACS.

Invisible cantilevered structures over a void

Roofs for cinemas and theatres were the most original structural geometric figures that Nervi & Nebbiosi built, again during the 1920s. These allowed Nervi to develop the creativity he had revealed in the *Gioco della Pelota* roof. This kind of structure, which in the coming years became even larger and without intermediate supports, ultimately won him international fame. The degree of finish of the concrete execution remained rudimentary because the structures were never intended to be visible from inside in the halls (even the visible parts on the roofs were made according to the same criteria and then plastered).

The structure of the roof for Bruno Banchini's theatre in Prato, built between 1923 and 1925, is significant in its quest for a virtuous geometry of radial beams, partly everted as in the *Gioco della Pelota* roof. Nervi came up with a design of acrobatic aspirations to cover the existing masonry walls of the circular auditorium and support the opening skylight. The ring within a square, onto which the rays of the flat roof's cantilever beams converged, is held suspended in the middle of the auditorium by a system visible above the roof and formed by four high perforated beams. Those beams are arranged in a square and anchored by four diagonal forks to the big corner pillars and the masonry in an interplay of opposing tensions that exploits the mechanical potential of reinforced concrete. The ring fitted at the top of the four perforated beams plays an essential role in this interplay of forces in joining the beams and holding them together in a rigid system that is able to react to the tensions set up by the diagonal forks. In designing the broken sections that typify the device on the roof, or in allowing the corner reinforcements in the joint between the forks and perforated beams to protrude slightly, Nervi must have experienced the pleasure of discovering a form dictated by the logical expression of moment diagrams, significantly adjusted to delineate the new beauty of a reinforced concrete body that has harnessed the forces at play. Only the keenest spectators would have been aware of the presence of the new structure buried in the ceiling of the auditorium – the enormous circular skylight open in the centre of a flat roof, not a dome.

8-9. P. L. Nervi, reinforced concrete structure for the Teatro Banchini (now the Teatro Politeama), designed by A. Chiti, Via G. Garibaldi 33, Prato, 1923-25.

8. Constructing the formwork for the roof (AFR).

9. View of the roof structure building site (AFR).

Nervi's aspiration to create a reinforced concrete structure that was not merely rational but endowed with an aesthetic nature of its own did not, however, become truly evident until he designed the structure for the *Cinema Teatro Augusteo* with a cable car station underneath, built between 1925 and 1929, along Via Roma in Naples on behalf of the Società Anonima Funicolare Centrale. The design of the structure followed the general lines of the design drawn up by Arnaldo Foschini and hinged on the stripping away necessary to create the grand void of the circular auditorium, some 30 metres in diameter, with its overhanging gallery. The ring-shaped frame, punctuated by 17 large pillars arranged along the auditorium (the eighteenth was removed to leave a gap for the stage), radiated with beams and slabs that stretch towards the perimeter walls to act as an attaching and stiffening device, while extending upward into an imposing truss of pillars, beams, slabs and diagonals. The result was almost a geometrical *mazzocchio* polyhedron with 18 spans, which extended through the height of one storey in order to create rooms and finish in an overhang. At the foot of the truss, 18 cantilever beams project even further forward with acrobatic grace to engage with the annular beam of the opening disc in the centre of the auditorium. The points of convergence of the diagonals, beams, and pillars in the polyhedral latticework extend into nodes that are crucial to the rigidity of the roofing system, forcing the most cantilevered section of the truss to assume the static function of a circular Vierendeel girder. The diagonals attached to the 17 large pillars (which are crucial to generating differential tensions in order to prevent the trussed roof from toppling over) and the annular beam that incorporates the cantilevered sections,

10-12. P. L. Nervi, reinforced concrete structure for Cinema-Teatro Augusteo, designed by A. Foschini, Piazza Duca d'Aosta 263, Naples, 1925-29.

10. View of the hall under construction, showing formwork for the tiers (photograph by T. and G. Vasari; BSTA, 024-2914).

11. View of the reinforced concrete structure of the hall (photograph by T. and G. Vasari; BSTA, 024-2912).

12. View of the hall roof level with its cantilevered radial structure (photograph by T. and G. Vasari; BSTA, 024-2913).

are the cornerstones of the mechanical system invented by Nervi to cantilever the entire roof. The principles determining the flow of forces under tension revealed similarities to the structure of the *Teatro Banchini* roof.

In the action of supporting the grand auditorium gallery over the void, Nervi again mastered the multiple forces at play, holding the cascade of parallel truss girders supporting the tiers, with nine angled paired truss girders of different lengths anchored to the great annular pillars (the lower and upper bars of each pair of truss girders are traversed by opposing tensions), stretching the broken girder that bolsters the descent to the lower edge of the gallery like a bowstring.

The design of the entire skeleton reveals the mechanical insight behind the flow of forces, expressed by trusses. They could be easily moulded into the desired lines due to the reinforced concrete, and the balance is always a permanent action of tensions caught in the only possible moment of pause in both the roof and the gallery. There is no doubt that the soaring acrobatic gesture of unleashing the weight of the reinforced concrete over the void of the auditorium is a visible reflection of all of Nervi's youthful passion for aviation and flight.

Despite being buried by cladding and decorated infill, the structure creates a pattern of raised ribs that is evident everywhere in the *Cinema Teatro Augusteo*, as well as in the cable car station. In this case, just as in the *Gioco della Pelota* or *Teatro Banchini*, Nervi demonstrated his ability to imagine geometries with complex static functions for extensive roof structures, even though he failed to affirm their identity in the *Cinema Teatro Augusteo*, and parts of the overall design show an overridingly instrumental use of reinforced concrete. This empirical need was determined by the requirement to create a building that looked as though it were built out of masonry. The building comes together best in the vanishing point of the void, and this coherence is then diluted only to disappear altogether when it comes into contact with the masonry. The only visual records of the geometry of the structure, showing the wholeness of its lines and networks, are the building site photographs, which also document the assembly stages of the metal reinforcements and formwork. The commissioning of the Roman photographic studio of Tommaso and Giorgio Vasari to photograph the spans of the structure before they were enclosed by masonry and cladding reveals Nervi's awareness, apparent in his careful choice of photographer, of the expressive power of his reinforced concrete creation, which was nevertheless destined to be buried beneath architecture.

13-15. P. L. Nervi, reinforced concrete structure for Cinema-Teatro Augusteo, designed by A. Foschini, Piazza Duca d'Aosta 263, Naples, 1925-29.

13. View of the tier beams (photograph by T. and G. Vasari; Maxxi - PLN, F06015).

14. View of the hall structure (photograph by T. and G. Vasari; BSTA, 024-2911).

15. View of the hall (Maxxi - PLN, F06024).

Fot. VASARI-Roma

POPOLARI
L.4

chapter two the *stadio berta* and the nature of form

The daring grandstand roof

During the 1920s, the debate over the role of reinforced concrete structures in the genesis of a new rationalist architecture reached Italy, fostering interpretations that went far beyond the brief statements set out in Futurist manifestos and amounting to a quest to identify the traits of fascist style. Journals focused increasing attention on the inroads made by reinforced concrete in defining the shape of the skeleton, even though, in the case of industrial buildings, they only examined the pouring of concrete without claddings.

The statement made by the young architects of *Gruppo 7*, who in 1927 wrote the first formal defence of the "general skeleton of the construction," resonates as a rebuttal of the widespread belief that reinforced concrete had to be buried beneath a scaffolding of traditional materials and styles.[1] Yet even they made no reference to the tenet of "truth to materials." Their critique of "concealing the sincerities of the material beneath fictitious claddings" left no shadow of a doubt that *Gruppo 7* promoted a reinforced concrete "aesthetic" in the manner of Le Corbusier, where the skeleton and its material were essential and invisible guides. The group's ideas about reinforced concrete crystallised around a quintessential example of a building whose structural material was there for all to recognise: the Lingotto building housing the Fiat factory, designed by the engineer Giacomo Matté Trucco. A work by Nervi was destined to take the place of the Lingotto as a new model for the visible reinforced concrete skeleton aesthetic.

Back at the beginning of February 1930, the City of Florence voted to build a "Campo Sportivo" (sports ground) on the land along Via Faentina, an idea that had been aired since November 1929 by the Gruppo Rionale

[1] Gruppo 7 [Ubaldo Castagnoli, Luigi Figini, Guido Frette, Sebastiano Larco, Gino Pollini, Carlo Enrico Rava, Giuseppe Terragni], *Architettura (IV). Una nuova epoca arcaica*, in, "Rassegna Italiana," X, May 1927, No. 108, pp. 467-472.

Fascista of the Cure neighbourhood. The work was to be dedicated to the memory of Giovanni Berta, a fascist militant of the Florentine Action Squads who died in 1921.[2] The celebratory intent of the *Campo Sportivo* design is something that must be kept very much in mind in order to understand the development of Nervi's plans for what were to become the symbolic structures of the *Stadio Berta*. The city Technical Department, led by the engineer Fiorenzo De Reggi, had already drawn up a brief design for a grandstand in reinforced concrete. This featured an area covered by a cantilevered canopy supported by a row of vertical pillars traversing the tiers and obstructing the view of the playing field.

We do not know what prompted Nervi to submit a design for the grandstand of the "Sports Facility" to the City of Florence on 16 March 1930, when the Gruppo Rionale Fascista and the City Hall were already involved in negotiations to take over a larger area of land on Campo di Marte (the plot was to be officially granted in August).[3]

In the draft contract awarding the work for the *Stadio Berta*, the Technical Department described a reinforced concrete grandstand "in one piece" and with a length of 110 metres, with a cantilevered roof limited to 70 metres and devoid of "intermediate pillars obstructing a free view."[4] In early May it was decided to immediately build a grandstand known as the *Tribuna Reale* or *Tribuna d'Onore* at the same time as the playing field and enclosures. The roof, it was stressed, had to be "cantilevered as far as possible to eliminate central pillars or supports" and hence had to be made out of reinforced concrete.[5] The structure described in the City Hall Technical Department documents corresponds to Nervi's design, even though this is never mentioned. In the design for the grandstand, drawn up in March and fine-tuned in the subsequent months, Nervi borrowed some guidelines from the Technical Department design for the sports ground in Via Faentina, but added his own special touch to the work due to the configuration of the roof, which no longer had any pillars and extended to partly shield the tiers (he maintained the original lengths stipulated in the Technical Department documents). The grandstand roof at last made it possible for him to design a structure that was not intended for warehouses, factories, or bridges and could remain visible without having to rely on building site photographs to record its existence, as with the *Cinema Teatro Augusteo*. The experience of constructing diagonals and forks extending like ties able to defy the gravity imposed by the weight of structures in the case of the *Teatro Banchini* and *Cinema Teatro Augusteo* was to prove crucial to his ability to imagine lines jutting into space and also destined to become significant. Nervi's challenge of flinging weights into the void like fantastical aircraft now took on epic dimensions as he knowingly set out to explore all the aesthetic potential of the lines and details of the structure.

From the earliest studies, to overcome the effects of thermal expansion, he aimed to break down the roof into two parts made out of different materials: a reinforced concrete structure, which took the form of "cantilever beams" (*mensoloni*), according to his own definition,[6] and a brick roofing layer. This separation gave rise to the appearance of depressed arches between the cantilever beams.

[2] Carlo Battiloro, *La vicenda architettonica*, in, Andrea Galluzzo, Carlo Battiloro, Francesco Varrasi, eds., *La grande vicenda dello stadio di Firenze*, Firenze, Edifir, 2000 (pp. 31-56), p. 31; Marco Piccardi, *Lo stadio di Pier Luigi Nervi*, in, Sandro Picchi, ed., *Firenze. Lo stadio racconta. Oltre 75 anni di storia cittadina nelle memorie del Franchi*, Firenze, Giunti, 2007, (pp. 38-57), p. 47.

[3] Piccardi, *op. cit.*, p. 49.

[4] The system of traffic and of public access to the tiers had not yet been specified, *Norme e condizioni alle quali dovrà ottemperarsi nello studio*

16. P. L. Nervi, *Stadio Comunale G. Berta* (now *Stadio Comunale A. Franchi*), Via M. Fanti 4, Florence, 1930-33. "Veduta prospettica delle tribune (perspective view of the stands)," delivered to Florence City Council on 29 September 1930 (ASCF, 028261).

All the aesthetic quality of the roof lies in the profile of the reinforced concrete structure that Nervi controlled with artistic sensitivity in every respect, to ensure that the moulded shape expressed the mechanics of the forces, and more.[7] In an early version of the roof, significantly explored through perspective in order to grasp the aesthetic effect, Nervi designed a semiarch that rose from the tiers to end miraculously without any vertical support. The semiarch is tapered at its end and thickens at the point of convergence of the bracing depressed arches with the straight section that anchors the overhang to the vertical structures at the base. This significant distinction between curved and straight lines, including the significance they impose on the parts – a semiarch soaring into space and held by a tie – reveals Nervi's source of inspiration to be the arches of bridges designed by Robert Maillart: the one he built over the Rhine at Tavanasa in 1905, and the one he had recently completed in the summer of 1930, at Salginatobel, both designed and geometrically modelled as two semiarches touching at the centre. Nervi saw the continuous beam running along the edge of the roof as an essential expedient for bringing together and bracing the tips of his cantilever beams, which were otherwise left too free. The acquired creative and technical ability that had come from designing cantilevered devices for the roofs of theatres in Prato and Naples, alongside his innate aspiration toward flight, now allowed him to build a half-bridge launching itself into space – and not merely a beam reinforced by a post. The bracing depressed arches themselves, juxtaposed as artificial props against the cantilever beams, indicate the extent to which he envisaged the semiarch as an unalterable profile: the intuitive expression of a symbolic structure that could be refined, but not changed, in the design process.

del progetto e nella costruzione delle Tribune in cemento armato per il Campo Sportivo "GIOVANNI BERTA," undated, p. 1, ASCF, pack No. 7467.

5 *Missiva dell'Uffizio Tecnico al Podestà*, 6 May 1930, ASCF, pack No. 7472.

6 *Descrizione e norme dei lavori*, undated, p. 1, ASCF, pack No. 7470.

7 This is borne out by his statement during a speech he delivered on 28 March 1963, when he was awarded the Mörsch medal, Maxxi, R3/13.

17-19. P. L. Nervi, *Stadio Comunale G. Berta* (now *Stadio Comunale A. Franchi*), Via M. Fanti 4, Florence, 1930-33.

17. Cross-section of the stand and roof plan, 1930 (ASCF, 027528).

18. Plan and cross-section of the covered stand, 1930 (ASCF, 7470-1).

19. Working drawing of the stand roof structure, showing the reinforcement ironwork arrangement and measurements (*The Works of Pier Luigi Nervi*, London, The Architectural Press, 1957, p. 8).

Nervi modified the intrados profile of his cantilever beam tie so that it also became imperceptibly curved, but without its line ever coinciding with that of the semiarch. This change had nothing whatsoever to do with the static trend, but became essential in order to perceive the functioning of the system. Only with the profile modified in this way could Nervi demonstrate the roles of the cantilever beam foot and tie as prongs of a fork with static modes of operation that were in opposition, yet both crucial to allowing the arched portion to overhang into space. The bracing arches between the cantilever beams still performed a role that was secondary to the final image. The disappearance of the edge beam and consequent rounding of the sharp ends are further confirmation of Nervi's mastery of structural identity and his informed aesthetic brilliance.

His removal of the edge beam joining the cantilever beams is not merely a detail: it was driven by the desire to draw the gaze of the spectator beyond the ends of the arch tips toward the wide open space of the sports field to create a feeling that nothing stood in the way of metaphorical flight. To unleash the full meaning of the arched tips, Nervi braced the ends of the cantilever beams by changing the resilient cross-section of all the members in order to form a T. The subtle flanges of the T, whose sizes he changed during the course of the design process – they were eventually built 90 centimetres wide – ran continuously across the spans to weave together the various members against the gusting wind. Nervi later returned to this kind of cross-section in other future designs as a way to join and stiffen the slender parts of his structures.

The variable profiles of the cantilever beams now rendered the different forces acting in the various cross-sections plastic, while the curved connections, which are highlighted in some sketches as essential details due to their shape, emphasise the dynamic flow and sudden direction changes in the bifurcations powerfully reinforced by metal brackets. The lines remain graceful, in the same way as those of Maillart's bridges, leaning significantly toward the sports field in order to express the weight of their material substance and the tension of the forces that animated them. Spectators observing them would have been aware of their vital struggle against gravity and reminded of the enterprise and technical knowledge necessary to achieve that daring gesture. For an intimate understanding of the significance and aesthetics of those lines, it is crucial the comparison between the design of the metal reinforcement of the shelter's structure and the photograph of the same part, comparison which Nervi repeated several times.[8] Only by looking at the metal bars and the way in which they are folded and intertwined is it possible to understand the invisible mechanical soul that made it possible for inert matter to hover in the air.

The cantilever roof made it necessary to invent posts, diagonals, and inclined tier beams with differential static functions to take up the weights of the cantilever beams and spread them evenly across the components of the skeleton, without concentrating stress yet relying on a resultant force approximately barycentric to the first, and widest, span. The reappearance of these devices, which Nervi had already used with static awareness to generate a given flow of forces, as in the *Cinema Teatro Augusteo*, contributed to determining

[8] See, for example, Giovanni Michelucci, *Lo stadio "Giovanni Berta" in Firenze dell'Ing. Pier Luigi Nervi*, in, "Architettura. Rivista del Sindacato nazionale fascista architetti," XI, March 1932, No. 3, (pp. 105-116), pp. 111-112, and Nervi, *Scienza o arte del costruire?* cit., plate X.

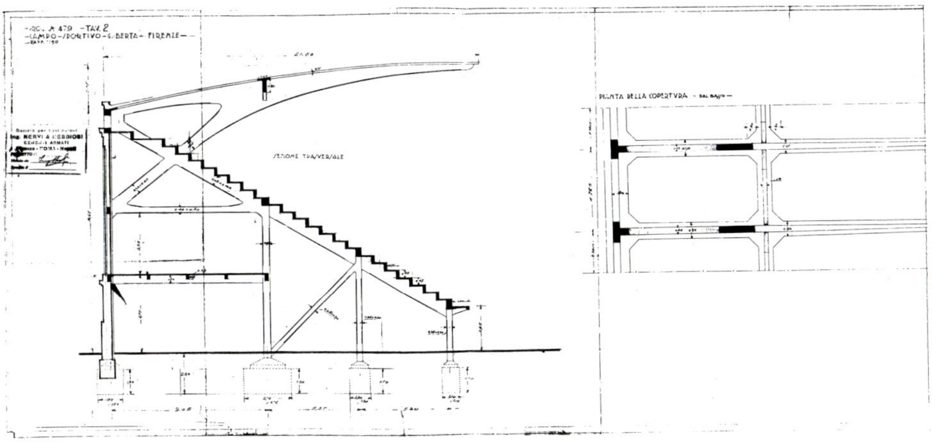
SEZIONE TRASVERSALE
PIANTA DELLA COPERTURA

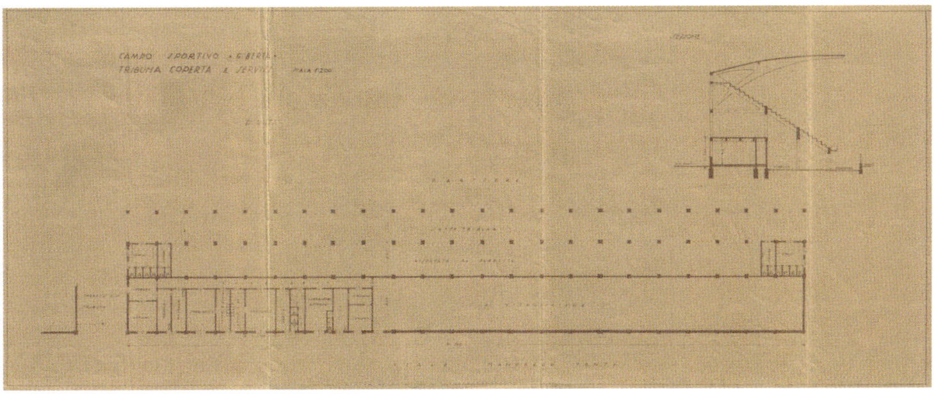
CAMPO SPORTIVO «G. BERTA»
TRIBUNA COPERTA E SERVIZI

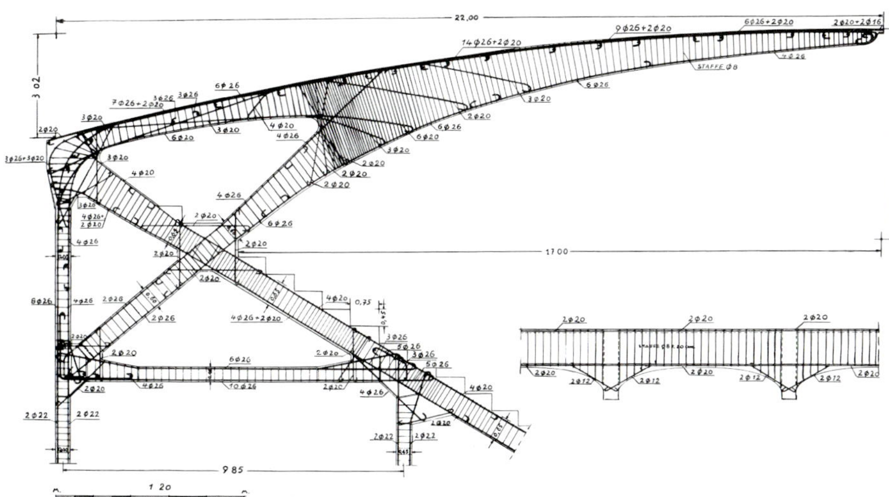
22,00
17 00
5,00
9 85
1:20

the complex shape of the skeleton onto which the stresses generated by the weight of the cantilever roof and the tiers converged. The spectators unwittingly participate by adding their weight to the mechanics of Nervi's structure, as though they were mounting an invisible scale already under tension.

The simple and effective explanation that Nervi provided of the functioning of the grandstand system reveals an intrinsic poetic side to his aesthetic vision of reinforced concrete. He moulds the material without giving rise to critical thickening, pursuing an intuitive flow of forces in which the structure must find the reasons for its shape, "without having to resort to a proper main joint cross-section."[9] Even though his design was still partly affected by the constraints inherited from the layout designed by the Technical Department, his words forcefully expressed his vision of the entire structure as a "bodily unit," with each component participating in the same system of forces and tensions and, ultimately, in the natural flow of weight to the ground.

There is nothing overtly extraordinary in the geometry of the pattern woven by Nervi under the tiers, particularly when compared to the taut lines of the cantilever beams. Yet it is crucial to the emergence of a more general view of the role of the skeleton in Italian architecture, at that particular juncture of the cultural debate, that Nervi planned to leave the structural device in view, including the part beneath the tiers. This makes it possible to gain an unblinkered view of the reinforced concrete frame and to understand the entire system, just as when one observes the structure of a Gothic cathedral firstly from the outside and then from within. The design of the connection profiles is therefore also explored in the structure beneath the tiers, and the only closed section is on the ground floor (housing washrooms, changing rooms, and storage rooms), where the pillars stand out from the infill walls, making a theorem of the visible structure. These offer evidence that the structure was intended to be left in view, the edges of each component cut in straight lines with exposed connections. It is significant that when designing the skeleton of the tiers, Nervi felt the need to resort to defining a "trussed structure."[10] When he had learned to master all manifestations of this form, even his trussed structures took on an original aesthetic appearance.

Starting in September 1930 with delivery of the design, including details of the cantilever beam reinforcement that accounted for the intrinsic quality and plastic force of Nervi's structure, administrative formalities followed and eventually led, on 1 December, to the signing of a contract for the work. This was to be completed by 7 July 1931, and it was awarded to Nervi & Nebbiosi. The text reflects the May draft, changing the description of some parts of the grandstand (the changes affected the roof, now extending across the entire 110-metre length, the height above the ground of the first shelf of tiers, which had been increased from 1.60 to 2.65 metres, and the stairs leading to the tiers).[11]

Converting the local sports grounds into a neighbourhood municipal stadium dedicated to Berta, and also available to the Associazione Calcio Fiorentina for *Serie A* games, meant that from submission of the design for

[9] *Missiva della Società per costruzioni Ing. Nervi & Nebbiosi al Podestà*, 16 September 1930, ASCF, f. 13623, quoted in, Battiloro, *op. cit.*, pp. 39-40.

[10] *Comune di Firenze. Direzione Terza. Servizi Tecnici. Lavori di costruzione delle tribune coperte in cemento armato al Campo Sportivo in Piazza d'Armi. Impresa Ing. Nervi e Nebbiosi. Registro di contabilità*, 14 January 1931 – 13 February 1932, p. 5, ASCF, pack No. 7470.

[11] *Missiva della Direzione III – Servizi Tecnici – Ufficio II al Podestà*, 1 October 1930, quoted in, Battiloro, *op. cit.*, p. 35, and *Atto tra Comune*

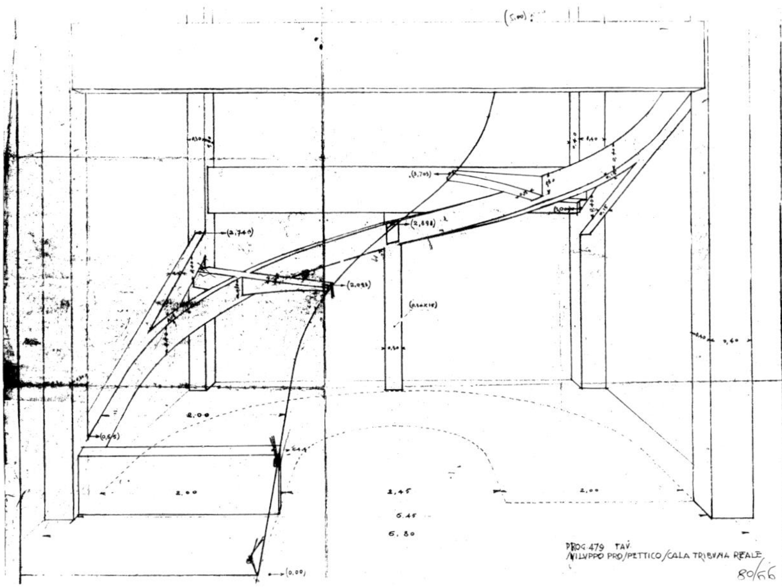

20. P. L. Nervi, *Stadio Comunale G. Berta* (now *Stadio Comunale A. Franchi*), Via M. Fanti 4, Florence, 1930-33. "Sviluppo prospettico scala tribuna reale (Perspective view of the grandstand stairs)," 1930 (ASCF, 027689).

the award of the work, the reinforced concrete trussing had to disappear behind the infill walls of the small volume occupied by the washrooms, which now reached the top of the structure. The result was the facade of the *Stadio Berta*, a veritable masonry screen organised into ridges to create a gigantic abstract pattern over the pillars, crowned by a cornice and decorated in the middle with blind niches and the Fascist *fasces lictoriae* insignia over the entrance doors. Apart from the section intended for the entrance, no rooms were planned in the upper part behind the blind infill walls. There is every reason to believe that the design of the facade was the work of Nervi. This view is supported by a Technical Department document stating that the "company itself drew up the architectural design chosen for the facade" – in other words Nervi & Nebbiosi.[12] Even at the operational stage, the Technical Department simply designed the layout of the rooms. We cannot rule out that Nervi's contempt, repeatedly stressed in his writings, for architecture that is about nothing more than building "facades," arose out of this very experience of having to apply a facade to the structure himself.[13]

Between signing the work contract and the end of December, Nervi altered the structure of the tiers and, in order to make the part under the grandstand more accessible, he removed the diagonals, replacing the "static effect" by "sturdy connections between the pillar attachment and the main beam of the tier" and by "increasing the cross-section of the beam and pillars."[14] The spectators now gained access to the tiers through open gates at the sides of the stand, and the whole area "under the stand" was "reserved for the spectators."[15] Nervi also planned to add stairs at the ends of the stand and to build

di Firenze e Società in nome collettivo Ing. Nervi e Nebbiosi sedente in Firenze, 1 December 1930, ASCF, pack No. 6415.

[12] *Missiva dell'Ufficio Tecnico al Podestà*, 29 January 1931, ASCF, f. 102, quoted in Battiloro, *op. cit.*, p. 39. See also Nervi & Nebbiosi Design No. 479, tav. 8, approved by Nervi, ASCF, pack No. 7470.

[13] Pier Luigi Nervi, *Arte e tecnica del costruire*, in, "Quadrante," June 1933, No. 2, p. 28.

[14] Pier Luigi Nervi, letter to the Florence City Hall Techical Department, 2 January 1931, ASCF, pack No. 7467.

[15] The design for this stage of the project documented a sequence of 23 spans, amounting to an overall length of 110 metres, see *Campo sportivo G. Berta. Tribuna coperta e servizi*, undated drawing, ASCF, pack No. 7470.

21. P. L. Nervi, *Stadio Comunale G. Berta* (now *Stadio Comunale A. Franchi*), Via M. Fanti 4, Florence, 1930-33. Covered stand under construction, photograph dated 25 March 1931 (photograph by F. Barsotti; S. Picchi, ed., *Firenze. Lo stadio racconta. Oltre 75 anni di storia cittadina nelle memorie del Franchi*, Firenze, Giunti, 2007, p. 45).

a spiral staircase instead of flights of stairs to access the restricted area of the grandstand. In order to make the spiral staircase, he developed a spiral beam with ledges to support the cantilever slab. This was the first appearance of a fragment of another virtuoso structure that was destined to take on very different proportions and significance in the development of the *Stadio Berta* grandstand design.

Nervi also increased the overhang of the cantilever roof beams, which now measured 22.60 metres, "for practical and aesthetic work purposes," with a consequent "knock-on increase in the entire load-bearing structure to the foundations."[16] This final tweak to the lines revealed not only his concern for protecting the spectators from the rain more effectively but also his pleasure in extending this flight into space to its limits and making the acrobatic appearance of the roof even more spectacular.

When the grandstand was in its final form with a facade, the roof's weights and stresses slipped beneath the tiers to be taken up by struts or ties no longer visible to the spectators; the lines and connections stiffened into figures that seem inexpressive, pragmatic, and brutal when compared to the eloquent stylishness of those visible in the roof. The system remained evident only on the two sides, but these were destined to disappear during the enlargement of the grandstand. The cantilever beams appear to be miraculously suspended in space, almost as if trying to break free and strain forward to the playing field in order to show the spectators, stunned by so much overhang, all the energy and flight of the fascist movement itself – or at least the most faithful interpreters of Mussolini's revolution might have thought so (the metaphor of outstretched arms?). When we reread the words of one of the first enthusiastic admirers of Nervi's stadium, Pietro Maria Bardi, we are left with no doubt about the extent to which the roof with its cantilever beams lent itself to representing the spirit of the fascist era: "hurled into the air with daring and with confidence."[17] Raffaello Brizzi described it as "a daring and solemn constructive body."[18] The Italian word *ardito* (daring) has great significance in the rhetoric of fascism or political futurism.[19]

There was undoubtedly an involuntary collision, one not foreseen by Nervi, between the inspiration that led him to invent the roof structure and the celebratory rhetoric of the fascist martyr that was manifest in the theatrical backdrop to the structure's imposing entrance. From then on – and not only in his development of the *Stadio Berta* design – Nervi set himself the task of creating a structure that was, in itself, rhetorical so that he could avoid having to apply facades, in a quest to resolve static and political forces in an aesthetic and rational whole.

The first stage of work

The work was awarded to Nervi & Nebbiosi without the publication of a call for tenders on the strength of the design submitted by Nervi. De Reggi was appointed construction manager. When the company opened the building site, the work for the stadium included, under the terms of the contract, only

[16] *Riserve al certificato di collaudo delle tribune coperte*, undated, pp. 4-5, ASCF, pack No. 7469, and *Collaudo tribuna coperta*, p. 2, ASCF, pack No. 7467.

[17] Pietro Maria Bardi, *Lo stadio di Firenze*, in, "Casabella," VI, April 1933, No. 4, p. 5.

[18] Brizzi used the expression in his opening address of the 1931-32 academic year see Mauro Cozzi, *La scuola del razionalismo*, in, Mauro Cozzi, ed., *Edilizia in Toscana tra le due guerre*, Firenze, Edifir, 1994, (pp. 81-106), note 15, p. 105.

[19] Translator's note: the *arditi* were special assault troops, many of whom joined the fascist movement.

[20] *Descrizione e norme dei lavori*, undated, p. 1, ASCF,

construction of "a single body," in other words the 110-metre long "grandstand in reinforced concrete" with a "cantilever roof" and the "slabs of the stairways leading to the tiers and to the grandstand."[20] Nervi & Nebbiosi agreed on the reinforced concrete structure finishes. The spectators were to sit on tiers clad in "smooth cement laid with a trowel," but those who wandered beneath the grandstand would enter an unfinished construction, where all the concrete surfaces were "left rough, exactly as when the timber formwork was stripped away."[21]

With a view to building the cantilever beams and spiral staircase, Nervi sent a request to the employment office to be able to use "workers who were not from Florence, but particularly skilled, given the difficulty of the work,"[22] while in order to check the type of steel bars and roof floor, he made a prototype in Rome, in the building site in Via Brofferio, where Nervi & Nebbiosi was putting up a residential building designed by Nervi.[23] After comparing the two hollow-brick blocks and concrete floors he tried out in Rome, he rejected the Adamoli type and chose the Perret version.[24]

After the work of digging the foundations, completed in January 1931, then building the pillars in February and the tiers in March and April, the "cantilevered roof spans" were built between April and May, with installation of the formwork, referred to as "timber reinforcements," and steel bars.[25] Pouring of the roof concrete was completed on 30 May. The hollow-brick

pack No. 7470.

[21] *Ibid.*

[22] Pier Luigi Nervi, letter to the Florence City Hall Techical Department, 2 February 1931, ASCF, pack No. 7471.

[23] *Riserve al certificato di collaudo delle tribune coperte*, undated, p. 2, ASCF, pack No. 7469.

[24] *Ibid.*

[25] *Comune di Firenze. Direzione Terza. Servizi tecnici. Lavori di costruzione delle tribune coperte in cemento armato al Campo Sportivo in Piazza d'Armi. Impresa Ing. Nervi e Nebbiosi. Registro di contabilità*, 14 January 1931 – 13 February 1932, ASCF, pack No. 7470.

blocks and concrete floor was separated from the cantilever beams by means of asphalted cardboard, to allow movement caused by thermal expansion.

Nervi was already able to appreciate, during this first phase of work, how designing structures along unconventional lines would prove costly because of the "very great amount of timber wasted and the formwork costs,"[26] even though it was rationally justifiable for shaping the reinforced concrete. His aspiration to building economy, which was evident in the solutions he came up with from his earliest building jobs for the Società Anonima per Costruzioni Cementizie, led him to reconsider the design of the tier structures for future *Stadio Berta* stands and also made him increasingly sensitive to the various practical and creative constraints imposed by the whole process from the carpentry work to the laying of the reinforced concrete.

Liquid mud and the stadium building site

The sequence of phases involved in the project and the *Stadio Berta* construction was dictated by considerations of the role of reinforced concrete in modern architecture, which cropped up again and again in Italy, albeit in different contexts. We cannot rule out that the building work, documented by photographs published before completion of the site and showing the grandstand skeletons still bare and free of infills and cladding, may have helped to fuel the debate around 1930 between those who were beginning to take sides in favour of a form of reinforced concrete architecture that, if it had not quite reached the point of leaving the formwork impressions on view, at least had all of its skeletons unmasked.

The first stages involved in erecting the skeleton of the *Stadio Berta* coincided with this taking of sides over the role of reinforced concrete in modern architecture. The republication in June 1930 by "La Casa Bella" of the *Manifesto dell'architettura futurista* to coincide with an exhibition of the works of Antonio Sant'Elia can be taken as a new start for the futurist idea of architecture based on the "truth to materials" that had been theorised about since 1914. The passage of the *Manifesto* on the "new beauty of cement and iron" that is "polluted by overlaid carnivalesque decorative incrustations,"[27] was bound to take on new urgency in view of what Nervi was about to do in the *Stadio Berta*, which went beyond the Corbusian "aesthetic" proclaimed by *Gruppo 7*.

The first photographs depicting the construction of Nervi's Stadium were published in the March–April 1931 issue of the journal "Eclettica," printed in Florence. The aerial photographs document the scaffolding of the formwork for the grandstand roof, and they may be the inspiration for the description of "reinforced concrete" poured "with ultra-modern criteria." The frame without infills visible in the photograph of the grandstand is termed a "reinforced concrete skeleton." Another photograph shows the bare structure of the atrium spiral staircase.[28]

During the very months when the grandstand was approaching completion, a debate started in Florence and then spread throughout Italy over the

[26] Pier Luigi Nervi, letter to the Florence City Hall Chief Engineer, 4 May 1931, ASCF, pack No. 7469.

[27] *Manifesto dell'architettura futurista,* by Antonio Sant'Elia launched in Milan on 11 July 1914, quoted in *Sant'Elia, oggi,* in, "La Casa Bella," III, June 1930, No. 30, (pp. 9-11), p. 9.

[28] Il discobulo, *Le opere del Regime. Lo stadio fiorentino al Campo di Marte,* in, "Eclettica," March-April 1931, No. 3, pp. 45-48.

[29] Ugo Ojetti, *Discorso di S. E. Ugo Ojetti per inaugurare la scuola fiorentina di architettura,* in, "Annuario della Regia Scuola, Scuola Superiore di Architettura," 1931-1932, Firenze, 1933, (pp. 15-23), p. 19.

[30] Cozzi, *La scuola del razionalismo* cit.

role of reinforced concrete structures in modern architecture. This may have prompted Nervi to develop his image of the stadium towards a statement skeleton without wall cladding as well as daring geometrical solutions that were an expression of the contemporary cultural movement. The journalist and critic Ugo Ojetti started a debate over the matter of the truth of reinforced concrete, which involved such luminaries as Giuseppe Pagano-Pogatschnig, Pietro Maria Bardi, Massimo Bontempelli, Piero Bargellini, as well as Nervi himself. Ojetti lived in Fiesole and was therefore able to keep an eye on Nervi's work, which became a secret model for the concrete architecture that formed the backdrop to his critical considerations.

In his inaugural speech to the first-year courses at the Regia Scuola Superiore d'Architettura, made on 1 March 1931, at the Accademia di Belle Arti in Florence, Ojetti described concrete as "liquid mud" and even equated it, when put together with a steel reinforcement, with the authoritarian state: "reinforced concrete, made up of two contradictory elements, namely iron and a liquid mud, will stand as a symbol of our civilisation, which strives to press the liquid, shifting, fleeting crowd around the more or less yielding iron of authority [...]."[29] Ojetti's reference to Nervi's half-built stadium was also evident in that at the same ceremony, the man who promoted the establishment of the School of Architecture and was to be its future President, Brizzi, proclaimed the beauty of the new *Stadio Berta*.[30]

Bare or clad structure

During the months when the grandstand was constructed, Nervi began to explore the "completion" project of the *Stadio Berta* with uncovered stands along the perimeter and to develop the design of the main facade. The stands initially appeared to be separated by four entrance gates, where it was planned to erect monumental arcades. A single uncovered straight stand ran facing the grandstand. Now the plan assumed the configuration of a colossal capital *D* that, despite stemming from the straight section of the athletic speed event track, was also bound to be seen as the first letter of *Duce*'s name in homage

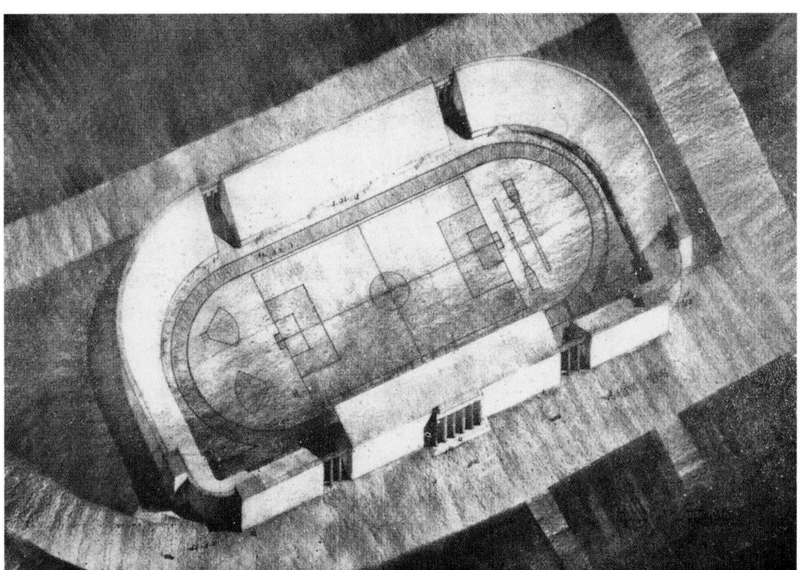

22. P. L. Nervi, *Stadio Comunale G. Berta* (now *Stadio Comunale A. Franchi*), Via M. Fanti 4, Florence, 1930-33. Axonometric drawing of the stadium completion design, 1931 ("Eclettica," March-April 1931, p. 45).

23-24. P. L. Nervi, *Stadio Comunale G. Berta* (now *Stadio Comunale A. Franchi)*, Via M. Fanti 4, Florence, 1930-33.

23. Perspective drawing of the uncovered stands with rooms, 1931 (ASCF, 027473).

24. Perspective drawing of the uncovered stands with exposed structure, 1931 (CSAC, n. inv. pra 31, coll. 148/6).

to the leader. Nervi also remodelled the imposing nature of the facade by adding Roman-looking statues in niches and increasing the thickness of the pillars, partly to make room for the ducts leading to the basement ventilation systems. This was probably the time when Brizzi proposed his plan to extend the stadium with continuous stands laid out along an elliptical plan.

Nervi also examined the possibility of an elliptical plan when he designed the second version of the stadium extension, with the stands still separated by openings but, this time, without any pillars. The monumental order of the grandstand appears to be extended along the perimeter, and the only revealed part of its structure are the pillars, which are nevertheless clad in brick bulked up by fins and accentuated with colour. In order to fill in the empty spaces beneath the tiers with walls and windows to make gyms, halls, and washrooms, Nervi designed the spans of the structure to obtain vertical lines of pillars supporting the infills along the perimeter. Uncertainties over the financial estimates also led him to examine a solution, treated as an interim solution by the authorities at the time, in which the reinforced concrete structure inevitably assumed a leading role. Thus the image of the imposing monument to popular sports is not evident in a drawing study by Nervi which depicts the exposed structure of the long stand facing the grandstand, captured with its lines bare and free of infills. The only structure preventing access to the stadium is a continuous stone boundary wall. A comparison between the two perspectives of the same stand, both designed from the same viewpoint, with the structure now buried by infills, now left completely on view, is emblematic of the alternative images of the *Stadio Berta* as seen by Nervi and by the Technical Department. Even if Nervi intended to represent the work in its rough state, the perspective of the bare skeleton nevertheless demonstrated his will to show the authorities the expressive power that reinforced concrete could wield by itself in characterising a modern stadium for the fascist nation. He used that very perspective to show his fascination with the aesthetic effect of the fast-paced rhythm of the variously angled bare bars that were interwoven in space. He characterised them with abstract sculptural accentuations of the trussed structure that was instead intended to be buried beneath the grandstand wall. For Nervi, that perspective was intended as a revelation of the formal power of a trussed structure.

Just at the time when people were talking about forms of "liquid mud," and Pagano was promoting an expression of reinforced concrete without wall cladding in order to allow the "bones," the "skeleton," the "slender solidity," or the "slender grid" to be appreciated,[31] Nervi seemed to want to go back and strip the *Stadio Berta* of any masonry device, as in the initial plans, in order to make a statement about what ultimately became the central theme of the Italian debate on modern architecture: the "reinforced concrete skeleton" photographed during the construction of the grandstand. He had drawn perspective views in 1930 to bring out the cantilever beams of the grandstand roof, and he now again used perspective with the intention of demonstrating the potential of the exposed trussed structure of the tiers. The perspective view of the bare skeleton was intended to express the idea of an architecture made up of "slender trusses." This series of plans for the stadium nevertheless marks

[31] Giuseppe Pagano – Pogatschnig, *I "materiali" nella nuova architettura*, in, "La Casa Bella," IV, May 1931, No. 41, (pp. 10-14), p. 14.

COMUNE DI FIRENZE
73/20

25-26. P. L. Nervi, *Stadio Comunale G. Berta* (now *Stadio Comunale A. Franchi*), Via M. Fanti 4, Florence, 1930-33.

25. "Sezione trasversale in asse con la tribuna d'onore (Cross-section aligned with the grandstand)," 1931 (ASCF, 027512).

26. Main facade along Via M. Fanti ("Architettura. Rivista del Sindacato nazionale fascista architetti," March 1932, No. 3, p. 110).

the first time that Nervi depicted the skeleton with the aim of emphasising its creative potential, as if to show what side he was on in the debate over the role of reinforced concrete in modern architecture. The perspective view of the skeleton is of particular significance in the process of designing the stadium as well as, more generally, in developing Nervi's essential concept of the skeleton. It reveals an unprecedented image of a fascist stadium that even hints at the development of design toward the invention of other, even more sculpted structural elements. Yet Nervi did not reveal the significance of that perspective view until June 1931.

In the version of the project for the extension to the *Stadio Berta* documented by a plan drawn up in June 1931, Nervi returned to the D-shaped plan and abandoned the open gates to allow the tiers to run continuously along the perimeter, perforated at their base by seven *vomitoria* (exits) to allow the audience to leave the stadium (two were added in the curves at the apices of the D). It was planned to wall in the structure, as in the case of the grandstand, given that on 4 May, when Nervi submitted his "design and quotation for the construction of uncovered straight stands," he specified that those works had been "developed in harmony with the general plan of the playing field that he had originally been given."[32] At the same time, the firm undertook to build the facade and the rooms in the grandstand in accordance with the "drawings" and "construction details" that "would be supplied and approved by the works management."[33] The drawing of the "cross-section aligned with the grandstand," where the subtle profiles of the structure are overlapped by imposing giant tiers and false segmental vaults, is more than expressive of the changes in the course of the project that reveal Nervi as a compliant interpreter of the various demands imposed on him.

The specifications for the uncovered stands described the finishes with details that again allow a glimpse of possibilities other than those specified in the design. These were again determined by the uncertainties over funding, demonstrating that the final work would partly be the outcome of chance and circumstance. This document states that the face of the tiers exposed to the elements would be clad in "cement plaster," while the reinforced concrete of all the visible structures beneath the tiers would be left unclad: "the remaining structures shall be handed over in their rough state after stripping off the timber formwork."[34]

In the case of the option of continuous stands with *vomitoria*, the idea of a bare structure was not therefore completely abandoned, although it ultimately assumed the status of an interim stage in the shape of the stadium. Certain clues help us toward a better understanding of all the meaning hidden in the perspective view of the bare skeleton that Nervi had drawn for the option of the stadium with stands separated by gates. That he was still harbouring hopes of leaving the work unfinished to reveal the lines of the structure seems apparent in the way he calculated the strength of the reinforced concrete used for the tiers, despite the impenetrable pragmatism that apparently guided his decision. In a memo addressed to De Reggi in June 1931, he specified the

[32] Nervi, letter to the Chief Engineer, 4 May 1931 cit.
[33] *Stadio G. Berta. Impresa Ing. Nervi e Nebbiosi. Atto di sottomissione,* 1 May 1931, p. 1, ASCF, pack No. 7471.
[34] Nervi, letter to the Chief Engineer, 4 May 1931 cit.
[35] Pier Luigi Nervi, letter to the Florence City Hall Chief Engineer, 16 June 1931, ASCF, pack No. 7471.
[36] Contract of 20 June 1931 signed by Rodolfo Nebbiosi, p. 1, ASCF, pack No. 7469.

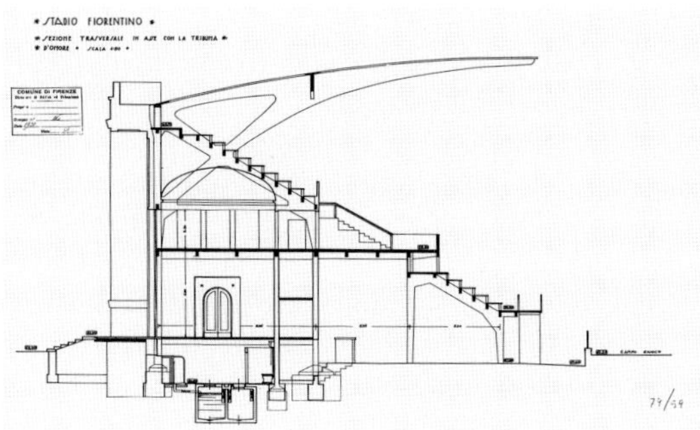

static functioning of a system made up of the two "sections," each 30 metres long, into which the two straight uncovered stands to be erected in front of the grandstand were to be divided. This system was designed to be "statically balanced," even without masonry: "This load option was considered in the eventuality that if the masonry walls of the grandstand are not built at the outset, it will remain built of Reinforced Concrete alone and must therefore be able to stably withstand wind action in every direction."[35] There is no doubt that the perspective view of the bare skeleton helped Nervi to show the Technical Department how the *Stadio Berta* would look if the funding for the work did not permit the construction of the masonry infill. Having drawn a perspective view of the reinforced concrete skeleton without any infill, and having calculated the strength of the reinforced concrete so that it could stand as the only building material used in the stadium, can only mean that Nervi believed in a new architecture based on structure alone.

Construction of the uncovered stands

Nervi & Nebbiosi's work on the *Stadio Berta* continued with the construction of an uncovered stand facing the grandstand, in two 60-metre sections separated by a central opening for a *vomitorium* under the tiers. The shape of the structure was different to that of the grandstand. This was not merely because the stand lacked a roof but because Nervi no longer had to adhere to the stand skeleton layout designed by the Technical Department. All the structural components were shaped in the manner that most closely reflected the static rules of moments and forces; the pillars were thinned out to create wide openings, tapered off lower down at their support hinges, and branched off into posts that braced (by means of stiffened joints) the sloping beams of the tiers whose ends were extended into cantilevers to control and reduce the moments. In this way, Nervi succeeded in creating a grand continuous opening punctuated by a repeated series of exceptional portal frames with beams combined in a spire formation. The structure was still designed to be infilled, as evidenced by the persistent presence of pillars and horizontal beams along the stadium perimeter along the height where it was planned to locate the windows.

The contract for the construction of the two uncovered stands, signed on 20 June 1931, specifies that "pillars, beams, inner and outer faces of the steps, and the structure in general shall be left rough in the same state as when the timber formwork is stripped away."[36] Nervi took care to reinforce the concrete partly with the aim of preventing cracking, using solutions that were to resurface in other building projects where the concrete was left exposed: "particular care was taken when designing the reinforcement of the tiers to prevent the formation of cracks in the step treads, including under the action

27. P. L. Nervi, *Stadio Comunale G. Berta* (now *Stadio Comunale A. Franchi*), Via M. Fanti 4, Florence, 1930-33.
View of the uncovered stand structure (photograph by F. Barsotti; Acm-EPFL, 0172.02.0003.2133).

of the extreme temperature fluctuations to which they will be subject, by arranging the irons in a lattice at a 45° angle to the stand centreline."[37] This invisible latticework was destined to assume very different configurations in Nervi's structures.

The two "straight uncovered stands" were built between July and September (the contract specified that they should be completed by 24 October).[38] The firm also constructed the boundary wall in the other sectors, which already included the reinforced concrete pillar stumps of the future stand structure.[39] September also saw the first official opening of the *Stadio Berta*.

Nervi & Nebbiosi continued to work in accordance with the same building site criteria, showing no concern if the formwork boards they fitted led to imperfect edges on pillars, beams, and struts. Nervi limited himself to touching up the most critical points of the reinforced concrete with cement plaster, even if defective workmanship was noted – "perhaps as a result of the excessive haste with which the work was carried out, some of the formwork was curved due to lack of reinforcement, which resulted in slight curvature in the joists forming the steps."[40]

The photographs taken by Ferdinando Barsotti, a Florentine photographer who specialised in architectural photography, document the strength of Nervi's structure with shots that sometimes accentuate the monumental scale of the sequence of spired portal frames and the ability of the lines of reinforced concrete alone to generate an architecture that was almost archaic in its lack of any finish.

Construction defects and inspection of the grandstand roof

While work on the two uncovered stands started, the reinforced concrete structure of the grandstand was inspected by the engineer Vittorio Tognetti, and the first construction defects were recorded in July 1931: the amount of iron in the shelter reinforcement was less than that specified in the contract; the curvature of the roof floor was not executed to perfection, meaning that rainwater pooled in some areas; the plaster on the tiers had become detached in many places. Another defect was identified when it was found that the concrete covers of the tier beams "fell at the slightest stress."[41] Plaster on the first-floor rooms, laid on laths and reinforced with a wire mesh, had also become detached. In order to make the plaster cladding adhere to the reinforced concrete tiers, Tognetti advised preparing a metal reinforcement made out of wire to be wound around nails.[42]

The grandstand roof was inspected in May, July, and September by subjecting some parts of the structure to loading 20, 40, and 135 days after pouring. De Reggi was able to observe that "under the test load the deflections were less than the theoretical deflections, and no cracks or failures of any kind arose […]."[43]

When he was about to undertake the construction of the uncovered stands in July, in an article entitled *Scienza o Arte dell'Ingegnere?*, Nervi took a stand against all kinds of "rules" and "equations" for the sizing of structures through calculation alone and also against the "rational" trend of modern

[37] Nervi, letter to the Chief Engineer, 16 June 1931 cit.
[38] Pier Luigi Nervi, letter to the Chief Engineer of the City of Florence, 27 June 1931, ASCF, pack No. 7467 and *Verbale di ultimazione dei lavori*, 19 September 1931, ASCF, pack No. 7469.
[39] Pier Luigi Nervi, letter to the Works Management, 27 May 1931, ASCF, pack No. 7471.
[40] Vittorio Tognetti, inspection report, 2 August 1932, p. 5, ASCF, pack No. 7649.
[41] *Missiva della Direzione III – Servizi Tecnologici all'Impresa*

architecture, in the name "of sense and constructive intuition," "feeling," and "inspiration."[44] His urging of the engineer to "feel the size and the intimate essence of the work" reflects the imponderable creative and aesthetic questions that guided him when outlining the structures of the *Stadio Berta*: variables everywhere, section by section, in a quest for the exact alchemy between forces, elasticity, the resilience of the reinforced concrete, and the aesthetic appearance of the lines. "One reaches the apparently absurd conclusion," he stated, "that in order to achieve uniform strength, you need to ensure that the element that should by all rights remain uniform to ensure uniform material use, in other words the safety load, actually remains variable – and to a significant extent. The true essence of engineering lies in measuring this variability in the choice of those coefficients, in the evaluation of immeasurable factors, and at this point it is no longer science but art."[45]

The structure of the stadium was indeed to change to match not only the changing safety loads but also other imponderable values that made Nervi's creative and artistic actions even more evident.

Ing. Nervi & Nebbiosi, 17 October 1931, ASCF, f. No. 116, quoted in Battiloro, *op. cit.*, p. 42.
[42] *Stadio "G. Berta." Collaudo Tribuna Coperta* cit., p. 5.
[43] Fiorenzo De Reggi, *Verbale di collaudo provvisorio*, 10 September 1931, p. 3, ASCF, pack No. 7469.
[44] Pier Luigi Nervi, *Scienza o Arte dell'Ingegnere?*, in, "L'Ingegnere. Rivista Tecnica del Sindacato Nazionale Fascista Ingegneri," vol. V, July 1931, No. 7, pp. 473-474.
[45] *Ibid.*, p. 474.

The *Torre di Maratona* and the *Monumento alla Bandiera*, obelisks to fascism

In his plan for the completion of the *Stadio Berta*, Nervi demonstrated his will to infuse his work with increasingly evident symbolic connotations that aimed to celebrate not so much the activist Berta as, more generally, the social and political foundations of fascism. The timing was significant for the regime, because Mussolini was then encouraging architects to seek a fascist style, and Bardi used Mussolini's words as a springboard for emphasising, at the beginning of 1931, in *Architettura, arte di Stato*, that architects should be responsible for the "task of illustrating Mussolini's deeds,"[46] thus announcing the rhetorical impetus that was to overwhelm rationalist architecture. Nervi became one of the leading proponents of the aspiration to a form of state architecture.

Nervi developed the *Stadio Berta* plan, expanding his repertoire of structures by wedging a figure at the point of the opening for the *vomitorium* left between the two straight stands: the resulting *Torre di Maratona* was erected facing the grandstand. A plaster model made in approximately July 1931 documented the official plan, which was intended to guide the new construction work, again entrusted to Nervi & Nebbiosi. The *Torre di Maratona* was shown on this plan. All of the *Stadio Berta* is invested with the imposing character introduced in the grandstand facade. The spans of the structure are all infilled with glass and masonry, and the perimeter is marked by the steady rhythm of the pillars that have been transformed into abstract orders.

In the version documented by the model, the *Torre di Maratona* features a massive pedestal and powerful stem, both stepped, and betrays no hint of a reinforced concrete construction. The structure's appearance, almost that of a stone obelisk, suggests that Nervi was directly inspired by the obelisk planned by Enrico Del Debbio, designed by Costantino Costantini, and being built in the *Foro Italico* in Rome in homage to Mussolini. This decision by Nervi followed an intuitive, yet specific, cultural strategy, intended to suggest that the reinforced concrete structure embodied the quintessential nature of the Fascist regime. His structure now had to demonstrate that it could retain all of its eloquence but do so without the old-fashioned Pharoanic pomp planned for Mussolini's obelisk – a monolith measuring 17,10 metres in length due to stand on more colossal blocks. The whole process had begun when the block was transported from the Carrara quarries in November 1928 on a sled pulled by buffalo to Carrara Harbour, at which point it was mounted on a special pontoon that took it to the Tiber. It was taken from here by boat in November of the following year to the Farnesina dock in Rome and then erected in September 1932 by means of a special machine. The work was eventually inaugurated on 4 November 1932.[47]

After building his jutting roof on cantilever beams, which had its main origin in engineering principles but could nevertheless also be interpreted as representing the *arditezza* (boldness) of the regime, the *Torre di Maratona* gave Nervi an opportunity to begin to consciously explore the rhetorical potential of reinforced concrete structure and its ability to address the desire to celebrate national and social values, although without the rhetorical trappings that surrounded the erection of Mussolini's obelisk, which must have

[46] Pietro Maria Bardi, *Architettura, arte di Stato*, in, "L'Ambrosiano," XLVI, 31 January 1931, p. 1.
[47] Maria Grazia d'Amelio, *L'obelisco marmoreo del Foro Italico a Roma, storia, immagini e note tecniche*, Roma, Palombi Editore, 2009.

appeared ridiculous to Nervi. The *Torre di Maratona* marks Nervi's first attempt to imagine a modern structure suited to the rhetoric of fascism.

In his transition from the model to the final version, Nervi succeeded in getting away from the mask of the obelisk and even infusing the *Torre di Maratona* with these same qualities as the grandstand roof through his solution of replacing a pointed terrace with a powerful overhang at the foot of the tower. This projects over the crowd in an epic dialogue with the cantilever beams of the grandstand, forming a structure that empathised with the fascist people, saluting with outstretched arms in the Roman style, as the footballers would do before the matches due to be played in the half-finished stadium. The *Torre di Maratona* ensured that the *Stadio Berta* began to become the most emblematic work in reinforced concrete, closely moulded in the fascist spirit.

We cannot grasp the full significance of the tower as the start of a quest for a reinforced concrete structure for fascism, without considering the other project that Nervi was managing around the same time (the end of 1931), which was also devoted to celebrating the regime: the *Monumento alla Bandiera*, to be erected on a hill in Rome where it would dominate the skyline with its eerie presence (Monte Mario, with the Foro Italico and Mussolini's obelisk under construction at its foot). The design seemed to spring from a vision by Nervi and was prepared with Rubens Magnani for to the tenth anniversary of the Fascist Revolution, which was celebrated in 1932: "My aim was to design and translate into a plan the most significant monument imaginable: that of our flag."[48] There can be no doubt that for Nervi, the design of this monument was an excuse to imagine an amazing device that he could use to erect a tall, slender structure in the background to his structure for the *Stadio Berta*. The comparison with Mussolini's obelisk was now no longer affected by the impulsive drive to imitate those prismatic shapes but had entered a domain of which Nervi had total mastery: technique. The main material was the same as that of the obelisk, but was planned to be broken down into pieces instead of as a monolith. The idea for the design is enclosed in a technical device that contains the very premises of what was to become, in Nervi's work, the prefabrication of parts to achieve the goal of a monolithic structure.

Nervi formed the slender tower with its imperceptibly truncated cone outline to rest on the base of a parabolic structure in reinforced concrete. It was clad in dark travertine by laying blocks of white marble with beds of cement mortar, again white and high strength, and then all the layers were compressed using a device that was likely inspired by the recent inventions of Freyssinet in the field of prestressing, Guido Fiorini in tensile structure, and above all the research of Eugenio Miozzi in the "system of induced stresses"[49]: a wedge at the tip in reinforced concrete and ties, also in reinforced concrete, anchored to the base of the monument. In a later version, Nervi left suspended the block, to adapt the device to possible settling of the masonry caused by temperature or yields. As with the grandstand roof miraculously suspended in mid-air, we again see Nervi's attraction for the void, now in the form of an exceptional "heavy mass freely suspended […] and capable of generating the required stress," as he explained to Miozzi.[50]

[48] Pier Luigi Nervi, *Monumento alla bandiera,* in, "Quadrante," December 1933, No. 8, p. 44.

[49] Pier Luigi Nervi, letter to Eugenio Miozzi, 20 December 1931, Maxxi, Epistolario.

[50] *Ibid.*

28. P. L. Nervi, *Stadio Comunale G. Berta* (now *Stadio Comunale A. Franchi*), Via M. Fanti 4, Florence, 1930-33.
Plaster model showing the *Torre di Maratona*, 1931 (ASBU).

29. C. Costantini, obelisk, Piazza L. De Bosis, Rome, 1932 (postcard, property of the authors).

30-31. P. L. Nervi and R. Magnani, design for the *Monumento alla Bandiera*, 1931.

30. Cross-section (BSTA, 024-6920).

31. Perspective drawing (Maxxi - PLN, F07308).

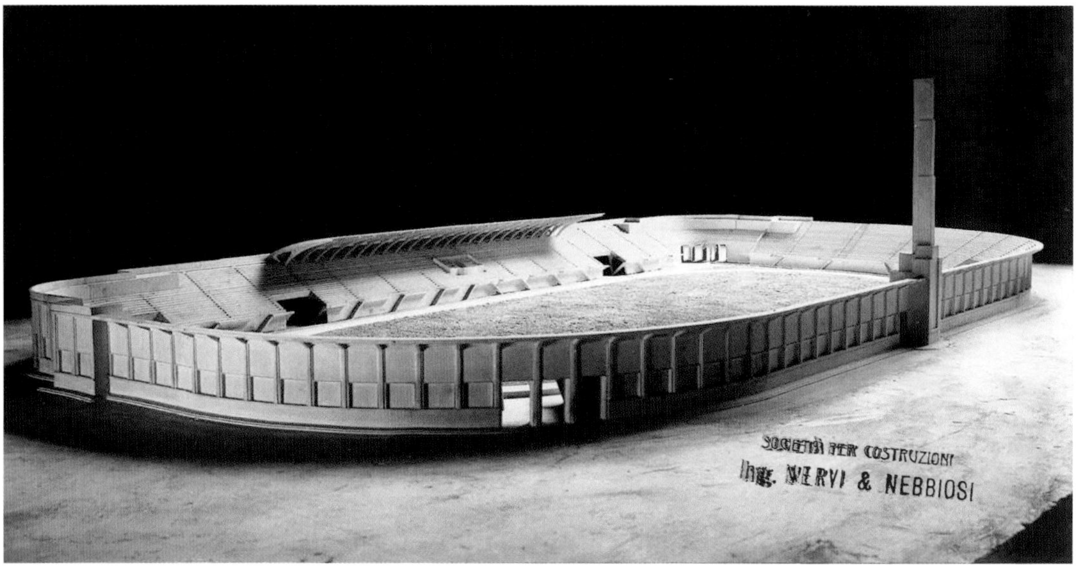

The choice of materials made it possible to create the *Monumento alla Bandiera* in spectacular colours. Its design included a massive base in dark travertine – the colour of the regime, since that base housed a room containing fascist memorabilia – a slender stem in white marble, as in the Mussolini obelisk, up to the glass and steel raising the *fasces* aloft and the waving flag, making them seem miraculously suspended at night, as if a religious apparition, due to special lighting trained on the top, as Nervi explained.[51]

Nervi's idea of using masonry reinforced with reinforced concrete hollows became an authoritative technical pretext for inventing a work that is outstanding for its expressive daring, perfectly in tune with the fascist rhetoric behind the creative impetus. Every detail of the project, including prestressing by means of cables and weights, and the extremely fine joints between the marble blocks, was designed to create an image of an "immense marble monolith" able to represent through its structure the symbolic unity of the *fasces*, to be conquered through sophisticated contemporary mechanical science and not through the technique used by the ancients for the monolith used in Mussolini's obelisk.

Nervi's monument did much more than raise the symbol of fascism to the heavens; it concealed within itself the potential of a military device, a control tower for air navigation and a transmitting station that Bardi had deliberately advised setting up on a promontory along the Adriatic coast. The words that Nervi used when he described the project in "Quadrante" seem to come from the pen of Bardi himself, when he described the "religious severity" of the room containing the "fascist memorabilia," the "progression of shapes and colours" the "*fasces* in gleaming steel," as well as the "climb to symbolic solemnity that the work upholds." While fascist rhetoric was still open to misinterpretation, as was the case with the *Stadio Berta* grandstand, the pure engineering virtuosity demonstrated in the *Monumento alla Bandiera* meant that any doubts were driven away by the eloquence of the distinguishing traits, almost as though it were Nervi's intention to strip away any lingering question about who was *Il Duce*'s most authoritative engineer. The invention of a device for pre-stressing marble allowed a glimpse of brand-new ways to take advantage of traditional Italian materials using state-of-the-art building rules derived from the mechanics of reinforced concrete. This helped the cultural debate over arches and columns to emerge from the sterile discussions of Ojetti or Piacentini and paved the way for an Italian technique that was still wide open to definition and which became the foundation of masterpieces by Terragni, helped along by Nervi's insight. From this viewpoint, the *Monumento alla Bandiera*, where the role of steel was limited to reinforcing the concrete of the ties and base, should also be seen as a self-critical version of the *Torre di Maratona*, for which Nervi & Bartoli had to use significant quantities of metal to reinforce the concrete. Nervi's intention when designing the structure was to react against the nostalgia for ancient Roman building that was embodied in the obelisk at the *Foro Italico* – more forcefully than he had done with the *Torre di Maratona*. With his *Monumento alla Bandiera* project, he was in his own way fighting the traditionalism of the regime in terms of its use of old-fashioned techniques, and demonstrating that fascist workmanship could be inventive and modern, fuelled by contemporary achievements, even if traditional materials were used. Only Terragni supported him in this belief.

[51] Nervi, *Monumento alla bandiera* cit.

Spiral staircases in the shape of *fasces*

The *Torre di Maratona* was defined following the invention of another figure, again designed by Nervi, to accentuate the rhetorical strength of the *Stadio Berta* structures without succumbing to an over-academic approach or the ancient Roman nostalgia of Ojetti, Piacentini, and a plethora of the regime's architects and leaders. Nervi introduced five spiral staircases, three of which were ultimately built, not only to improve the egress of spectators from the various uncovered stands but also to create further opportunities to celebrate the aesthetics of exposed reinforced concrete and to act as a counterbalance to the roof of the grandstand and the *Torre di Maratona*. The way in which the structural system of these staircases was designed reveals a will to overcome the conventional architecture that was stifling the stadium with its colossal regularity.

After planning stairways in straight ramps, each to be contained within two spans of the stand and enclosed within the perimeter wall of the stadium, Nervi decided to transform the pedestrian traffic system into a visible manifestation of spatial energy and the plastic strength of reinforced concrete. Thus, with the same creative impulse with which he shaped the cantilever beams of the grandstand or the overhanging terrace of the *Torre di Maratona*, he freed the stairways from the cage of the spans, arranged them beyond the perimeter of the stadium, and sought to create them using shapes that symbolised the flow of spectators and the symbolic strength of the structure. Unsurprisingly, given the success of this particular motif within the expressive arsenal of reinforced concrete, even in the monumental shapes employed in the Lingotto, Nervi, the engineer and builder, chose a spiral ramp to demonstrate how the material could flow freely once plucked from the geometrical rigidity of the stand spans. The genealogical descent from the spiral staircases of the house in Lungotevere Arnaldo da Brescia and the atrium of the *Stadio Berta* to the stairs of the stands was, however, anything but linear. As with the *Torre di Maratona*, Nervi's insight was primarily motivated by the will to create an exemplary fascist structure that could suggest a modern approach to *romanità*,[52] due to the critical distance that was possible in Florence. Once again, the Foro Italico in Rome offered Nervi crucial inspiration for the symbolic completion of this Florentine stadium where the innovative features of the Roman complex celebrating the fascist cult of sport become iconic. As he was engaged in designing the stadium, work proceeded on the spectacular spiral staircase ramp featuring a cantilevered slab on a single beam, within one block of the building of the Scuola Superiore Fascista di Educazione Fisica (subsequently renamed the Accademia) at the Foro Italico, designed by Del Debbio in 1927 (it was completed in October 1932). The structure was designed by the engineer Aristide Giannelli.

The studies for the *Stadio Berta* staircases show how Nervi's design developed from a spiral ramp with a cantilever slab, borne by a midline structure of four pillars with a beam which boldly descends from the version by Del Debbio to a pillar structure consistent with the monumental order to which a spiral edging beam was anchored to support the cantilever slab, then becoming a slab that hung freely in space before ultimately achieving a novel way of anchoring the spiral edging beam to another spiral beam with opposing geometry – both spiral beams being welded at a single mid-point. The result

[52] Translator's note: *romanità* is the essence of ancient Roman civilisation espoused by the Italian fascists.

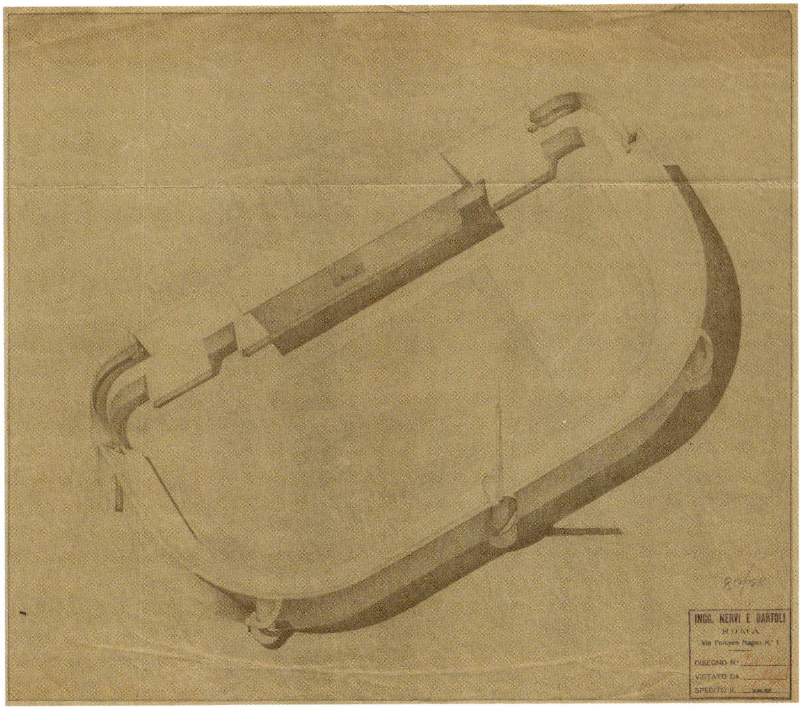

32. P. L. Nervi, *Stadio Comunale G. Berta* (now *Stadio Comunale A. Franchi*), Via M. Fanti 4, Florence, 1930-33.
Axonometric drawing showing the spiral staircases, 30 May 1932 (ASCF, 027681).

is nothing less than a sculptural engineering feat with its unprecedented set of spiral staircases made out of reinforced concrete. It acted as a counterpoint to the soaring roof of the grandstand and laid the foundation for freeing the entire structure of the stadium from the confines of masonry, even that of the tiers – though they were not meant to be exposed.

The grid of the two spiral beams with opposing lines harnessed the space through the flow of forces at play. It expressed the changing vitality of the structure and its material; it hurled it into the void and held it in one acrobatic movement, heralding other spatial grids by Nervi. The system of the two spiral beams crossing to form a *fasces* became even more complex – a veritable lattice – in the *Torre di Maratona*, where Nervi introduced a sequence of pillars that were not infilled with masonry to support the tower, almost as a hymn to a structural *fasces* made out of reinforced concrete. In this way, the combination of tower, spiral beams, and pillars formed a shape that was ideally suited to the quest for rhetorical character that drove the masterpieces of fascist architecture: a stylised fascist *fasces* in reinforced concrete. The structure marked the beginning of a quest for a symbolic virtuosity that was destined, at least for a certain time, to be the theme of Nervi's development of the idea of structure toward increasingly complex solutions, with geometries based on something more than principles of rational and economical construction.

The technical drawings show that the interwoven spiral beams of the staircase were destined to become a thick and dense tangle of steel reinforcing bars. The ramp, which projected by 3 metres, took the form of a slab that tapered along the edge to follow the shape of the cantilever and look even more slender. It is traversed by invisible metal reinforcements that are woven together to form the slender beams that anchor it, at regular intervals, to the spiral edging beam. The same diamond meshed reinforcing bars used on the

Fot. VASARI-Roma

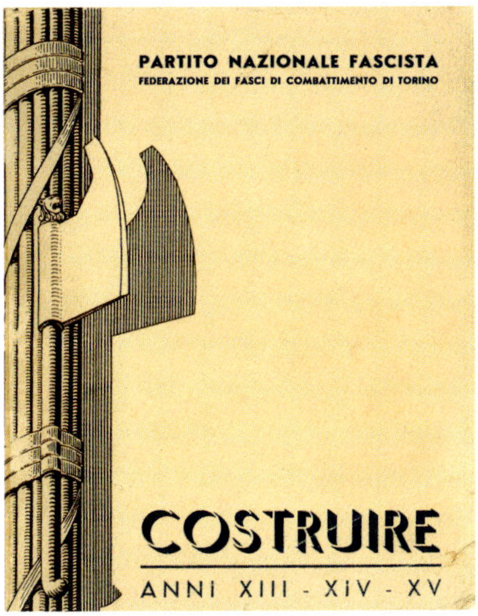

33. E. Del Debbio, Accademia Superiore Fascista di Educazione Fisica, Piazza L. de Bosis 15, Rome, 1927-32. Spiral staircase under construction (photograph by T. and G. Vasari; Maxxi - EDD, F02134).

34-35, **37-38**. P. L. Nervi, *Stadio Comunale G. Berta* (now *Stadio Comunale A. Franchi)*, Via M. Fanti 4, Florence, 1930-33.

34. Spiral staircase. Study drawing (CSAC, n. inv. pra 31, coll. 148/6).

35. Spiral staircase. Study drawing (CSAC, n. inv. pra 31, coll. 148/6).

36. *Fascio littorio* (Partito Nazionale Fascista. Federazione dei Fasci di Combattimento di Torino, ed., *Costruire. Consuntivo degli anni XIII-XIV. Preventivo per l'anno XV*, Torino, Società Poligrafica Editrice, 1936).

37. Spiral staircase with the *Torre di Maratona*. Perspective drawing (CSAC, n. inv. pra 31, coll. 148/6).

38. Spiral staircase. Study drawing, 17 November 1932 (ASCF, 028270).

39-41. P. L. Nervi, *Stadio Comunale G. Berta* (now *Stadio Comunale A. Franchi*), Via M. Fanti 4, Florence, 1930-33.

39. Spiral staircase under construction (photograph by F. Barsotti; S. Picchi, ed., *Firenze. Lo stadio racconta. Oltre 75 anni di storia cittadina nelle memorie del Franchi*, Firenze, Giunti, 2007, p. 44).

Torre di Maratona terrace are woven among the slender beams and anchored to the main ribs by virtue of their geometry. Technical drawings of the reinforcements and building site photographs showing the metal bars document Nervi's vision of a reinforced concrete ribbed everywhere like *sidero-cemento*. Nervi described the spiral staircases as though the elements made up an organic whole without main and secondary parts: "the load-bearing structure of the cantilever external spiral staircases is made up of two spiral architraves that meet to form a kind of triangulated system with curved sides, two of which are rendered non-deformable in a plan view by the slab of the staircase."[53]

Nervi's identification with the revolutionary and corporatist fascist spirit was now total. This was the same spirit he had felt in Terragni and simultaneously staged with the rhetorical and symbolic reinforced concrete structure he created for the *Stadio Berta*. At this juncture, Nervi must have begun to recognise in that very physical make-up of the reinforced concrete – that web of reinforcing rods bound and welded by the concrete – a metaphor for what was beginning to become a fascist civilisation, of which he felt he was able to represent the essence.

The invention of the spiral staircases reopened the question over the structure of the tiers, exposed or infilled, which Nervi revisited with perspective views depicting the two different possibilities. At this point, however, he dared to imagine that these spans could be infilled with large windows and side sectors, both set back from the edge of the skeleton in order to expose the reinforced concrete warping without any cladding to increase its thickness. By means of another perspective drawing of the same type as that proposed in 1931, Nervi set out the relentlessly spare outlines of his reinforced concrete structure for the *Stadio Berta*; the lattice structure of the straight tiers and the spiral structure of the staircase intertwined in a space that they defined and enhanced. It was in visions such as this that fascism seemed to have finally disappeared from reinforced concrete structure, as was to happen with the grid-like skeleton of Terragni's *Casa del Fascio* – these mirages of the transubstantiation of fascist rhetoric into matter and geometry achieved lyrical heights in both cases.

[53] Typescript, 28 November 1936, p. 3, ASCF, pack No. 7467.

40. Spiral staircase under construction (photograph by F. Barsotti; S. Picchi, ed., *Firenze. Lo stadio racconta. Oltre 75 anni di storia cittadina nelle memorie del Franchi,* Firenze, Giunti, 2007, p. 58).

41. Spiral staircase (photograph by F. Barsotti; Acm-EPFL, 0172.02.0059.0350).

104

42-43. P. L. Nervi, *Stadio Comunale G. Berta* (now *Stadio Comunale A. Franchi*), Via M. Fanti 4, Florence, 1930-33.

42. Spiral staircase and *Torre di Maratona* (photograph by F. Barsotti; Acm-EPFL, 0172.02.0003.2131).

43. General view of the three spiral staircases and of the *Torre di Maratona* (photograph by F. Barsotti; Acm-EPFL, 0172.02.0003.2126).

Curved stands with entrances, or archaic concrete

Nervi designed so many other structures to complete the *Stadio Berta* stand circuit that the finished work became a catalogue of different solutions, a reflection of the project's progress through different stages. Nervi drew billowing lines for the structure of the uncovered stands backing onto the grandstand, intended to free up enough space for gymnasia. This resulted in mighty formless arched members on which the tiers were built. These rested on the row of pillars around the perimeter, all tapering toward the base and braced by horizontal beams arranged at a height designed to allow the fitting of large windows once the work had been completed by infilling. It is significant to the identity of the figures designed by Nervi that he now followed his description of a "lattice structure" by writing of a "flying buttress structure."[54]

When he designed his flying buttress structure, Nervi also altered the shape of the two entrances in the curved stands. Without further ado, he finally freed the load-bearing structure of the tiers and changed their geometry, abandoning the frame with the outer row of pillars and replacing it with mighty pillars that were set back with a curved, cantilevered crown to follow the shape of the tier, as in the adjacent "flying buttress structure." The perspective drawings emphasise the plastic and monumental effect of this type of structure. The fact that structures with similar lines to some of Terragni's architectural works appear alongside the entrances is confirmation that the shape of Nervi's reinforced concrete was lending itself more and more to exemplifying *Gruppo 7* theories on the structure of "rationalist" architecture.

For the design of the large pillars, Nervi also sought a rhetorical expression manifested in the continuity of the curved line impressed on the profile facing the city, down to the bases of the pillars, in order to give those mem-

[54] *Stadio Comunale Giovanni Berta Firenze. Progetto di completamento. Computi Metrici*, p. 10, ASCF, pack No. 7467.

44-45. P. L. Nervi, *Stadio Comunale G. Berta* (now *Stadio Comunale A. Franchi*), Via M. Fanti 4, Florence, 1930-33.

44. Perspective drawing of the unrealised gymnasium beneath the uncovered stand along Via M. Fanti, 1931 (ASCF, 027688).

45. View of the area beneath the stand along Via M. Fanti (photograph by F. Barsotti; Acm-EPFL, 0172.02.0003.2134).

bers a sophisticated monumental quality despite the harshness and simplicity of the figure. Because the areas under the tiers with entrances on the curved section were not intended to house any rooms, the design of both the stands with entrances reveals a quest for a harmony between the curve of the tiers and that of the large pillars that goes beyond the simple juxtaposition of pillars, beams, and posts. The configuration of that structure may be taken as the end-point of Nervi's reflections on the ideal shape of the structure to be left exposed for a stadium. Those large pillars with their arched outline reveal all the monumental character of the reinforced concrete structure, ideally suited to a stadium that was destined to become an emblematic work of the regime. Those large curved pillars enabled Nervi to raise triumphal entrance porticos to the *Stadio Berta*, expressions of modern fascist monumentalism made possible by a quintessentially Italian reinforced concrete engineering culture, iconic figures of the "new archaic age" invoked by *Gruppo 7*.

As with the cantilever beams, whose ramifications cannot be seen in the skeleton beneath the tiers, the large pillars of the stands with entrances in the curved section are apparently endowed with their own special nature because part of their structure is hidden. Their stately soaring flight from the ground like ancient columns, the evident sense of safety they convey through their weight alone, and the miraculous effect of being able to support the eccentric weight of the overhanging tiers without risk of tipping over are expertly researched effects. They are based on a psychology of form diametrically opposed to that of the cantilever beams or spiral beams and made possible by arching the stem of the pillars beneath ground level to balance the eccentric loads. Only Gothic builders have been able to achieve such miraculous effects in their structures, hiding their underlying mechanisms from sight, as Nervi did with the cantilever beams and pillars of his stadium, and thereby venturing into the theoretical excesses of "*architecture mystèrieuse*." He was to repeat the same use of large pillars to achieve the opposite effect: a structure that leaned so far forward that it left its observers dumbfounded. The design was for two small diving boards built in 1934 for the Ugolino Golf Club swimming pool in Florence, whose upsurge was restrained by the invisible twisting effect of the underground foundation structure.[55]

Building site of the completion

After a halt to building due to a lack of funds, and after the first sporting events, hosted in a stadium made up of the grandstand and the two straight uncovered stands, the final building work was resumed in summer 1932 and ended in December of the same year. Before building his latest virtuoso works on the *Stadio Berta*, Nervi broke off his partnership with Nebbiosi. Starting in June, with an official agreement signed on 27 August, he established with Bartoli the Società Anonima per Costruzioni Ingg. Nervi & Bartoli, based in Rome.

On 13 June, Nervi & Bartoli were commissioned to build the three spiral staircases, the *Torre di Maratona*, and the remaining stands, including those with the entrances on the curved section, with instructions to finish pouring the concrete by 15 October and finish the work by 4 November 1932. It was significant that the date set by the contract coincided with that of celebrations

[55] Pier Luigi Nervi, *Il Golf dell'Ugolino a Firenze*, in, "Casabella," VIII, April 1935, No. 88, pp. 8-15.

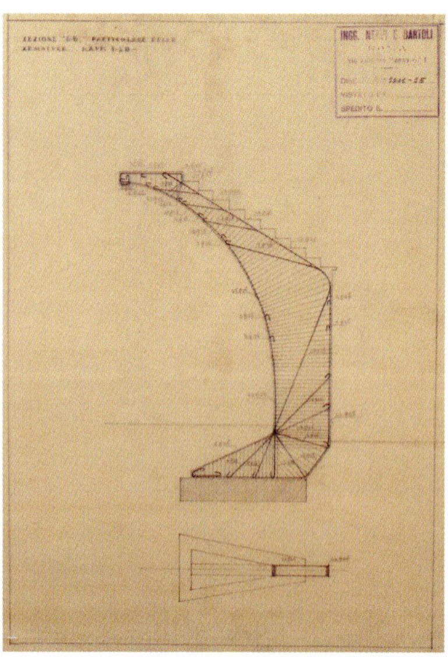

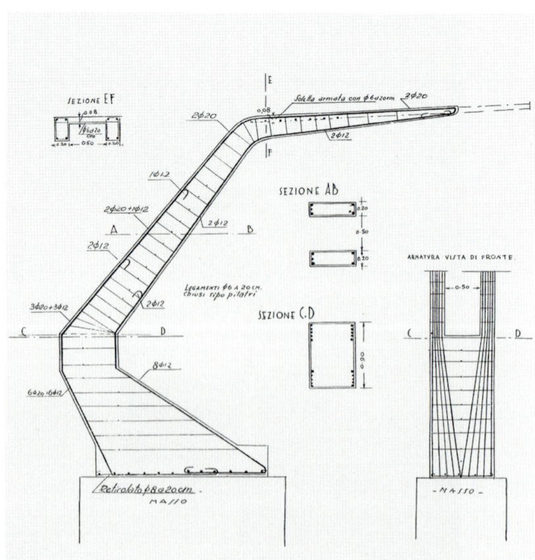
SEZIONE EF
E
F
Soletta armata con φ6 a 20 cm
2φ20
2φ12
1φ12
SEZIONE AB
2φ20+1φ12
A
B
2φ12
ARMATURA VISTA DI FRONTE
Legamenti φ6 a 20 cm.
Chiusi tipo pilastri
SEZIONE C.D
3φ20+3φ12
C
D
8φ12
6φ20+6φ12
Ritirolato φ8 a 20 cm.
MASSO
-MASSO-

46-47. P. L. Nervi, *Stadio Comunale G. Berta* (now *Stadio Comunale A. Franchi*), Via M. Fanti 4, Florence, 1930-33.

46. Perspective study drawing of the stand with entrance on the curved section, 1932 (CSAC, n. inv. pra 31, coll. 148/6).

47. "Particolare delle armature (Detail of the reinforcement)" of the structure of the stand curved section (CSAC, n. inv. pra 31, coll. 148/6).

48. P. L. Nervi, Structure for the diving board of the Ugolino golf club swimming pool, designed by G. Bosio, Via Chiantigiana per Strada 3, Impruneta, 1934.
Reinforcement of the structure ("Casabella," April 1935, No. 88, p. 15).

held in Rome to mark the first decade of the fascist revolution, in which the *Stadio Berta* was apparently to play a crucial role. The firm organised the building work in two stages (building the curved stands firstly in the south-west sector, then in the north-east sector), estimated the number of workers necessary for each one (60-65 carpenters, 45-50 steel fixers, 50-60 hodmen), the number of cubic metres of concrete (daily output of 80-90 cubic metres), and the setting up of 120 quintals of iron and 600 square metres of formwork. The works were subdivided into nineteen pouring stages. In July, the formwork was set up, and the metal reinforcements of the south-west sector were prepared. The firm imposed a "surcharge for the curved formwork of the tiers and covered stands, the ticket office and spectators' hall and the spiral staircases."[56] The concrete pouring began on 10 August and ended on 23 October. All the structures were completed on 20 November, a few weeks later than expected. No finishing was required for the concrete skeleton.

Construction of the spiral staircases must have posed some problems for the carpenters in setting up the formwork; Nervi stipulated at the contractual stage "regularisation of formwork with a layer of plaster."[57] Because their curves made them difficult to mould with wooden boards, the spiral staircases became an object lesson for Nervi on the conflict between the formal possibilities of reinforced concrete and the limits set on the manufacture of the timber moulds required to model the structures – the formwork. Although he had already encountered difficulties in building the structures of the grandstand, the spiral staircases marked a starting point for an aversion to timber formwork that was to recur in his future writings and would ultimately lead him to invent a cementing technique freed of the constraints of carpentry. Regarding the spiral staircases of the *Stadio Berta*, he said: "which made me realise for the first time, the extent to which the full development of reinforced concrete is linked up with the problem of timber formwork and its inability to adapt itself to curving or spiralling surfaces."[58] No other creator of reinforced concrete structures laboured as Nervi did to seek the expressive essence of a combination of steel and concrete, beyond the provisional mould necessary to form it into a figure.

The concrete in the "completion" structure was intended to be left exposed, with certain exceptions specified in the bill of quantities. The steps of the uncovered stands were finished with "cement plaster," while in the sectors backing onto the grandstand, "lime plaster" was laid on the faces of the brick walls, infilling the skeleton overlooking the sports ground.[59] For the spiral staircases, Nervi & Bartoli agreed with its customer that they would continue to plaster the annular beams in order to correct the impressions left by the boards of the formwork,[60] and they would paint the intrados of the ramps, pillars, and beams.[61]

It is relevant to an understanding of the idea of the stadium's concrete structure that Nervi decided to erect the looming *Torre di Maratona* with its cantilevered platform, also clad in plaster, in front of the grandstand, which

[56] *Atto di Accollo*, 4 August 1932, p. 24, ASCF, pack No. 6420.
[57] *Ibid.*, pp. 24-25.
[58] Quoted in, Huxtable, *op. cit.*, p. 24.
[59] *Stadio Comunale Giovanni Berta Firenze. Progetto di completamento. Computi Metrici* cit.
[60] *Allegato 9. Intonaco a cemento travi scale elicoidali*, in, *Importo lavori eseguiti in più e in meno allo stadio "Giovanni Berta,"* pp. 1-2, ASCF, pack No. 7467 and *Analisi nuovi prezzi da combinare per il lavoro del campo sportivo "Giovanni Berta" in Firenze*, p. 1, ASCF, pack No. 7471. The work also included "imitation stone architraves" at the entrances, *ibid.*, p. 2.
[61] *Allegato 7. Tinteggiature*, in, *Importo lavori eseguiti in più e in meno...* cit., pp. 1-2, and *Verbale di ultimazione dei lavori*, 15 December 1932, ASCF, pack No. 7472.

49-50. P. L. Nervi, *Stadio Comunale G. Berta* (now *Stadio Comunale A. Franchi*), Via M. Fanti 4, Florence, 1930-33.

49. *Torre di Maratona*, cantilever platform under construction (photograph by F. Barsotti; S. Picchi, ed., *Firenze. Lo stadio racconta. Oltre 75 anni di storia cittadina nelle memorie del Franchi,* Firenze, Giunti, 2007, p. 54).

50. *Torre di Maratona* (photograph by F. Barsotti; Acm-EPFL, 0172.02.0003.2127).

112
ARELLI VIA PUCCI · 6
CALZOLERIA
CENTRALE
VIA MARTELLI · 4
ZE

was covered in lime mortar. The intrados of the platform at the foot of the tower was the only area of concrete not clad in plaster, in order to avoid detachment. Terranova plaster with a "special coloured mortar" was used to clad the *Torre di Maratona*.[62] The choice of this plaster, which was widely used in the late 1920s as a replacement for quality coatings, could be explained by its technical qualities of strength and adhesion to the surface.

In the majestic pillars of the stand, with entrances on the curved sections, the reinforced concrete was left without any cladding. The formwork boards were installed with Nervi's usual pragmatism, without any particular attention to design. At the end of the work, a carpet of protruding nails that had been positioned during the stripping stages could be seen on the faces of the pillars. Even though the concrete remained exposed until approximately the 1960s, Nervi likely planned to clad the pillars with a layer of plaster. The function of the nails on the pillars could have been to apply a metal mesh or wire weave, as Tognetti had advised to repair plaster that had become detached from the grandstand tiers.

The *Stadio Berta* construction stages continued through concrete finishing solutions that ultimately made up a kind of enigmatic and random sample of the expressive possibilities of that material and might even be seen as Nervi's gradual coming to terms with the theoretical implications inherent in his structure. It was almost as though he were responding to the various stances taken by critics and architects, between 1930 and 1932, on the question of whether to leave the skeleton exposed or clad. The *Stadio Berta* ultimately showed the two faces of reinforced concrete: one courtly and polished, achieved through different devices; the other monumental and raw. Much of its expressive force comes through having been left unfinished, with the bare lines of its structure, without architecture.

One reference that is significant in this regard may be read in documents of the Florence City Hall in 1936, where the *Stadio Berta* is described as "not [...] yet completely finished," and the works to be carried out were listed, such as the rooms beneath the uncovered stands ("changing rooms, gymnasia, meeting rooms, exhibition rooms etc."). These were to be enclosed by means of "the facade along the perimeter of the Stadium, a facade that is currently only present at the grandstand."[63]

A work of brutal engineering

The links between Ojetti's comments about concrete being liquid mud and Nervi's work were revealed in February 1932 by the Journal "L'Ambrosiano" in a short article entitled *Lo stadio di Firenze* by Bardi. In the next column started a polemical *Guida dell'Italia Fascista* (Guide to Fascist Italy), which listed the buildings worthy of the regime, beginning with the stadium.[64] Bardi wrote the article during a visit to the building site, when he climbed onto the canopy together with some students from the Florence School of Architecture, defining it as "an extremely daring challenge to the laws of gravity"[65] – he did not yet know Nervi personally. Bardi believed that the "construction" of the *Stadio Berta* was "lacking in any decorative element, naked and veined with its essential ribs," and that the work was a "magnificent monument made out of 'liquid mud'." He used the stadium raised by Nervi with its reinforced concrete shapes as an example of the new fascist architecture, in opposition to "Roman arches" at the height of the critical debate over "arches and columns." "[...] a fan of immobile, slender, agile arms, almost fragile when considering the vastness and weight of the protective pouring," wrote Bardi. "[...] the

Florence stadium stands as a clear example of contemporary architecture. These are, in response to the critic who attempted to boycott 'liquid mud', the monuments of Fascist Italy: certainly not Roman arches. Yet the stadium is dedicated to one of the martyrs of the Revolution. Therefore: fascism, sports, fresh air, and reinforced concrete. The first verses of a new poem."[66]

Unsurprisingly, given the evolution of the *Stadio Berta* toward that "general skeleton of construction" that the young Italian rationalists declared to be the foundation of the new architecture, Nervi was invited to take part in the Terza Mostra Nazionale di Architettura Razionale, organised in Florence in March 1932 by Giovanni Michelucci and Adalberto Libera. In an article published during the same month and based on a report by Nervi, Michelucci hailed the *Stadio Berta* as "one of the most notable and daring modern structures in Europe." In his eyes the stadium possessed a "striking modern quality."[67]

Bardi showered more praise on the *Stadio Berta* when he stated that it crowned his list of Italian naval, aeronautical, and civil engineering masterpieces that began with the Lingotto. These were examples of what he called "living architecture." "The charges brought against the new rationalist architects of elementarism, primordialism, bareness, and ultimately of 'engineering' are the best praise that can be given to 20th-century builders," he affirmed at the beginning of 1933, his gaze fixed on Nervi's then completed work.[68] The debate over arches and columns, liquid mud, and the role of concrete in

51. P. L. Nervi, *Stadio Comunale G. Berta* (now *Stadio Comunale A. Franchi*), Via M. Fanti 4, Florence, 1930-33. View of the playing field (photograph by F. Barsotti; Acm-EPFL, 0172.02.059/348).

[62] *Stadio Comunale Giovanni Berta Firenze. Progetto di completamento. Computi Metrici* cit., p. 20, and *Atto di Accollo*, 4 August 1932, cit., p. 11.
[63] Report *Stadio "Giovanni Berta" in Firenze*, 28 November 1936, p. 1, ASCF, pack No. 7467.
[64] Pietro Maria Bardi, *Guida dell'Italia attuale* and *Lo stadio di Firenze*, in, "L'Ambrosiano," XLVII, 9 February 1932, No. 34, p. 1.
[65] B. [ardi], *Lo stadio di Firenze* cit.
[66] *Ibid.*
[67] Michelucci, *Lo stadio "Giovanni Berta" in Firenze dell'Ing. Pier Luigi Nervi* cit., p. 105. Michelucci's article was based on a report that was probably written by Nervi himself, see Typescript, 28 November 1936, cit.
[68] Pietro Maria Bardi, *Edilizia e Ingegneri*, in, "L'Ambrosiano," XLVII, 2 February 1933, No 28, p. 3.

99

52. P. L. Nervi, *Stadio Comunale G. Berta* (now *Stadio Comunale A. Franchi*), Via M. Fanti 4, Florence, 1930-33. Curved stand (photograph by F. Barsotti; BSTA, 024-2057).

architecture ultimately had an impact on the material's visibility, when Ojetti observed the extent to which the qualities of "sincerity," "bareness," "truth," and "simplicity" upheld by the proponents of reinforced concrete jarred with the widespread practice of stone cladding: "not even a fingernail of cement is left uncovered."[69] Publication of the exposed concrete surfaces of the uncovered stand structure was the very action that helped to turn Nervi's stadium into an icon of "reinforced concrete in modern architecture,"[70] a "work of brutal engineering," according to Pagano's pithy description.[71]

Building on the success of the *Stadio Berta*, Nervi became a leading name in architectural criticism, and he outlined his vision of reinforced concrete from the pages of "Casabella." His views went beyond proclaiming the need to bare the structure, claiming that a new aesthetic could be achieved by enhancing all the technical characteristics of the new materials.[72] More than bareness and truthfulness, the thing that interested him was defining a concept of *technique* as a blend of scientific knowledge and static intuition achieved through the experience of construction.[73]

Nervi also controlled the publicising of photographs of the *Stadio Berta*, beginning to develop a shrewd strategy that he was to refine through subsequent publications.

He used various strategies when presenting the photographs in order to accentuate the aesthetic effect of these structures. In addition to cladding the pillars, beams, and slabs of the spiral staircase with plaster or white paint, he made further corrections to the photographic prints, retouching them to remove any building blemishes that were still visible, evident in the photograph published in "Casabella" in April 1933.[74]

The photographs are significant for grasping the value that Nervi attached to certain structures of his stadium. When the work was completed, it was clear that the *Stadio Berta* had undergone a non-linear design process, and the various hold-ups over the finishes had ultimately become a series of mixed messages that had been changed and refined over the months, sometimes solely concerning the engineering, sometimes more openly rhetorical. The cantilever beams of the *Torre di Maratona* and the spiral staircases can be traced back to the same creative impulse, variously expressed with different and growing awareness, corresponding to a certain image of the regime that was nostalgic for the original revolutionary drive of fascism. The two spans with entrances on the curved sections at the tips of the straight section of the plan in the shape of a massive letter D, on the other hand, display an expressive quest for a structural design that lends itself, more than the other parts of the *Stadio Berta*, to the noble and grandiloquent aspirations of fascist rhetoric. The large pillars of those stands, with their arched silhouette, were made famous by Barsotti's photograph, which records all of the imperfections of construction, the unfinished state, and the monumental strength, accentuated by the rawness of the surfaces. Nervi was particularly attracted by that

[69] Ugo Ojetti, *Ancora le colonne e gli archi. Risposta a Marcello Piacentini...*, in, "Pègaso," V, March 1933, No. 3, (pp. 358-363), p. 361.
[70] G. S., *Il cemento armato nella moderna architettura*, in, "L'industria italiana del cemento," V, February 1933, No. 2, pp. 42-44.
[71] G. P. P. [Giuseppe Pagano Pogatschnig], *Ing. Pier Luigi Nervi: stadio comunale Giovanni Berta a Firenze*, in, "Casabella," VI, April 1933, No. 4, p. 40.
[72] Pier Luigi Nervi, *Problemi dell'architetto*, in, "Casabella," VI, May 1933, No. 5, p. 34.
[73] Nervi, *Arte e tecnica del costruire* cit. We may also glean more information on Nervi's ideas about 'intuition' in Pier Luigi Nervi, *Pensieri sull'ingegneria*, in, "Quadrante," October 1933, No. 6, p. 20, and Pier Luigi Nervi, *Ancora sulla sorveglianza delle costruzioni in beton armato*, in, "L'Ingegnere. Rivista Tecnica del Sindacato Nazionale Fascista Ingegneri," December 1932, No. 12, p. 883.
[74] See "Casabella," VI, April 1933, No. 4, p. 10.

photograph, borne out by the fact that it was published as a full-page spread at the end of his article, *Considerazioni tecniche e costruttive sulle gradinate e pensiline per stadi*,[75] and on the back cover in the short set of photographs collected in the booklet on the *Stadio Berta*, published by Nervi & Bartoli.[76] Nervi must have recognised in those pillars a specific feature of the work, expressing values other than those informing the spiral staircase with the *Torre di Maratona* and the grandstand roof with its cantilever beams. In Barsotti's photograph, the curved reinforced concrete pillars stand as a statement against the opposing positions voiced by critics and architects over the fate of marble arches and columns. It is no coincidence that those massive pillars became Nervi's starting point when he built other stadia for the regime.

The fact that Barsotti's photograph acknowledged blemishes in the concrete also implies that Nervi was predisposed to consider the visibility of the material to be part of the structural expression.[77] Bardi, who had been Nervi's mentor since his visit to the stadium in Florence, wrote some meaningful words about the matter of reinforced concrete cladding and its camouflage with "arches and columns." If he had not had the stadium before him, despite its contradictory and sometimes even equivocal nature, Bardi might not have been able to make such telling statements about the visibility of the material, designed pehaps as a reply to the disguises Ojetti lamented: "everyone thinks that a reinforced concrete construction skeleton is a thing of beauty: when this skeleton is clad, padded, and tattooed, it disappears under an inadequate, unbearable cloak."[78]

The *Stadio Berta* reveals that Nervi believed that intuition, truthfulness, and the static possibilities of structure could be used to invent figures able to replace the rhetoric guaranteed by "arches and columns" and other "formal elements." In the design of the cantilever beams, the spiral staircases, or the *Torre di Maratona*, he was able to instill in the reinforced concrete a fundamental degree of rhetoric. When the stadium was completed, this become the foundation of his concept of structure, beyond any rules of calculation: "an idea with its origins in feelings," giving rise to what he defined the "nature of form," a blend between "inspirational lyrical feeling" and "implementing a technical event."[79] There is no doubt that the "nature of form" instilled in the reinforced concrete structure made it possible to identify the *Stadio Berta* as an icon with the revolutionary and modern face Mussolini desired for his regime. "[…] the aesthetic result is closely dependent on the expressive capacity of purely static and constructive forms, each of which has its own particular nature." Nervi's apodictic words foreshadowed the crucial role he was to confer on the concept of form.[80]

Cracks in the stadium and Brunelleschi's dome

Following initial load tests on the completion structures, carried out from 22 October 1932, Tognetti confirmed that the "set of works examined shows indisputable properties of stability."[81] After the inspection of the grandstand in December, the firm of Nervi & Bartoli was forced to consolidate defective plas-

[75] Pier Luigi Nervi, *Considerazioni tecniche e costruttive sulle gradinate e pensiline per stadi,* in, "Casabella," VI, December 1933, No. 12, (pp. 10-13), p. 13.
[76] *Opere di completamento dello Stadio Comunale Giovanni Berta, Firenze, Ingg. Nervi & Bartoli, Anonima per Costruzioni, Roma*, Milano, Treves-Treccani-Tumminelli.
[77] For one view of the role of Nervi as an early admirer of the exposed concrete technique, beginning with the *Stadio Berta*, see Mario Desideri, *La tecnica del costruire*, in, Gabriele Milelli, ed., *Eredità di Pier Luigi Nervi*, Ancona, Istituto marchegiano Accademia di scienze-lettere ed arti, (pp. 39-47), p. 42.
[78] Pietro Maria Bardi, *Architettura nuova materiali nuovi,* in, "Il Lavoro Fascista," VIII, 4 January 1934, p. 4.
[79] Pier Luigi Nervi, *Problemi della realizzazione architettonica,* in, "Casabella," VII, February 1934, No. 74, (pp. 2-3), p. 3.
[80] Nervi, *Considerazioni tecniche e costruttive sulle gradinate e pensiline per stadi* cit., p. 12.
[81] *3° Verbale delle prove di carico*, 19 November 1932,

tering on the roof and, in some rooms, at its own expense.[82] During the final inspection, performed in July 1933, Tognetti recorded the imperfect execution of the structural form in some stands, various cracks in the plaster on the tiers, and not enough sealant in the expansion joints.[83] Nervi & Bartoli considered the structural imperfections to be an "inevitable" consequence of the settling of the formwork reinforcements, accentuated by the summer temperatures, and they attributed the phenomenon of cracking to the inspection work having been carried out in summer before the plaster had finally set, preventing sufficient watering of the surfaces.[84] After referring to a German standard on the manner of constructing "shuttering" in order to "rigidly bear the weight of the pouring," Tognetti told the firm to carry out repairs by "hammering the surface of the reinforced concrete structure, wetting it, and immediately plastering it with high-strength cement or wet cement, then keeping it constantly wet for about 10 days."[85] These repairs did nothing to prevent the incident, which occurred on 10 July 1934, when a layer of plaster became detached from the roof.[86] Another report described the malfunction of the expansion joints during abrupt temperature changes: "the expansion joints of the various reinforced concrete did not work properly, and [...] the detachment caused by expansion of the concrete did not take place along the planned line of least resistance but along a broken line that moved away from the original line, causing the detachment of large pieces of plaster and concrete mixture."[87] In the wake of the defect reports drawn up during the inspection, Nervi stated his opposition to expansion joints in reinforced concrete structures: "I am not convinced of the efficacy and usefulness of expansion joints."[88]

In the context of his reflections on the stadium defects and the matter of heat expansion, Nervi reconsidered the technical properties of his *sidero-cemento* and proposed a concrete that was reinforced not only to withstand weight but also that possessed special "secondary reinforcements" distributed to achieve a "structure capable of elastically withstanding the expansions and contractions caused by temperature changes."[89] This led to him stating what was to become the technical – and not the aesthetic – goal of his exposed concrete: "monolithic and waterproof surfaces."[90] As far as Nervi was concerned, achieving this technical quality was a necessary condition for doing away with the concrete cladding.

Nervi's participation in the committee of experts appointed in 1934 to examine the reasons for the cracks that had occurred in Brunelleschi's dome in Santa Maria del Fiore represented another important chapter in his understanding of the phenomena of expansion and contraction in a large structure, as well as cracking and the role of joints, since the testing concluded that the cracks were caused by temperature fluctuations.[91] Until the end of the 1930s, Nervi continued to believe in the possibility of building large structures without expansion joints and being able to counter movements through the technical improvements in the thermal insulation, which was what he also suggested for Brunelleschi's dome.

p. 18, ASCF, pack No. 7612.
[82] Piccardi, *op. cit.*, p. 53.
[83] *Relazione del collaudatore alle osservazioni delle impresa al certificato di collaudo*, 9 October 1933, ASCF, pack No. 7467.
[84] *Riserve alle risultanze di collaudo delle tribune in c° a° dello stadio G. Berta di cui all'atto di accollo 4 August 1932*, 12 August 1933, p. 2, ASCF, pack No. 7469.
[85] *Relazione del collaudatore alle osservazioni delle impresa al certificato di collaudo*, 9 October 1933 cit., p. 16.
[86] Piccardi, *op. cit.*, p. 53.
[87] Report of 20 September 1934, ASCF, pack No. 7612.
[88] Nervi, *Considerazioni tecniche e costruttive sulle gradinate e pensiline per stadi* cit., p. 11.
[89] *Ibid.*
[90] *Ibid.*
[91] Pier Luigi Nervi, *Considerazioni sulle lesioni della Cupola di S. Maria del Fiore e sulle probabili cause di esse*, in, Opera di Santa Maria del Fiore di Firenze, ed., *Rilievi e studi sulla cupola del Brunelleschi, eseguiti dalla Commissione nominata il 12 gennaio 1934-XII*, Firenze, Tipografia Ettore Rinaldi, 1939, pp. 44-49.

chapter three
reticular structures for the regime

Modern shapes and *romanità*

The *Stadio Berta* was the first prestigious work by Nervi, started by Nervi & Nebbiosi and completed by Nervi & Bartoli. Nervi's firm continued to be involved in various building sites for factories, wool mills, warehouses, roofs, bridges, tanks, reservoirs, and various buildings, from houses to churches, in Lecce, Rosignano Solvay, Florence, Prato, Genoa, Rome, and elsewhere. Some of the main projects came from government agencies such as the Italian Royal Air Force, the Monopoli di Stato and two companies from the emerging Italian chemical industry, namely the Società Chimica dell'Aniene, based in Rosignano Solvay, and the Azienda Nazionale Idrogenerazione Combustibili, based in Leghorn and Bari. In the wake of the success of the *Stadio Berta*, Nervi began a fruitful collaboration with architects and became an influential columnist in prestigious journals, including "Casabella," "Quadrante" and "Architettura."

Nervi's joint ventures with architects often took the form of projects with special structural and technical requirements, sometimes carried out for presentation in competitions or exhibitions. After Coppedé, Capponi, Giovannozzi, and Magnani, his main partners became Cesare Valle and Luigi Carlo Daneri. His projects continued to be driven by the creative impulse he had shown in the *Stadio Berta* and by the discovery of bold structures, often using sophisticated techniques that were always dependent on his firm's iconic material: reinforced concrete. Sometimes his projects were also designed to give visible form to the rhetorical demands of the regime, like the more experimental parts of the stadium. As with the stadium, everything appeared to be focused on defining the salient features of a modern style that tended to differ from rationalist architecture due to the part played by structure. Nervi was supported in this aspiration on various occasions by Bardi and his journal "Quadrante."

His design for a floating hotel on a circular plan, drawn up with Magnani and presented at the Terza Mostra Nazionale di Architettura Razionale, held in Florence in 1932, set out to demonstrate the possibility of transferring techniques tested in reinforced concrete ships and boats to rationalist archi-

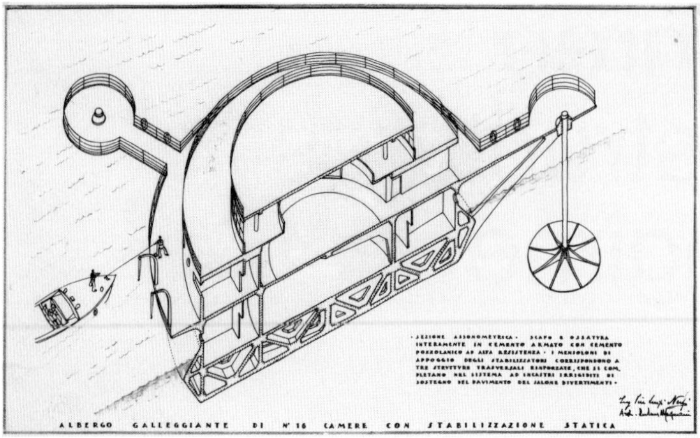

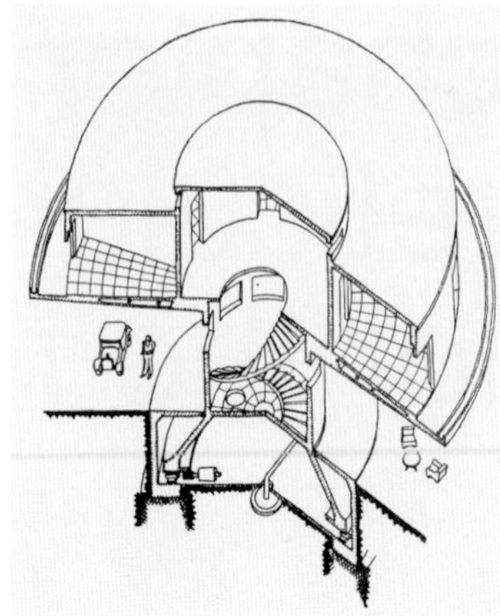

53. P. L. Nervi and R. Magnani, design for a floating hotel, 1932 (Maxxi - PLN, F07301).

54. P. L. Nervi, design for a revolving house, 1934 ("Quadrante," May 1934, No. 13, p. 27).

55. P. L. Nervi, structure for the church of San Marcellino, designed by L. C. Daneri, Via Bologna 8, Genoa, 1934-36 ("Architettura. Rivista del Sindacato nazionale fascista architetti," October 1939, No. 10, p. 627).

56. P. L. Nervi, phosphate salt warehouse, Via Barletta, Margherita di Savoia, 1935-36 (Maxxi - PLN, F08519).

tecture. For the construction of the whole structure, it was necessary to use "high-strength pozzolanic cement"[1] whose hydraulic and physical properties Nervi valued for the manufacture of waterproof mortars that were resistant to the action of sea or sulphated waters. The oscillations produced by wave motion were reduced through the use of tentacle-like extensions with discs submerged at depth in still water and known as "stabilisers." These became the subject of a patent filed by Nervi on 4 March 1932.[2] The floating hotel project and indeed the building of the stadium allow us to discern the fascist policy of ensuring facilities for sports, education, and the health of the people. This is evident in the fact that Nervi and Magnani envisaged mooring the hotel in locations not far from urban centres, to allow inhabitants who were unable to travel for their summer holidays access to a resort. It is impossible to ignore the suspicion that, as with the *Monumento alla Bandiera*, the hotel may have been designed to demonstrate the potential of reinforced concrete for creating floating military devices that could be deployed for coastal defence.

The will to win over the building trade to the special properties of reinforced concrete also led Nervi to imagine, in his 1934 design for a "revolving house" – again with a circular plan – a mechanical device that would make it possible to rotate the reinforced concrete structure according to the position of the sun or the landscape to be admired.[3] The engineer Angelo Invernizzi had built a prototype of this kind of construction with his Villa Girasole at Marcellise, near Verona.

The work carried out by Nervi & Bartoli for the San Marcellino Church in Genoa, designed by Daneri and built between 1934 and 1936, reveals certain distinguishing structural features.[4] Despite their white Finale stone cladding, the tall, slender pillars of the portico reveal from the ratio between height and cross-section that they were erected in reinforced concrete (in response to unrealistic avant-garde assumptions, criticised by Nervi).[5] The thin masonry divided into niches makes the cylinder shape able to withstand the roof's pres-

[1] *Ing. Pierluigi Nervi e Arch. Rubens Magnani: Albergo galleggiante di 16 camere con stabilizzazione statica*, in, "Architettura. Rivista del Sindacato nazionale fascista architetti," XI, May 1932, No. 5, p. 241.
[2] Pier Luigi Nervi, patent IT 305134, *Dispositivo per la stabilizzazione di galleggianti nei riguardi delle oscillazioni prodotte dal moto ondoso*, 4 March 1932, ACS.
[3] *Una casa girevole*, in, "Quadrante," May 1934, No. 13, p. 27.
[4] Pier Luigi Nervi, *Chiesa di San Marcellino a Genova*, in, "Architettura. Rivista del Sindacato nazionale fascista architetti," XIX, October 1939, No. 10, pp. 627-630; APCSM, *Relazione finanziaria*.

sure, and slender reinforced concrete beams with an elliptical profile were created in order to support the floor of the nave. Lastly the 24-metre diameter shell-shaped dome was the "only example" of its kind, according to Nervi.[6] As with the spiral staircases of the *Stadio Berta*, constructing that shell made Nervi aware of the difficulties of preparing formwork for pouring concrete.

After the flying buttress structures designed for some of the *Stadio Berta* stands, Nervi began to take a closer look at the parabolic arch shape that was beginning to attract increasing interest, even from rationalist architects, due to its static qualities. Nervi introduced a sequence – monumental in parts – of parabolic arches in reinforced concrete in order to build (in 1935-36) the structure of a sophisticated salt warehouse for the Monopoli di Stato in the Margherita di Savoia saltworks in the province of Foggia (the salt was dropped from an upper gallery to form parabolic mounds). He explored a structure jointed into three hinges in order to build the arch, as he also did in 1932 for the pedestrian walkway over the Bisenzio in Prato. The design of the skeleton for the warehouse head sections and the facades of the machinery building also reveals a certain quest for expressiveness. The concrete was worked in the rough, as was the norm for Nervi & Bartoli buildings.

Nervi continued to offer his services as a specialist engineer and builder in the stadia that Mussolini intended to build in various cities throughout Italy and in the capital itself. After proposing a steel roof for the stand of the Leghorn stadium, which would have to be clad because he felt that the metal structure could not remain exposed without arousing "a feeling that the construction was excessively utilitarian,"[7] he designed football and athletics stadia for Turin and Rome between 1932 and 1935, which were planned in reinforced concrete.

In his plan for the design of a concourse for the Turin Stadium, the capital D shape became even more evident than in the *Stadio Berta*, made monumental by single enclosed spaces at the apices of the straight bar framing the main entrance.[8] The design demonstrated all the figurative potential of

[5] Nervi, *Scienza o arte del costruire?* cit., p. 12.
[6] Pier Luigi Nervi, *Aesthetics and Technology in Building. The Charles Eliot Norton Lectures, 1961-1962*, Cambridge, Massachusetts, Harvard University Press, 1965, p. 102.
[7] Nervi, *Considerazioni tecniche e costruttive sulle gradinate e pensiline per stadi* cit., p. 12.
[8] *Ing. Pier Luigi Nervi. – Progetto di stadio per 50,000 spettatori*, in, "Architettura. Rivista del Sindacato nazionale fascista architetti," XII, 1933, special issue, V Triennale di Milano, p. 166.

57-59. P. L. Nervi, competition design for Turin stadium, 1932.

57. "Stadio per 50.000 spettatori. Prospetto sul fianco (Stadium for 50.000 spectators. Side elevation view)" (Maxxi - PLN, F07290).

58. View of the stand from the playing field (Maxxi - PLN, F07289).

reinforced concrete that had not been fully expressed in the Florentine work. From the outset, Nervi imagined the structure of the covered and uncovered stand tiers without infill, allowing the stark lines to define the aesthetics of the fascist stadium and unwaveringly proposing a view of architecture as an expression of the static strength of the new construction system within the context of the contemporary debate over architecture and engineering so dear to Bardi. The design for the Turin stadium ultimately took on all the connotations of a manifesto for modern architecture, or fascist style, reduced to reinforced concrete alone. The lack of infill meant that Nervi designed the trestles supporting the tiers as powerful strips capable of creating monumental arcades from a perspective view, while their mighty bars were variously angled to allow an excellent view of the playing field from the stands. Nervi achieved the apotheosis of the new structure's static power and creative force in the bold overhead tier that runs continuously over the uncovered stands, poised on a single row of angled pillars, where the spectators raised to the sky would have the intoxicating aesthetic experience of hanging vertiginously above the crowd – hovering in space on an overhead platform in reinforced concrete instead of staring in amazement at a canopy projected into space. Nervi was now able to take his place as a brilliant choirmaster whipping up the vital intoxication and exaltation that the regime intended to instill in its people. His structures informed the feelings of the crowd by acting on the perception of the spectators' lived experience through a sophisticated rhetoric without visible signs.

Nervi came up with a variant of the *Stadio Berta*'s cantilever beams in order to correct the bracing configuration, and the entire system became an artificial mechanism. The structure arose out of a static approach that immediately incorporated the bracing. This laid the ground for a fork-shaped bifurcation with an angled post converging into its node and giving rise to an overhanging beam coplanar to the forks. Two V-shaped ties were attached to the ends of the fork to anchor the weight of the roof. This entire interplay of structures leaping into space and of counterweights, effected by tall, slender beams, straddled a series of archaic portal frames that stood beneath the acrobatic shapes like ungainly Telamons. This contrast of shapes was to become one of the features of Nervi's bold and intuitive creativity.

When the reinforced concrete structure had to be moulded for a stadium intended to celebrate sports in the capital city, and thereby interpret the ex-

59. "Stadio per 50.000 spettatori. Vista della 2ª tribuna (Stadium for 50.000 spectators. View of the second stand)" (Maxxi - PLN, F07288).

60-61. P. L. Nervi and C. Valle, competition design for the *Stadio Massimo,* Rome, 1933-35.

60. Model (BSTA, 024-7589).

61. Perspective drawing of the covered stand (BSTA, 024-7592).

pressive values of the regime's rhetoric, Nervi drew lines ubiquitously reflecting veiled memories of ancient stadia. This device imbued the structure with an unmistakable aura of *romanità* that neither the *Stadio Berta* nor the design for the Turin Stadium possessed, apart from in occasional glimpses. Nervi increasingly showed himself capable of reflecting the individual "nature of form" of his structures, determined by the type of programme and the desired level of grandeur required of the work.

The design for the stadium in Rome, again drawn up with Valle, and ambiguously referred to as *Stadio Massimo* due to its dimensions and historical impact, marked the apotheosis of *romanità* in the development of Nervi's idea of structure. This was evident in the elliptical plan, designed as an unmistakable homage to the Coliseum, with its three tiers of galleries. This created a spectacular capacity that exceeded the expectations of the programme (150,000 spectators instead of the planned 100,000) and also achieved an air of majesty due to the ostentatious rhythm of pillars, soaring up to the top and ending in a curve, as in the pillars built in the *Stadio Berta*'s curved stand incorporating entrances. These now encircled the entire work to form a new reinforced concrete level, embodying the majesty of Roman amphitheatres. The fact that the large pillars are the only visible part of an infilled structure is significant in terms of achieving the effects of substantial volume and monumental infill that Nervi desired, in contrast to the transparency – with all lines visible – that he envisaged as a feature of the stadium in Florence and then put into practice in the Turin stadium. The disappearance of any roofing helps to determine the "nature of form" of the structure for the *Stadio Massimo*. As confirmation of its *romanità*, it was planned to clad the reinforced concrete and infills with travertine marble slabs at base level and "unalterable light plaster,"[9] with the aim of making the other surfaces look like stone. However, as Nervi was preaching in his articles, from the stand-

[9] Agnoldomenico Pica, *Progetto per uno stadio a Roma*, in, "Architettura. Rivista del Sindacato Nazionale Fascista Architetti," XII, November 1933, No. 11, (pp. 705-712), p. 705.

62. P. L. Nervi, design for a bridge over the Biedano stream, Viterbo, 1934. Model (*The Works of Pier Luigi Nervi*, London, The Architectural Press, 1957, p. 19).

point of a theoretician of a reinforced concrete-centred view of architecture, this was not the "true Architecture"[10] – a form of *romanità* achieved with masking columns and arches. Instead it was an intrinsic shape conferred on the structure itself, sublimated in its lines and cadences to the point of suggesting engineering harnessed for the purposes of a regime aiming to restore the glories of the Roman Empire through the design of the stadium. Due to its unmistakable fascist rhetoric, the work is symptomatic of the imponderable margins that Nervi decided to instill in his concept of "form" because this still had an "emotional origin" for him at this particular stage of his structural creativity.

In order to appreciate the decisive departure from the structures designed for Florence and Turin and thus understand the specificity of the "form" of the *Stadio Massimo*, we need to pay attention to the few words Nervi expended to explain the main features of this design. He believed it important to stress the "structural simplicity" and the "balance of dimensions" that he sought "between the various parts," which are formal features antithetical to those that typified the other two stadia.[11] Nervi himself offered the key to appreciating the rhetorical significance of his structure, when he admitted with nationalistic pride that he had taken contemporary fascist taste as "an idea of emotional origin" for the "nature of form" or the "nature of *romanità*" which, as he wrote,

[10] Nervi, *Problemi della realizzazione architettonica* cit., p. 2.

[11] Pier Luigi Nervi, *Idee sulla costruzione di uno stadio per 120.000*, in, "Quadrante," August 1933, No. 4, p. 36.

[12] *Ibid.*

"may be summarised in the more rigorous uniformity and symmetry of the whole and the imposing impact of a single rhythmic sequence of very stark structural elements."[12]

The symbolism that breathed life into Nervi's structures in the mid-1930s is confirmed by his 1934 design for a viaduct over the Biedano stream near Viterbo. For this he created two slender trusses of right-angled mesh leaning against one another as in a house of cards, with pillars that look like cut-off strips with ribbing on the back (reminiscent of the cantilever beams of the *Stadio Berta*). These are overlapping and graceful, conjuring an image of a Roman aqueduct reduced to a skeleton, almost as though Nervi were again taking sides in the debate that set reinforced concrete structures against columns and arches. The downward tapering of the pillars in the three tiers forces each beam to act as base strip for the vertical reinforcement while also contributing to the structural slenderness that is evident across the trusses. The viaduct's right-angled bar geometry should also be seen as a device for avoiding having to erect grand centrings to construct the arches.

Designs for aircraft hangars with an expressive nature

Nervi continued to enthuse about exploring the possibilities of applying reinforced concrete to various kinds of structures, in increasing rivalry with other material, particularly steel, in the wake of his success with building sheds, tanks, and stadia. A decisive chapter in the rivalry between concrete and steel opened up for Nervi, again during the first half of the 1930s. From that time, a kind of geometry began to evolve that was destined to undergo significant changes and result in a structure of absolute technical, formal, and symbolic originality that was to distinguish Nervi's output during the 1940s and 1950s. Yet again, the special geometry of his structure owed much to the regime's drive to create public works that needed to possess a clear aesthetic and rhetorical value for the national propaganda of a modern fascist state.

After celebrating the many faces of the regime with the overhead structures of the Florence and Turin stadia and the imposing structure of the Roman stadium, Nervi and his firm became involved in building some of the emblems of the other side of a regime that was now aiming for military expansion to carve out a new empire. Nervi undertook the construction of some works for the Ministry of War against the backdrop of the international political scenario marked by war in Ethiopia, which broke out in October 1935, and the signing of the military alliances between Italy, Germany, and Japan in November 1936. These works were destined to establish themselves as symbols of the regime's military might and were masterpieces of reinforced concrete engineering. The significance of the structures designed for the works leading up to those masterpieces begged the question of whether or not the project had been guided by a "first fundamental idea,"[13] as confessed by Nervi with regard to the *Stadio Massimo*. It should not be forgotten that in 1934, Nervi issued an exhortation to "*think* like an engineer," in other words to come up with a "synthetic overview of the work" and then do the structural calculations.[14] He also reiterated that the key ingredients for "doing architecture" included, first and foremost, having "an idea" originating in "poetic inspiration" and then the "ability to express it and translate it into bodies that

[13] *Ibid.*
[14] Pier Luigi Nervi, *La riforma delle scuole di ingegneria e il prestigio dell'ingegnere*, in, "Il Lavoro Fascista," VIII, 4 January 1934, p. 4.

63. P. L. Nervi, patent IT 304282, *Capannone circolare per aeroplani o veicoli in genere con piattaforma anulare rotante*, 8 January 1932 (ACS).

64. P. L. Nervi and C. Valle, design for a square hangar, 1933 ("Architettura. Rivista del Sindacato nazionale fascista architetti," XII, 1933, special issue, V Triennale di Milano, p. 167).

65. P. L. Nervi and C. Valle, design for a round hangar, 1933 (BSTA, 024-6006).

obey the laws of technique (stability, economy, and durability) while simultaneously achieving the practical purposes for which the work is created." For the latter, he invoked a "perfect knowledge and absolute mastery of construction techniques."[15]

The works that the regime ordered to improve its military hardware, and which Nervi worked on from the beginning of the 1930s, were hangars housing Italian Royal Air Force aircraft. The first models were built in Orvieto and then replicated in other strategic locations. This was the beginning of research into the configuration of structures that could cover increasingly large spaces. The work was destined to bring about major developments in the output of Nervi & Bartoli, not merely in the designs of the structures themselves but also in the way of using concrete, which was cast in place or prefabricated.

The beginning of Nervi's new interest in military structures dates back to January 1932, when he filed a patent for a device that would improve the functioning of hangars by means of a turntable to make it easier to manoeuvre the aircraft.[16] Because the aircraft were positioned in front of the large opening through rotation of the turntable, the patented device did not include a cantilever structure along the perimeter ring designed to create a mobile wall that could be opened at any point, since the text described a circular or polygonal roof that was wholly cantilevered on pillars in the centre of the hangar. Nervi's intention to create a practical device, the turntable, is evident, alongside his desire to explore the possibilities of a cantilever roof to achieve large openings for the aircraft. At this stage of his studies, Nervi seems not to have called into question the most popular materials for hangar construction, such as iron and wood, although he mentioned among the possibilities for achieving a cantilever structure a type that had become established following recent German experiments conducted from the mid 1920s by Franz Dischinger and Ulrich Finsterwalder, namely "thin reinforced concrete vaults."[17] This detail is very important for understanding what led Nervi down a path that would result in him inventing his own type of thin vault.

One of the first outcomes of Nervi's studies into the structure of hangar roofs was a schematic design for a hangar, again with a circular plan, drawn up together with Valle in 1933. It shows a jutting roof around the entire perimeter, as already foreshadowed in the patent. Two variants exist of the roof: one with a metal structure and the other in reinforced concrete (variants with diameters of 57, 67.50, and 70.80 metres were explored).[18] The matter of the "thin reinforced concrete vaults" was now couched in original terms, different from those of the German techniques. The engineer who had trained in Muggia's firm, which held the licence for the Hennebique skeleton, continued to work in terms of a structure of ribs and infills along the same lines as the roof of the *Stadio Berta* and the other skeletons created for the stadium. An impression of strength and the ability to withstand bombardment was therefore made manifest in a mighty truss branching from the centre of the hangar, where it assumed a shape similar to that of the trestles of the stadium stands, before branching out into diagonal grids and long beams that reached the round perimeter beam. The radial geometry of the grid explained Nervi's subsequent

[15] P.L.N. [Pier Luigi Nervi], *Corsivo No. 134*, in, "Quadrante," August-September 1934, No. 16-17, p. 43.
[16] Pier Luigi Nervi, patent IT 304282, *Capannone circolare per aeroplani o veicoli in genere con piattaforma anulare rotante*, 8 January 1932, ACS.
[17] *Ibid.*, p. 2.
[18] *Ing. Pier Luigi Nervi e Cesare Valle. – Progetto di aeroporto per squadriglia civile di turismo*, in, "Architettura. Rivista del Sindacato nazionale fascista architetti," XII, 1933, special issue, V Triennale di Milano, p. 167.

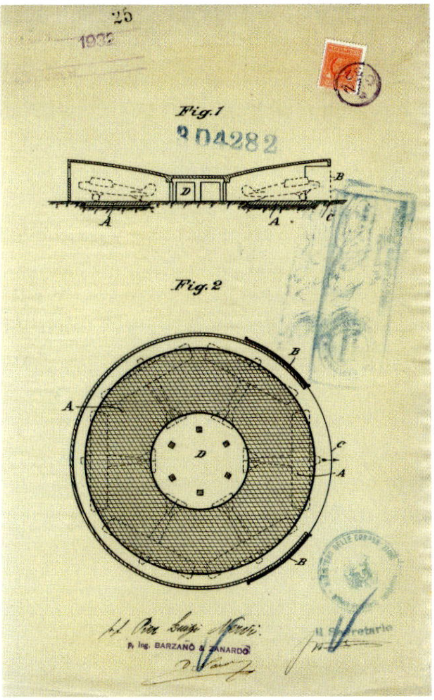
Fig.1
Fig.2
P. Ing. BARZANÒ & ZANARDO

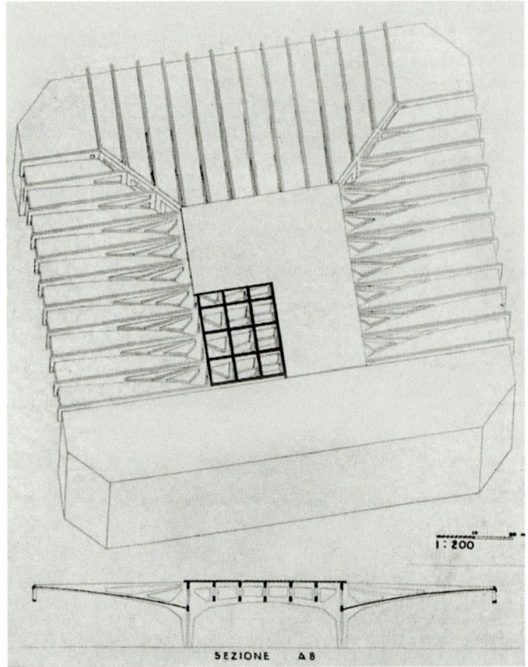
1 : 200
SEZIONE AB

66-68. P. L. Nervi, design for a hangar, Ciampino airport, 1934-35.

66. "Aviorimessa in cemento armato per l'aeroporto di Ciampino (Reinforced concrete hangar for Ciampino airport)." Option with three-hinged arch (CSAC, n. inv. pra 7, n. id. 12610, coll. 270/4).

67. Cross-section (CSAC, n. inv. pra 7, n. id. 12610, coll. 270/4).

68. "Schizzo prospettico con variante nella parete posteriore (Perspective sketch with variant in rear wall)" (CSAC, n. inv. pra 7, n. id. 12610, coll. 270/4).

69. P. L. Nervi, design for a covered swimming pool, about 1933 (Maxxi - PLN, F07276).

references to Gothic umbrella vaults. Other projects for square hangars by Nervi and Valle followed the same pattern, with the aircraft sheltered by bundles of overhanging beams that branched out from a dense grid of Vierendeel girders, as in contemporary monumental buildings erected by the regime (a similar structure was built in the Accademia Superiore Fascista di Educazione Fisica). The structure of the hangars, like those of the stadia in Florence, Turin, and Rome, possessed a specific iconic value, and it was impossible to escape the impression that the virtuoso shapes were generated with the idea of creating an iconic and mighty state-of-the-art war machine.

Even elsewhere, where the hangars were designed with more conventional structural geometries, certain traits nevertheless revealed hints of Nervi's experimentalism. The 1935 design of a "Hangar and services for an aerial reconnaissance squadron" (*Aviorimessa e servizi per una squadriglia O[sservazione]. A[erea].*) is significant, because the roof of the massive area for protecting the aircraft features an arched structure with three hinges, but Nervi planned a "reinforcement crossed and constructed so that the joists are angled at 45° to the arch centreline" for the curved slab.[19]

Studies for rectangular hangars at the military airport of Ciampino, which were implemented in 1934–35, demonstrate Nervi's creative drive and his determination to invent a device made out of reinforced concrete that could compete with the normal steel structures – avoiding, under all circumstances, any type of shell with a continuous surface after an initial proposal. Nervi firstly proposed a reinforced concrete skeleton that developed into depressed arches spaced 3.6 metres apart, to be covered by a Danusso floor.

Subsequent study sketches for the Ciampino hangars bear witness to his various solutions for weaving together arched truss girders, almost all asymmetrical, in order to fully open up one side for the movement of aircraft. Some cases saw the re-emergence of his idea of an extrados truss girder for a thin vault implemented in the *Gioco della Pelota* roof. In some versions, the structure became a sequence of three-hinged arches with an asymmetrical parabolic profile that was stiffened by beams to create a geometrical overhead truss to be infilled with panels and concrete asbestos cladding (*Eternit*). Nervi tended toward the option of an asymmetrical roof that allowed a large opening on the long side of the hangar. It was created using a few pillars spaced back from doors that slid closed on the cantilever truss girders. Sometimes the pillars were shaped like huge, archaic trellis-work, of the same type as the

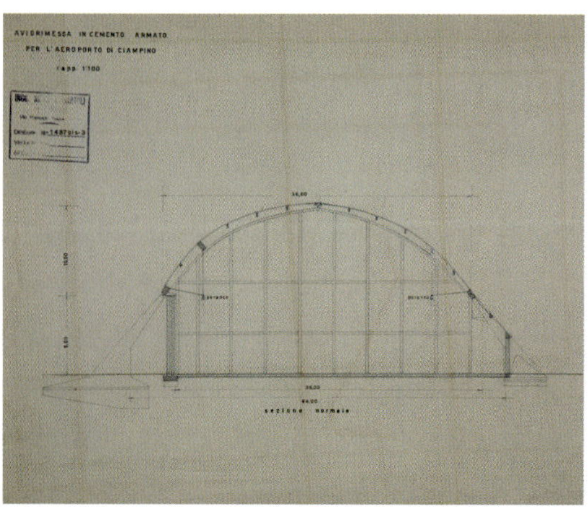

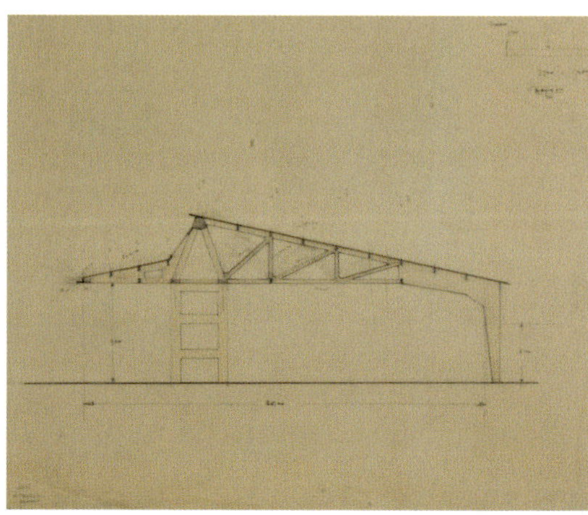

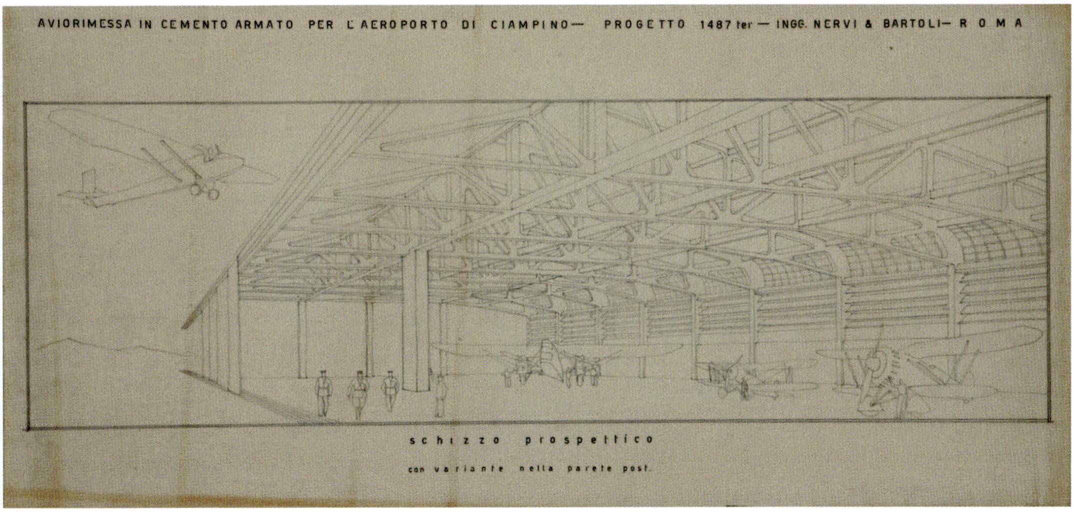

portal frame created in the canopy for the Turin Stadium, with the slender-barred roof truss girders resting on them. In the perspective view of the final version, the truss girder grid highlights the cutting-edge features of the reinforced concrete shelter for the Italian Royal Air Force. The draftsman also depicted *Il Duce* posing proudly during his visit to the hangar. He was followed by two soldiers and greeted with an extended arm salute by a military aircraft maintenance attendant.

The designs for the branched skeleton of the circular hangar with its slab with diagonal ribs and the various grid patterns for the Ciampino roofs lay the foundations for defining the kind of symbolic reticular structure for hangars of the regime that Nervi had been seeking since the early 1930s. The fact that the proposed roofs always included a bundle of exposed ribs reminiscent of *fasces* can be explained by the value that Nervi attached to that particular form of reticular structure. To fully grasp the extent to which this value was rooted in his entirely lyrical idea of eloquent structure, it is enough to read some of his statements about the "particular expressive nature" of any system of static forces:

"The same problem may almost always be resolved in different ways. Arches, straight girders, simple or complex reticular structures can all equally be used to achieve a capacity and support loads, but each of those solutions has its own specific intrinsic character that must be deeply felt in order to reach the result you wish to achieve. The evidence of effort in the ribs of a reticular structure can, in certain constructions, achieve a quality of dynamism and lightness in perfect harmony with the whole work, while in others, that are conceived in a different spirit, an irreconcilable conflict would result. It may be added that while some static balances are in themselves harmonious, other equally possible balances give a feeling of effort or of unnatural balance that no decorative or superficial artifice can remove."[20]

[19] *Aviorimessa e servizi per una squadriglia O.A. Calcoli di stabilità*, p. 2, CSAC, n. inv. pra 40, n. id. 12643, coll. 270/1.
[20] Nervi, *Arte e tecnica del costruire* cit.

The Orvieto hangars or the power of hyperstatic *fasces*

The work of Nervi and his firm Nervi & Bartoli became famous during the fascist regime, in his new guise as an engineer and constructor of military works, due to the first two hangars he built in Orvieto between 1935 and 1938 on behalf of the Italian Royal Air Force. Nervi & Bartoli was to continue building other military constructions, while Nervi also went on to study the application of cement to the manufacture of floating bombs. There is no doubt that the structure designed for Orvieto should be seen as the logical outcome of the studies on hangars undertaken for the Ciampino military airport designs. In order to truly grasp the expressive value of the structure, we must not forget the cultural and political climate behind the project.

It is no exaggeration to say that the nature and meaning of Nervi's structure for the hangars were foreshadowed by events such as the Esposizione Aeronautica Italiana, held in 1934 at the Milan Triennale building. The dynamic and modernist face of the regime and its military campaigns were celebrated with parabolic arches plastered with sheets of writing and photographs to make them even more representative, staged by Luciano Baldessari in his arrangement of the room devoted to *Aviazione e Fascismo* (Aviation and Fascism). What turned out to be the symbolic space of Nervi's hangars was identified by Baldessari's arrangements as well as, very aptly, by the presence of a *fasces* of lozenge-shaped strips among the artistic fragments of infill trusses between the arches. These prefigure the very meaning of a structural pattern that was apparently designed to pay homage to the *fasces* insignia. Libera had also designed a roof with a bundle or *fasces* of lines in one of the versions of his reinforced concrete structures, as proposed by Baldessari, but intended to create a barrel vault structure.[21] Terragni had suggested a truss in exposed reinforced concrete moulded into a parabolic vault as the roof for a hangar.[22]

As part of his research into a rational and symbolic structure for fascism, Nervi invented a reinforced concrete grid for the Orvieto hangars that represented the contemporary potential of aviation. Exactly as the quality of *romanità* expressed by the *Stadio Massimo* resides in the symmetrical rhythm of the strong, spare lines of its structure, in the same way the military might of the regime was expressed by the dense and ordered grid of ribs, with each part identical and combined in a single sturdy *fasces*.

Nervi's intuitive imagination was therefore driven by a set of technical, cultural, static, figurative, constructive, and poetic inspirations toward his decision to use a diamond-shaped grid geometry for the vaulted roofs, as opposed to systems of trusses and purlins. From 1931, Gaetano Minnucci had documented the kind of diamond-shaped mesh "trusses" or "tissues" that were very popular in Germany, obtained by installing wooden boards or light, prefabricated steel lamellae to create the vaulted roofs of buildings used for different purposes, namely exhibition buildings, church naves and even hangars – the latter built using a system designed by Hugo Junkers involving steel lamellae.[23] He concluded that "this type of roof undeniably has deep meaning from an aesthetic point of view."[24] The company Costruzioni Industriali Metalliche built a roof out of steel lamellae for a hangar at the Venice Lido and this emphasised the "aesthetic value" of the "pleasing design" and the "decorative" effect of the grid pattern.[25]

[21] *Progetto di atrio di stazione, 1927-1928*, in, *Adalberto Libera. Opera Completa*, Milano, Electa, 1989, p. 128.

[22] See Claudio Poli, *1930-1931. Progetto di hangar per idrovolanti a Como*, in, Giorgio Ciucci, ed., *Giuseppe Terragni. Opera completa*, Milano, Electa, 1996, p. 355.

[23] Gaetano Minnucci, *Notiziario tecnico. Coperture a lamelle,*

70. Società Costruzioni Industriali Metalliche, steel hangar, Venice Lido, 1933 ("Casabella," August-September 1933, No. 8-9, p. 51).

71. E. Torroja y Miret with S. Zuazo, roof *Frontón Recoletos* playing field, calle Villanueva, Madrid, about 1935. Demolished (property of the authors).

Nervi himself revealed the source of his structure when, in order to explain his works in Orvieto, he referred to the "iron hangars" and was able to offer a version that could assuage the fascist regime's concerns about saving steel by using his favourite material.[26] His decision to transfer the technique employed for diamond-shaped mesh trusses made out of timber boards or steel lamellae – valued for cheapness, lightness, speed of assembly, and, not least, the aesthetic value of its design – to the technique of cast-in-place reinforced concrete indicates the significance of his structure. Nervi's attraction to a diamond-shaped mesh truss was virtually foreshadowed by the grid of beams in the *Stadio Berta* staircases. The other feature of a timber or steel truss, namely that it is obtained by assembling prefabricated parts with nodes that are particularly sturdy because they are destined to become the points onto which the stresses converge, contains the premises for the development of Nervi's structure toward the precast reinforced concrete that Nervi & Bartoli was to develop immediately following completion of the first hangars.

Right in 1935, in Madrid, another engineer, Eduardo Torroja y Miret, used a similar cast-in-place reinforced concrete spatial grid for the structural frame of the openings obtained in the thin-shell vault of the *Frontón Recoletos* playing field.

The roof of the Orvieto hangars was strategically configured in the form of a pavilion vault instead of the barrel vault built at the Venice Lido, in order to reduce the shadow cast and therefore make the hangars less visible to enemy aircraft as well as attenuating the impact of wind loads, as Nervi himself explained.[27] The diamond-shaped mesh truss, altogether similar to that of Torroja, did not develop from the foundation beam, as in the metal lamellar structure, but was raised on rows of angled posts in reinforced concrete, known as spurs and arranged along the perimeter (Nervi in any case examined a solution with the grid extended to the ground). The pavilion-shaped vault prompted the solution of a grid raised off the ground to obtain a symmetrical reticular structural geometry while at the same time meaning that the aircraft no longer entered at the head, as in the Venice Lido hangars, but on one long side, where Nervi created an exceptional truss girder supported at the entrance and centre on colossal spurs, again in reinforced concrete. Yet the three spurs with the truss girder of the large opening, oppos-

in, "Architettura e Arti Decorative. Rivista d'Arte e di Storia. Organo del Sindacato Nazionale Architetti," X, April 1931, file VIII, pp. 408-412.

[24] *Ibid*, p. 412.

[25] *Le lamelle di acciaio,* in, "Casabella," VI, August-September 1933, No. 8-9, (pp. 50-51), p. 50.

[26] Pier Luigi Nervi, *Un'aviorimessa di cemento armato,* in, "Casabella-Costruzioni," X, April 1938, No. 124, pp. 4-9.

[27] *Ibid.*, p. 4.

72-73. P. L. Nervi, design for a hangar, 1935.

72. "Aviorimessa in cemento armato (Reinforced concrete hangar)." Cross-section (CSAC, n. inv. pra 7, n. id. 12610, coll. 270/4).

73. Option with structural grid extending to the ground (BSTA, 024-7326).

ing the sequence of twenty-one smaller spurs, generated a differential static behaviour in the symmetrical meshes of the structure. The asymmetry of the supports gave rise to a complex load distribution in the first hangars, which Nervi subsequently corrected.

The diamond-shaped mesh hangar grid meant that the reinforced concrete assumed hitherto unknown configurations and sizes. The design of the structure became a continuous grid with an original plastic strength. As in the case of the various structures of the *Stadio Berta*, it was able to give shape to the regime's aspirations for an avant-garde image that was now also autarkic. Once again placing his trust in the idea of overcoming heat expansion by using protective coatings instead of expansion joints, Nervi made his lamellar structure independent of the roof coating, the latter being intended to protect the grid of bars from temperature fluctuations.[28] The recognisable look conferred by the diamond-shaped mesh truss geometry ultimately also informed the design of the corner spur bars housing the hangar portal frame panels – another triangular truss.

During the process of designing the lamellar structure, Nervi did not neglect the building site and construction stage. His thoughts about how to construct a hangar that still had a three-hinged arch configuration may have led him to change the geometry of the structure. Once building work was complete on the first hangar, Nervi explained that the qualities of a diamond-shaped mesh lamellar truss constituted an advantage in a monolithic reinforced concrete construction: that of achieving "better distribution

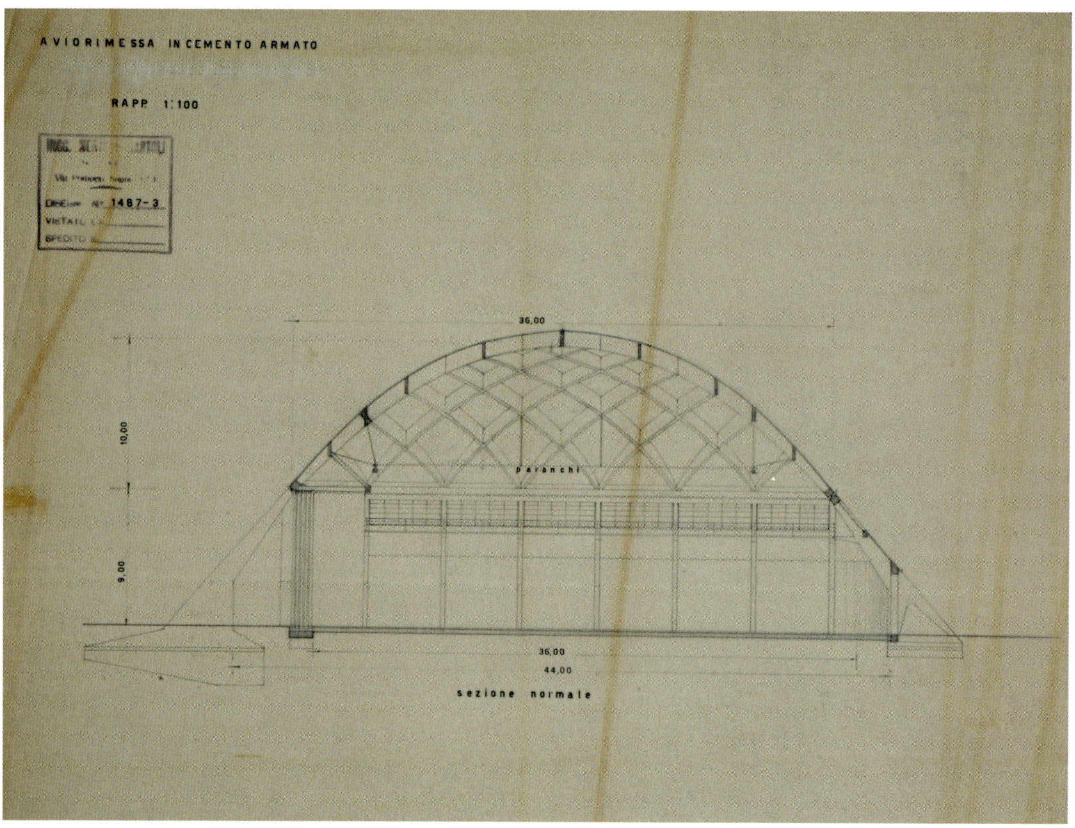

of internal stresses, involving as many structural elements as possible in the resistance."[29] From then on, his main roofs for large spaces were to be based on the same principle, which offered the possibility of dividing the structure into small parts to be prefabricated and installed to obtain a huge, strong, monolithic unit. In 1938, Nervi used the description "reticular structure," already used for the *Stadio Berta* grandstand, to describe this type of lamellar truss.[30]

The invention of the reticular structure of the hangars marked a fundamental chapter in the development of the concept and the shape of the reinforced concrete skeleton. It entered the universe of hyperstatic structure with authority due to those works and by virtue of its exceptional aesthetic appeal. The aesthetic appeal of the work is crucial to an understanding of the original creative and formal drive that continued to guide Nervi in the design of his structures; he always exceeded the permissible limits according to contemporary technical canons. The drawback of the reticular structure of the hangars lies in its hyperstaticity, which prevented any expansion movement due to heat stress, giving rise to states of internal constraint and tensions between the reinforced concrete structure and the brick covering. This was the stumbling block that made Nervi's invention a transitional stage in a process of research that was still ongoing and aimed at overcoming such limits.

Viewed from the perspective of the scientific implications involved in using reinforced concrete to build the hyperstatic reticular structure of the hangars, the roof and spiral staircases of the *Stadio Berta* appear as virtuoso

[28] *Ibid.*, p. 6.
[29] *Ibid.*, p. 4.
[30] *Ibid.*, p. 5 (captions to the photographs).

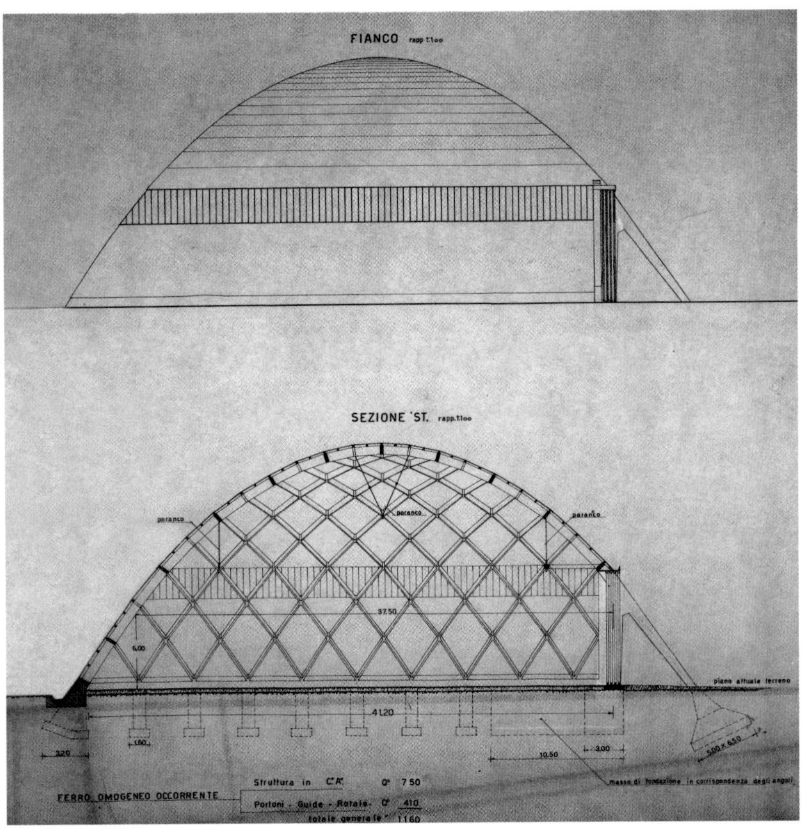

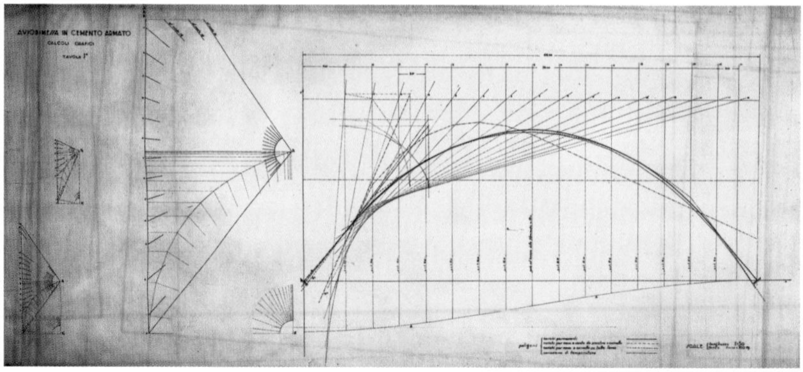

74. P. L. Nervi, design for a hangar, 1935. "Aviorimessa in cemento armato. Calcoli grafici (Reinforced concrete hangar. Graphic calculations)" (BSTA, 024-7333).

examples of an aesthetic and a symbolism that had not yet generated a founding principle of a future vision, since they were still dependent on "an idea of emotional origin."

Mathematical confirmation of the validity of a particular structure was sought not by calculation, but by using a new verification tool that was unique for the time. This involved being able to evaluate with scientific accuracy the behaviour of a hyperstatic structure through direct observation of its physical state using a model. In this way, after the initial calculations, which were approximate and difficult because of the diamond-shaped grid and its hyperstaticity, the static properties of the Orvieto hangar reticular structure were verified using an innovative method that was destined to be applied in the design and construction process of Nervi's future structures. Through his friendship with Danusso, Nervi was able to make use of the Milan Polytechnic Testing and Model Laboratory. There, in 1935-1936, Danusso and his colleague Guido Oberti produced a celluloid 1:37.5 model of the reticular structure with spurs that they could subject to stress using strain gauges and calibrated weights attached to wires and positioned at the nodes.[31] Nervi personally monitored the testing, which lasted for several months, at the end of which the original structural configuration was approved.

Nervi's action in turning to Danusso and Oberti's laboratory became the foundation of his original scientific method in the field of building science, which was then dominated by mathematics. Through this, he elevated his static insight about form, albeit hyperstatic and not fully verifiable by calculation, to become the cornerstone of his approach to structural design. The partnership between Nervi and the Danusso and Oberti laboratory, for the experimental checking of hyperstatic form using a model subjected to physical stresses similar to the real ones, added creative and scientific authority to a line of research that was still latent in 20th century experiments on new materials and new figures (we need only think of the tests carried out on models by Dischinger and Finsterwalder, or by Torroja).

Work on the Orvieto building site began in 1935. In 1938, by the time the work ended, Nervi had erected two hangars using the same construction system. Over the next few years, these prototypes were destined to lead to a new type of hangar made up of prefabricated elements. Nervi & Bartoli accumulated experience in manufacturing the formwork and reinforced concrete that proved crucial to the realisation of Nervi's masterpieces up to the early 1960s.[32] For the construction of the grand pavilion vault, the firm used a centring made out of timber boards. This was not set on a typical timber superstructure, but on a scaffolding of metal posts that had recently been designed by the Roman firm of Fratelli Innocenti.

Even though they were made out of cast-in-place reinforced concrete, the first two hangars possessed certain specific traits of the prefabrication identified as one of the characteristics of a diamond-shaped mesh truss in timber or steel. Because he was inspired by the metal lamellae, Nervi transferred the prefabrication criteria to the construction of the most complex work on his project building site, the timber formwork. He therefore prepared all the timber side planks necessary to manufacture the reinforced concrete lamellae of the diamond-shaped mesh. The side planks were fitted along timber guides arranged above the sparse wooden boards of the centring to identify the interwoven lines of the diamond-shaped mesh (the thickness of the lamellae to be built was nine centimetres, and they were approximately one metre high). The decision to prefabricate the side timbers of the formwork instead of the lamellae themselves can be explained by the difficulty of achieving particularly strong nodes between prefabricated lamellae. The discovery of a technical solution for the construction of those nodes was the breakthrough that made it possible to change over to the direct prefabrication of the structural lamellae in the second series of hangars.

The type of concrete used by Nervi & Bartoli was *supercemento*. Once the grid structure was complete, a floor of Perret reinforced perforated brick blocks was laid, and this was covered by waved concrete asbestos slabs. All the vault material was installed without coatings or paintwork.

The building site of the first hangar became the subject of a photo shoot, commissioned from Vasari's studio, which also took the photographs of the completed work. Nervi advertised his exposed grid of reinforced concrete ribs forming a reticular structure, enhanced by Vasari's symmetrical shots, as the ideal means of withstanding aerial bombardment.

The geometry of the grid made up of reinforced concrete lamellae, joining to form a pavilion that was as graceful as the aircraft it sheltered and held on the ground by spurs, generated an idea of space that was embodied in the very action of structural invention. The space depicted in Vasari's photographs is defined by a tissue of lines but went beyond Nervi's original intention to render the space proportionate and illusory. Despite the order imposed by the repetition of a few identical modules, the structure was also elusive because of the way it ceaselessly reverberated with no stable centre or defined perspective. Nervi's reticular structure gives substance to lines that are no longer bound by a historical view of architecture, even though they are based on exposed ribs reminiscent of Gothic vaults, Hennebique's right-angled mesh frames, or Anatole De Baudot's flowing bundles of ribs. After the *Stadio Berta*, Orvieto's hangars offered the regime and the debate new evidence of lyricism in modern structure. His scientific verification, his programmed hyperstaticity, the strength of his nodes and his corners, and the indissoluble unity of his *fasces* of lamellae ensured that Nervi's grid was becoming established as a means of overcoming any conventional structure and assumed, in the current debate, the same formal and cultural authority

[31] Gabriele Neri, *Capolavori in miniatura. Pier Luigi Nervi e la modellazione strutturale*, Mendrisio, Mendrisio Academy Press – Silvana Editoriale, 2014, pp. 37-43.

[32] Alberto Bologna, interview with Vittorio Nervi, 10 October 2011.

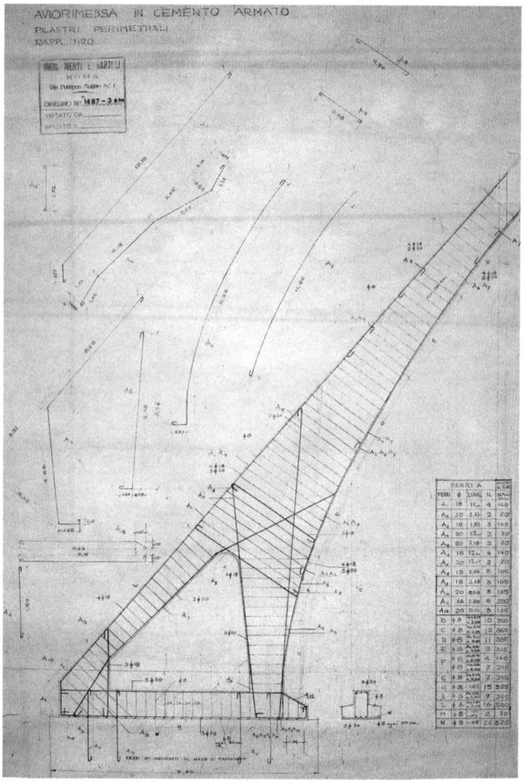

75-79. P. L. Nervi, hangar, Orvieto military airport, about 1935.

75. "Aviorimessa in cemento armato. Pilastri perimetrali (Reinforced concrete hangar. Perimeter pillars)," 1935. Reinforcement drawing (BSTA, 024-7316).

76. Building site (photograph by T. and G. Vasari; Maxxi - PLN, F04209).

77. Model built in the *Politecnico di Milano* Testing and Models Laboratory, 1935 (Maxxi - PLN, F05383).

78. Building site (photograph by T. and G. Vasari; Maxxi - PLN, F04210).

79. View (photograph by T. and G. Vasari; Maxxi - PLN, F04268).

80-81. P. L. Nervi, hangar, Orvieto military airport, about 1935.

80. View (photograph by T. and G. Vasari; Acm-EPFL, 0172.02.0059.0353).

81. View (photograph by T. and G. Vasari; Acm-EPFL, 0172.02.0059.0354).

82-83. P. L. Nervi, water tank for Società Solvay, Via Piave 6, Rosignano Marittimo, 1937.

82. Interior (Maxxi - PLN, F08537).

83. Exterior (Maxxi - PLN, F08533).

as the grid of isostatic lines taught by Oberti and intuitively assumed by Terragni as the cornerstone of the political and cultural meaning underpinning his most recent architecture. It was no coincidence that, at the end of the 1940s, Nervi became heir to Terragni's intuition about the creative potential of the isostatic lines that were to effectively replace the scientific meaning of a grid of lamellae.

A few months after completion of work on the first hangar, probably in 1937, some of the perforated brick blocks began to become detached due to thermal expansion, and Nervi was forced to accept the groundlessness of the beliefs he gained during his examination of Brunelleschi's dome, namely the possibility of counteracting such phenomena with roof coverings. He therefore had to add an expansion joint along the floor of the hangar.[33]

The construction of the roof grid took the use of cast-in-place reinforced concrete to a level of executive complexity that transcended the principles of economy and rules of execution normally applied by Nervi. After the symbolic grid of the *Stadio Berta* spiral staircase, Nervi continued to stress his formal taste for a structural geometry that makes the pouring contrived, if judged by the parameters of an approach that takes into account building site constraints, but which becomes, precisely because of that bundle of lines and its aesthetics, highly symbolic of a desire to appear at the forefront of technology against the backdrop of Italy's imminent entry into the war.

The construction of the first hangars did not close the chapter of Nervi's research into the dynamics of hyperstatic structure. The switch to prefabrication was an action that also affected the very nature of reinforced concrete. It was Nervi's intention, at least according to statements he made in the mid-1940s, to drastically reduce all the unpredictable phenomena due to atmospheric and compaction conditions when pouring in situ, which gave rise to different elastic moduluses in the various sections of a structure.

Underground anti-aircraft tanks

The first hangars in Orvieto were not the most striking and iconic works built by Nervi & Bartoli for the regime's military hardware. During the second half of the 1930s, a considerable number of achievements, designs, and patents by Nervi and his firm concerned constructions of a military nature and particularly tanks for fuels and oils. The series of tanks began with the great open water tanks built by Nervi & Bartoli between 1935 and 1937 on behalf of Società Solvay in Rosignano. Nervi broke down the tank structure

into a skeleton of reinforced concrete pillars and vaulted roofs made out of traditional brick masonry, almost as though he were transforming a contemporary model of Brunelleschi's dome, with its frames with small vaults, into a feat of contemporary engineering. The shape of the vaults reacted to the water pressure to create a circle of great cavities between the pillars. To prevent the ingress of water, Nervi created a durable waterproof layer designed to be independent from the structure and built using a metal reinforcement embedded in cement applied using a cement gun technique – "wired gunite plaster."

From 1936 to 1943, Nervi & Bartoli was engaged in designing and building underground fuel tanks along Italy's coastline, in Palermo, Cagliari, Siracusa, Taranto, Piombino, Gaeta, Trapani, Pozzuoli, and Portoferraio, for use by ports and military bases. Most of them were built on behalf of the Italian Royal Navy and Royal Air Force. The tanks were characterised by devices to prevent fuel spillage in the event of bombardment. This was achieved by special tank coating systems, by wall construction techniques that reduced the need for formwork, and by innovative methods of consolidating soils and shapes specially calculated to reduce the impact of bombs. The studies were accompanied by a set of patents filed by Nervi and his firm between 1936 and 1940.

One of the first underground tanks to be designed by Nervi & Bartoli was one created for the Società Nafta of Genoa. As before, this was produced with a massive ribbed wall structure made out of cast-in-place concrete, protected by a layer of "wired cementitious plaster." Small brick vaults were built against this to create a cavity with water; the vaults were in turn protected by a layer of gunite cement, as in the Solvay tanks. The roof took the form of a reinforced concrete structure made of columns widened at the base and a slab supported by a grid of beams with variable cross-sections. The cavity became the subject of a patent in January 1936 for fuel and oil tanks; the invention of special prefabricated blocks made out of concrete or brick allowed it possible to reduce the use of formwork.[34] In January 1937, Nervi & Bartoli patented a thin wall for tanks, measuring approximately 20 centimetres and in three layers, to be made out of gunite cement.[35]

In 1937, for the construction in Custolo, near Gaeta, of three circular naphtha tanks for the Italian Royal Navy, Nervi invented a system of making walls out of hollow blocks of concrete, woven together by means of a metal reinforcement and covered with a layer of reinforced gunite cement (he filed a

[33] Nervi, *Aesthetics and Technology in Building* cit., p. 28.
[34] Pier Luigi Nervi, patent IT 338800, *Serbatoio per liquidi più leggeri dell'acqua e con essa non miscibili, specialmente nafta, olio, benzina e simili,* 16 January 1936, ACS.
[35] Società Ingg. Nervi & Bartoli, Anonima per Costruzioni, patent IT 349384, *Procedimento di costruzione di una parete cementizia composta, formata da strati alternati di conglomerato cementizio e sottili strati di cemento puro per tenuta di liquidi o gas e prodotto relativo,* 23 January 1937, ACS.

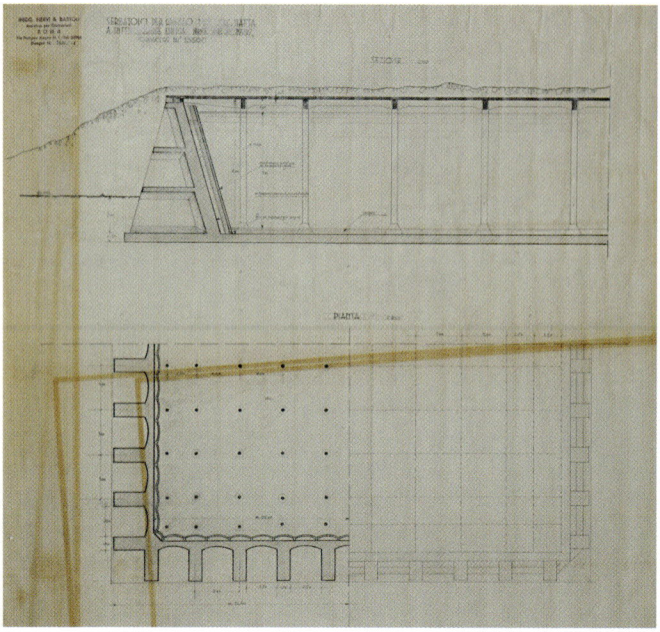

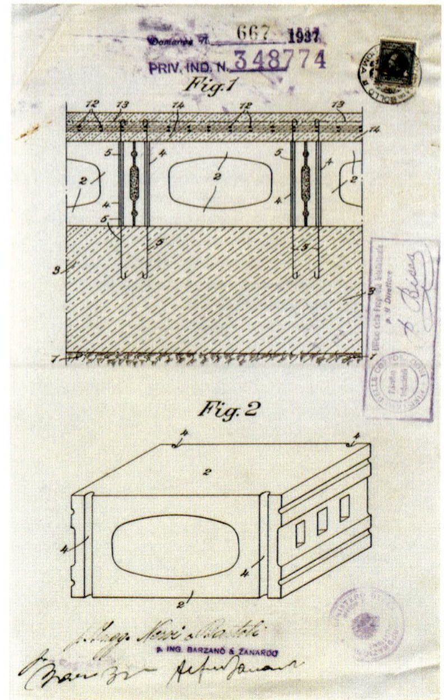

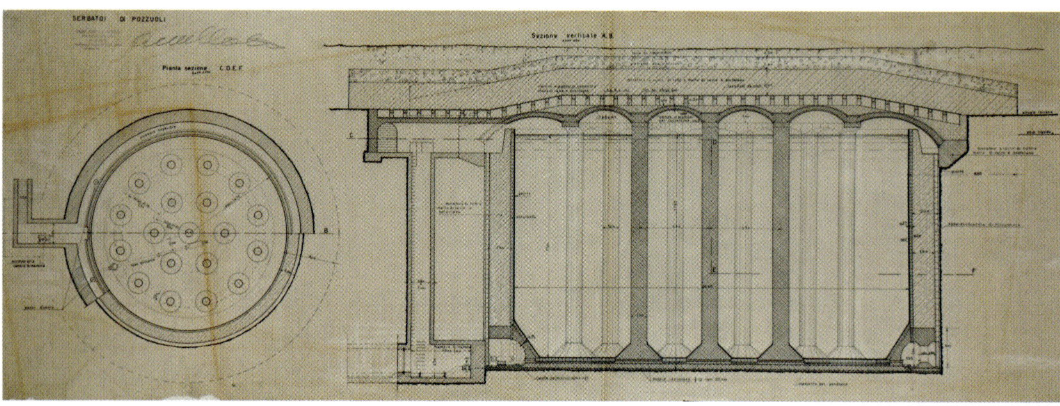

84. P. L. Nervi, Design of a "serbatoio per grezzo per Soc.[ietà] Nafta a intercapedine idrica (crude diesel oil tank with a water interspace for Società Nafta)," 1937 (CSAC, n. inv. pra 150, n. id, 13333, coll. 269/6).

85. Società Nervi & Bartoli, patent IT 348774, *Doppia parete cementizia di tenuta e controllo per grandi serbatoi interrati,* 6 February 1937 (ACS).

86. P. L. Nervi, underground tank, Via A. Artiaco, Pozzuoli, 1939 (CSAC, n. inv. pra 68, n. id. 13143, coll. 269/2).

patent for this type of masonry in February 1937).[36] The reinforced concrete slab covering the tanks was protected against bombs by means of concrete vaults produced using permanent forms. Nervi & Bartoli came up with another method of protecting the tanks against bombardment for which a patent was filed in October 1937. This was a truncated cone ending that was able to withstand explosions, as with the German bomb shelters designed by Leo Winkel.[37]

Nervi's other inventions, again intended to improve the tanks and patented before Italy entered the war, were designed with the aim of reducing the use of metal and making it possible to use local materials. They ranged from special products such as paints or layers of plaster to waterproof the walls made out of hollow blocks of concrete, instead of gunite cement with metal reinforcement,[38] to hollow walls to be filled with porous materials such as pozzolan, pumice or kieselguhr, to be impregnated with water,[39] sheet coatings,[40] and impervious clay walls.[41]

Nervi was able to improve some of his systems in a series of underground anti-aircraft fuel tanks built in 1939 in Pozzuoli, at the Air Force Academy, and in Palermo, for the Italian Royal Navy. His roofs were now made out of segmental annular barrel vaults, often with a protective coating, intended to produce a "vaulted armouring" against the bombing. It is likely that he adapted the system to local building conditions on each occasion, as evidenced by the Pozzuoli tanks, where the volcanic materials used since Roman times were within easy reach. For these cylindrical tanks, which measured 34 metres in diameter and 10 metres in depth, Nervi designed permanent forms made up of five-centimetre-thick brick vaults. A pozzolan-based layer approximately 2 metres in depth was poured over the reinforced concrete vaults ("core walls of tuff and lime mortar and pozzolan"). A cavity with arches made out of cement *tavelloni* (thin prefabricated slabs) and bricks was created to vent the gas. Another two layers were applied over the core walls, one of "tuff stone" and a camouflaged layer of "covering soil." The sequence of construction stages for the Pozzuoli anti-aircraft tank, as described by Nervi, involved digging an annular trench about ten metres deep and two metres wide in order to construct the cylindrical wall, which also acted as a soil containment structure (the wall was made out of the prefabricated blocks described in the 1937 patent, with cavities to be filled in the "core walls"); removal of soil from the pit, construction of pillars and vaults and the subsequent "forcing and loading of the walls"; lastly, the tank pit was coated with gunite. In the Palermo tanks, produced using the same procedure as those in Pozzuoli, a tunnel was also added at the top for inspection and in order to carry the suction tubes.

[36] Società Ingg. Nervi & Bartoli, Anonima per Costruzioni, patent IT 348774, *Doppia parete cementizia di tenuta e controllo per grandi serbatoi interrati*, 6 February 1937, ACS.
[37] Società Ingg. Nervi & Bartoli, Anonima per Costruzioni, patent IT 355466, *Procedimento di costruzione di una parete cementizia composta, formata da strati alternati di conglomerato cementizio e sottili strati di cemento puro per tenuta di liquidi o gas e prodotto relativo*, 8 October 1937, ACS and Società Ingg. Nervi & Bartoli, Anonima per Costruzioni, patent IT 364978, 28 June 1938, ACS.
[38] Società Ingg. Nervi & Bartoli, Anonima per Costruzioni, patent IT 364418, *Perfezionamento ai serbatoi in conglomerato cementizio per combustibili liquidi. Caso "B,"* 1 August 1938, ACS.
[39] Società Ingg. Nervi & Bartoli, Anonima per Costruzioni, patent IT 364519, *Perfezionamento ai serbatoi in conglomerato cementizio per combustibili liquidi. Caso "A,"* 5 August 1938, ACS.
[40] Società Ingg. Nervi & Bartoli, Anonima per Costruzioni, patent IT 377827, *Perfezionamento nella costruzione di serbatoi murari con foderatura in lamiere metalliche specialmente per combustibili liquidi*, 29 September 1939, ACS.
[41] Società Ingg. Nervi & Bartoli, Anonima per Costruzioni, patent IT 384152, *Perfezionamento nella costruzione di serbatoi per liquidi e particolarmente per idrocarburi mediante l'adozione di fondi o pareti in argilla plastica*, 27 March 1940, ACS and Società Ingg. Nervi & Bartoli, Anonima per Costruzioni, patent IT 388212, *Perfezionamento nella costruzione di serbatoi per liquidi e particolarmente per idrocarburi mediante l'adozione di fondi e pareti in argilla plastica*, 8 May 1940, ACS.

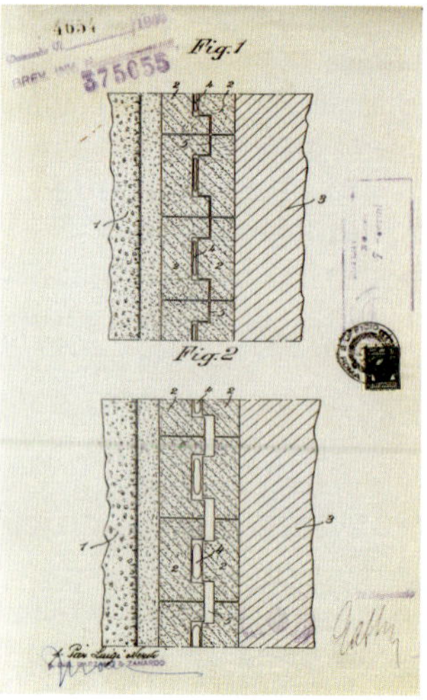

87. P. L. Nervi, patent IT 375055, Certification supplementing Industrial Property Right No. 363646 filed 5 July 1938 with the invention entitled: *Procedimento e dispositivo per creare e mantenere uno strato di pressione, determinato in valore e direzione, fra due corpi, ad esempio fra una struttura edilizia e il terreno di appoggio*, 19 May 1939 (ACS).

88. P. L. Nervi, underground tank, unidentified location. View of the building site (Maxxi - PLN, F08539).

89. P. L. Nervi, tank for lubricant oils, Parco della Favorita, Palermo, about 1940-41 (Maxxi - PLN, F02688).

Following the construction of the tanks in Pozzuoli and Palermo, in September 1939, Nervi & Bartoli filed an adaptation of the previous patents, which related to the method for constructing the "lining" of the petrol tanks, to be produced using a coating of cement, rows of bricks spaced to leave channels, a thick layer of a cementitious compound soaked in oil and small pebbles known as "moniglio," and lastly a covering of zinc sheets welded and nailed on special wooden strips embedded between the bricks.[42] The concrete structure of the roof was protected against any spillage of petrol by means of a coating of sheet iron and celluloid sheets.

In 1940, Nervi & Bartoli built other tanks for lubricating oils and liquid chemicals and resumed the study of protective coatings for tanks containing fuels and other products. Again in Palermo, for the Italian Royal Navy, alongside the tanks made the previous year, the firm built two circular tanks for naphtha, with a capacity of 15 tonnes. Their perimeter walls were protected by a coating that was a variation on the type patented in 1939 (wired glass slabs), with drainage, and a structure made out of tall, mushroom-shaped reinforced concrete pillars, with a base in the same configuration, stiffened approximately halfway along the stem by a frame of crosses, again in reinforced concrete, as in the previous rectangular naphtha tanks.

The construction of the hangars and underground tanks must have raised some technical issues relating to the foundations, reflected by the fact that Nervi & Bartoli filed some patents for the consolidation of unstable soils. The main device designed by Nervi involved injecting concrete or other similar compounds under pressure by means of hoses and a special machine into waterproof bags previously positioned between the soil and the masonry.[43] In a variant of this device, bags to be filled with water according to the same procedure were positioned between the special solid blocks in concrete or brick, in a male-female configuration, again with the aim of compacting the soil. Nervi planned to pour the concrete into the cavity created between the blocks.[44]

[42] Società Ingg. Nervi & Bartoli, Anonima per Costruzioni, patent IT 377827, *Perfezionamento nella costruzione di serbatoi murari con foderatura in lamiere metalliche specialmente per combustibili liquidi*, 28 September 1939, ACS.

[43] Pier Luigi Nervi, patent IT 363646, *Procedimento e dispositivo per creare e mantenere uno strato di pressione, determinato in valore e direzione, fra due corpi, ad esempio fra una struttura edilizia e il terreno di appoggio*, 5 July 1938, ACS.

[44] Pier Luigi Nervi, patent IT 375055, Statement supplementing Industrial Design Right No. 363646 filed 5 July 1938 for the invention entitled: *Procedimento e dispositivo per creare e mantenere uno strato di pressione, determinato in valore e direzione, fra due corpi, ad esempio fra una struttura edilizia e il terreno di appoggio*, 19 May 1939, ACS.

chapter four
designs for the rome *esposizione universale*, shells and autarky

The bar and the spiral

Building the hangars in Orvieto opened up to Nervi a whole universe of research into structures for grand reinforced concrete roofs. After the lamellar grid, he set himself the task of discovering other kinds of shapes that he could mould in reinforced concrete to enable forces to flow in accordance with understandable and appropriate directions, always in accordance with moment diagrams and a minimal use of materials, particularly the steel used to manufacture the reinforcements. Nervi & Bartoli's military studies and constructions went hand-in-hand with designs drawn up by Nervi for the great enterprise, which had been underway since 1935 for a new district in Rome to house the E42, the *Esposizione Universale* planned for 1942. His projects always featured the invention of structures that probed the potential of reinforced concrete to create extraordinary symbolic projections and imposing atmospheres, or the reuse of tried and tested figures that had become iconic.

Nervi's projects for E42 concerned a symbol to be erected at the foot of Via dell'Impero, the Building of Water and Light, a set of proposals for generic pavilions and the pavilion, or building, for the *Civiltà Italiana*. They were drawn up between 1936 and approximately 1940. His collaboration with Bardi was probably limited to the last pavilion. The fact that some designs for the pavilions were signed by Nervi and dated 11 July 1938 suggests that these were counter-projects for works that were already under construction during that month, such as the *Palazzo della Civiltà Italiana*. This was emblematic of the entire theoretical conflict that had been under debate since the early 1930s between a right-angled mesh frame in reinforced concrete, which was used to construct the work, and the arches, which added decoration in the travertine slab cladding.

The symbol of E42 that Nervi envisaged as crowning Via dell'Impero took the form of a design, submitted in February 1936, for an extremely high structure: the lyrical manifestation of fascist autarky and the "nature of *romanità*," to be erected as a new triumphal column made out of marble drums

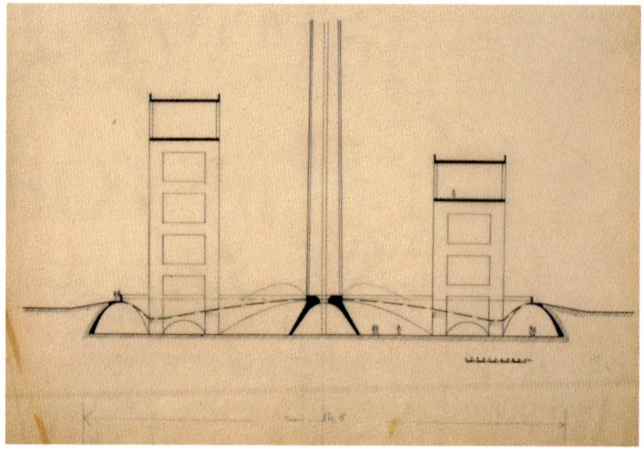

90-94. P. L. Nervi, designs for the *Palazzo dell'Acqua e della Luce* for the 1942 *Esposizione Universale*, Via dell'Impero, Rome, 1937-38.

90. Perspective drawing (BSTA, 024-6946).

91. Cross-section (CSAC, n. inv. pra 1364, coll. 267/3).

92. Plan and cross-sections (BSTA, 024-6034).

93. Model (BSTA, 024-6942).

94. Perspective drawing (CSAC, n. inv. pra 1364, coll. 267/3).

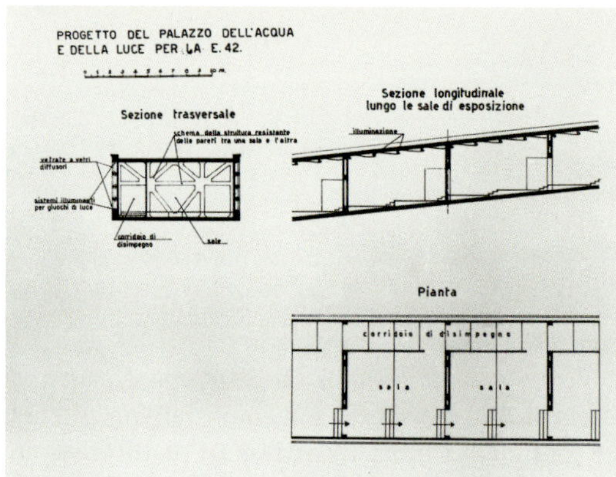

reinforced by a device described as an "axial stabilisation,"[1] directly derived from the one on the design for the *Monumento alla Bandiera*.

When it was decided to build a colossal triumphal arch during 1937, as a symbol of E42, Nervi began to develop the design for the Building of Water and Light, to be built at the foot of the arch. The version submitted in October 1937 and completed in early 1938, when a model was also built, as dictated by the circumstances of the Esposizione, indicates the extent to which his conception of structure was still influenced by poetic aspects, that "idea of emotional origin." The two elements to which the building was dedicated were captured and represented in the form of a water jet and a lightning bolt, which generated two different structures, which would both be capable of informing – always through the sheer virtuosity of his technique – the nature of the work in their own right. The first was a high, slender rod rising in the centre of a pool of water that challenged the proportions of the *Torre di Maratona* and the *Monumento alla Bandiera* (this was intended to contain the axial stabilisation device); the second was a colossal hollow beam, bent in accordance with the spiral shape so dear to Nervi, grooved with glass diffusers and lighting systems to create a play of light, which could contain the exhibition rooms and be cantilevered for 40-metre stretches. In some versions, it ended in a terrace such as the one at the foot of the *Torre di Maratona*. Nervi was unsure about the shape of the radial pillars supporting the spiral. They were first vertical, then inclined and also imbued with the dynamism of the box structure. The "strangeness of the construction," as he put it, literally interpreted his aspirations that his initial proposals for the *Esposizione Universale* should show "great pomp."[2] In subsequent versions dating from June 1938, Nervi transferred the hollow beam into a fluorescent halo that rested either on eight inclined pillows with the fountain stand in the middle, or on a colossal tripod of curved pillars, but always ensuring that the structure assumed the characteristics of form and symmetry appropriate to the cultural approach imposed by Piacentini.

Autarkic concrete

Nervi's ideas in February 1938 about reinforced concrete were destined to undergo changes following the entry into force of a circular from the Ministry of Public Works imposing steel savings in the building industry and ratifying Mussolini's declarations that steel should be directed into armaments manufacture as a matter of priority.[3] Architects and critics voiced their opinions of the consequences of the circular on the construction of building structures. The autarky helped Piacentini to find a constructive approach to the figures evoked in the debate over arches and columns. In addition to reconsidering the possibility of using bricks and stones, in the case of reinforced concrete, he proposed "semi-reinforced" concrete both to economise on steel and make the best use of concrete in addition to standardising the structures used in the same building so that the formwork could be used several times over, and the concrete could be manufactured using materials such as pozzolan to save on

[1] "Torre alta e sottile a struttura marmorea e stabilizzazione assiale," proposal No. 115 of 23 February 1937, in, *Elenco proposte in ordine cronologico al 31 gennaio 1940*, p. 13, envelope 1039, ACS.
[2] Nervi, *Scienza o arte del costruire?* cit., p. 122.
[3] Tullia Iori, *Il cemento armato in Italia. Dalle origini alla seconda guerra mondiale*, Roma, Edilstampa, 2011, p. 160, and *L'applicazione del decreto che vieta le costruzioni in cemento armato e l'impiego del ferro*, in, "Il Messaggero," 8 September 1939, envelope 378, ACS.

cement.[4] Another kind of autarky was expressed in the design submitted by Terragni, Cattaneo, and Lingeri at the second stage of the competition for the Reception and Congress Building for E42 that was drawn up in December 1937. This provided guidelines for a new use of traditional materials integrated in the reinforced concrete construction instead of used to clad the finished product. Granite was to be used in monolithic blocks to be set alongside the pillars of the framework to reabsorb compression stresses and therefore save steel in the reinforced concrete. Terragni and his colleagues were not, however, willing to give up the foundations of their concept of structure, and the same applied to Nervi.

The most authoritative Italian engineer and builder responded to Piacentini's proposals by defending what he saw as the unavoidable role of reinforced concrete in contemporary construction, although he accepted the challenge of reconsidering the techniques in light of the ministerial circular. Noteworthy in Nervi's reasoning about the logical use of traditional materials were certain remarks that lead in the direction indicated by the design of Terragni, Cattaneo, and Lingeri. A statement he made in 1938 against wasting stone, by cutting it into cladding slabs, could even be seen as a literal interpretation of the technical qualities of that design: "due to its superior strength, cladding stone almost always lends itself to performing a static function more effectively than concrete."[5]

Nervi did not fail to point out that Piacentini's proposal for autarkic cement, referring to the state of the art and Roman structures, involved significant cost for both transporting materials and the necessary weight of masonry. "I am almost certain that buildings like the Pantheon or the Bath would today be much less autarkic than their equivalents in iron or reinforced concrete,"[6] he stated from his standpoint of pragmatic entrepreneur and prudent builder; and, possibly also after assessing the stance taken by Terragni, Cattaneo, and Lingeri, Nervi suggested addressing the question of autarky by introducing the significant concept of "degrees of autarky" but nevertheless always remaining, like Terragni, within the sphere of a contemporary use of concrete. He was unequivocal: "Maximum product autarky corresponds to its lowest cost in foreign currency or gold."[7]

Empathy of the shell structure

As the debate raged over autarky, with significant consequences for reinforced concrete structures, and the events involved in designing and building E42 were complicated by the imminence of war and of competing cultural and political factions, Nervi seemed to seek a detachment when working on some of these projects that was almost Olympic, as though he intended to freely explore some new matters that were cropping up in the field of reinforced concrete shapes, partly with the aim of reducing material quantities and meeting the demands of autarky in his own way. The result was a series of designs for pavilions intended for E42, most of which were dated July 1938 and signed by Nervi. The crucial question in all the pavilions was that of the shape

[4] Marcello Piacentini, *Per l'autarchia. Politica dell'Architettura. III. – Riforme concrete*, in, "Il Giornale d'Italia," 17 July 1938, p. 3.
[5] Pier Luigi Nervi, *Per l'autarchia. I problemi economici delle costruzioni e la politica dell'architettura*, in, "Il Giornale d'Italia," 23 July 1938, p. 3.
[6] *Ibid.*
[7] *Ibid.*
[8] Pier Luigi Nervi, *Le strutture portanti del Palazzo per le Esposizioni al Valentino*, in, "Atti e Rassegna Tecnica della Società degli Ingegneri e degli Architetti di Torino," II, July 1948, No. 7, (pp. 118-122), p. 121.
[9] For works created by Baroni and Miozzi for E42, see *Relazione sulle cupole. Sistema*

of the roof, giving the space its own special connotation that was different on each occasion.

The structural system proposed in the various projects for circular pavilions does not differ from that of the circular hangars he designed with Valle. Yet even in the structural configuration, Nervi showed the extent to which he believed that every solution possesses its own specific eloquence or expressive rhetorical value, which determines its "nature of form." No ribs were in evidence in the circular pavilions, since they were not intended to represent the military might of the Fascist regime but rather to instill in the spectator a feeling of wonder that might even spill over into the rhetoric of the sublime. The designs for those pavilions demonstrate the uniqueness of Nervi's creative processes, which tended towards empathy, evidenced in the modelling of light and space, the latest structures for shells with a double curve, with a geometry derived from the hyperboloid, the hyperbolic paraboloid, the parabolic cylinder and the conoid, developed and tested by Eduardo Torroja, Bernard Lafaille, Fernand Aimond and – in the case of Italy – by Miozzi and Giorgio Baroni. Nervi may have also been thinking of the theoretical matters and mathematical formulae for shells thar was published by Dischinger in 1935, to which he referred in later years.[8] In any case, his initial certainties over the characteristics of those structures came about because he was fresh from having completed – albeit not without difficulties – the construction of the shell on the Genovese church cupola.

It was almost as though Nervi were engaged in a clash of ideals with his compatriots Miozzi and Baroni, who between 1936 and 1937 were already experimenting with and patenting construction systems and techniques for three-centimetre thick autarkic roofs in hollow-brick blocks and concrete or reinforced concrete, and who were also consulted in 1938 about the idea of building their own special products for E42.[9] Nervi also created thin shells with a double curve that were able to stand up by virtue of their own specific geometry, to reduce the amount of iron and concrete and to show his ability to use those materials even under autarky with surprising results in terms of form and space. A quest for spectacular effects of light and immensity led Nervi to imagine, for the design of a 70-metre diameter pavilion (classified as "*Padiglione Tipo B*"), a shell that originated in the centre of the plan, from a series of inclined radial columns, and then sprouted like a mushroom to the suspended edge of the circumference, to infuse a feeling of levity, and surrounded it with uninterrupted continuous glazing. Nervi examined the profile of the shell and its connection to the ground in order to express the fairy-tale levitational quality of the pavilion through structural design alone. The light that filtered in among the pillars from the grounds outside and emanated from the windows separated the shell from the ground to conjure up the surreal essence of a fantastic aircraft, with an even fabric-like quality resembling a parachute or an umbrella. That shell marks the re-emergence of Nervi's attraction to making structures hover in the void, the "idea of emotional origin" although its geometry reflects scientific research into new reinforced concrete

Brevettato Ing. G. Baroni, envelope 326, ACS and Eugenio Miozzi, letter to Vittorio Cini, 27 September 1939, envelope 195, ACS. On these aspects, see Claudio Greco, Tullia Iori, *Sprimentazioni autarchiche in Italia per le coperture di grande luce. I casi di Eugenio Miozzi e Giorgio Baroni*, in, Pier Giovanni Bardelli, Elena Filippi, Emilia Garda, eds., *Curare il moderno. I modi della tecnologia*, Venezia, Marsilio, 2002, pp. 101-112.

95-96. P. L. Nervi, designs of exhibition buildings for the 1942 *Esposizione Universale*, Rome, 1938.

95. *Tipo B* exhibition building, dated 18 July 1938. Interior (BSTA, 024-6022).

96. *Tipo B* exhibition building, dated 18 July 1938. Plan, elevation, cross-section (BSTA, 024-6024).

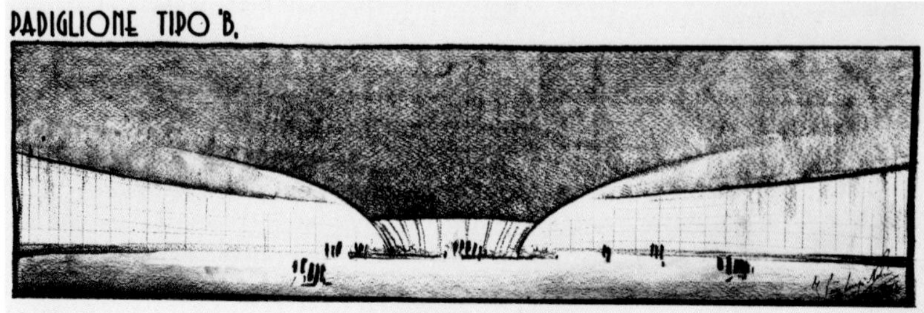

structures. "[...] the roof is made out of a thin vault with a hyperboloid cross-section in reinforced concrete *with only central supports*," Nervi went on to explain, stressing in italics the aspect that made the levitation effect possible.[10]

Nervi envisaged other shells, again with a "hyperboloid cross-section" but with a very different nature, for the *Civiltà Italiana* pavilion, for which he confirmed a circular plan (corresponding to "*Padiglione Tipo D*"). The imposing nature of this kind of pavilion is immediately evident in the four entrance porticos with its tall, slender pillars, similar to those of the church in Genoa. The roof is a continuous annular barrel vault or semi-toroid, supported in the centre on a ring of pillars delimiting an open-air garden. Two versions of the pavilion were proposed, differing in degrees of structural magniloquence: the one closest to rationalist taste allows a glimpse of the roof profile even from the outside and visitors always walk at the same level even when they enter the garden (the diameter was 105 metres and the design dates from 11 July 1938); the design that is more rhetorical in nature, and more likely to appeal to Piacentini (drawn up in 1940, according to Nervi's account),[11] depicts a pavilion more than doubled in size (with a diameter of 252 metres) and surrounded by a ring of pillars that creates a monumental abstract order, continuous with the pillars of the four porticos and powerful dramatic device.

While the shell for the 70-metre diameter leaves the support invisible, when the shell became the roof of the *Civiltà Italiana* pavilion, Nervi (particularly in the more overblown version) designed its profile to achieve greater magniloquence, in the style of Boullée, with the aim of celebrating the cultural and historical foundations of Italy, to which the pavilion is dedicated. The previous curved lines of pillars and the shell in the middle of the pavilion give way to a ring of tall pillars erected vertically to evoke the image of a circular temple accessed by means of monumental ramps, while a series of massive inclined structures around the circumference take up the weight of the roof and bend downward to reproduce the effect of a dome thereby reflecting the seriousness of the space and its solemnity. There is no doubt that the structure of the second version of the design is modulated to interpret the guidelines for the new master plan regulating the *Esposizione*, "based on a more rigid, heavy, rhetorical, and classical symmetry of masses and forms," as Nervi wrote.[12]

[10] Pier Luigi Nervi, *La tecnica ed i suoi orientamenti estetici*, in, "Lo Stile nella casa e dell'arredamento," July-August 1942, No. 19-20, (pp. 9-15), p. 15, captions to photographs.
[11] Pier Luigi Nervi, *Costruire Correttamente. Caratteristiche e possibilità delle strutture cementizie armate*, Milano, Hoepli, 1955, plate LI.
[12] Nervi, *Scienza o arte del costruire?* cit., p. 122.
[13] Nervi, *Per l'autarchia* cit.
[14] Roberta Martinis, *Pier Luigi Nervi e Pietro Maria Bardi: un'amicizia, due continenti*, in, Sergio Pace, ed., *Pier Luigi Nervi. Torino, la committenza industriale, le culture architettoniche e politecniche italiane*, Cinisello Balsamo, Silvana Editoriale, 2011, (pp. 69-75), p. 70.
[15] Nervi, *La tecnica ed i suoi orientamenti estetici* cit., p. 15, caption to images.

Both versions of the *Civiltà Italiana* pavilion still nevertheless reflect the "Italian style" debated by critics and that Nervi set out to achieve in July 1938 by pursuing "our sincere, powerful, and indestructible expression" as a spontaneous affirmation of the "spiritual atmosphere of our lives [...]."[13] Given that they were the only pavilions allocated a specific purpose in the series of designs by Nervi and since the guidelines for the floor plan described a circular, chronological exhibition space on the theme of Italian art, culture and history through the centuries, it is clear that these proposals were intended as a criticism against the *Civiltà Italiana* pavilion under construction at the E42.

After his difficulties in constructing the shell of the church roof in Genoa, Nervi became convinced of the need to find a way to build the kind of structures able to overcome problems created by wooden formwork, its deformability and its installation in accordance with complex surfaces. Nervi's series of designs for the E42 pavilion fell during the period between establishing the difficulties involved in constructing shells in accordance with normal reinforced concrete building site practices, and inventing a special technique and an unusual shape for a practical and economical shell structure. He developed them with the aim of establishing the creative potential of those complex structures. What was to become Nervi & Bartoli's next undertaking, involving a kind of cement construction that was thin and strong because of its special curvature, first came about as a result of those designs, which almost directly addressed early research by Miozzi and Baroni.

One comment written on the pictures of the circular pavilions opened up a decisive chapter in Nervi's research into structure for reinforced concrete: "a typical example of strength through form."[14] The concept of this kind of resistance conceals an idea of structure that no longer set out to entrust a skeleton derived from a Hennebique frame with a load-bearing role but to identify a load-bearing capacity in the shape itself. It could not be ruled out that the reinforced concrete shells of the circular pavilions concealed an invisible ribbing, as in the slabs of the spiral staircases of the *Stadio Berta*, or that they could be planned with stiffening ribs on the extrados, even though these were not present in the cross-section drawing. The suspicion comes from reading the caption published by Nervi in 1942 as a comment on the 70-metre diameter pavilion: "it is like an inverted umbrella supported by slats alone: a slender mushroom shape."[15] The fact that Nervi had also wondered about other ways of constructing those shells using re-

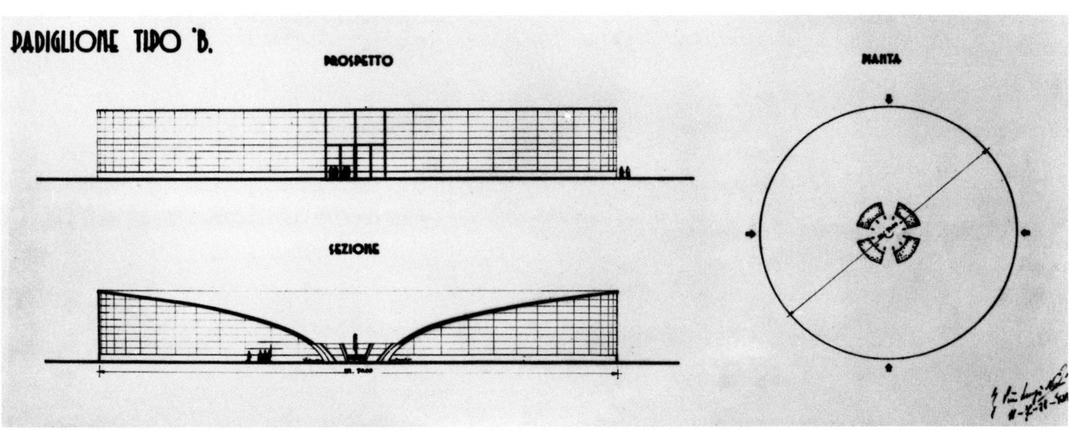

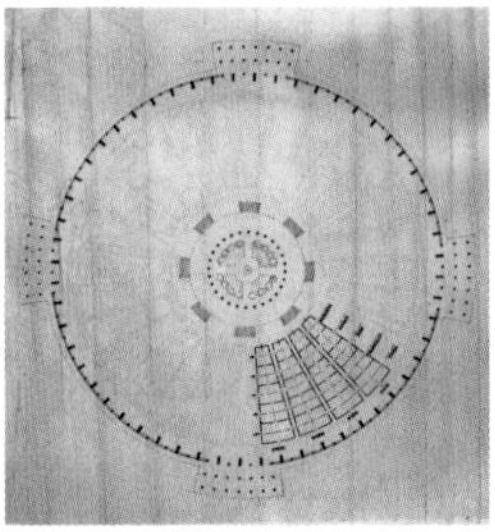

97-101. P. L. Nervi, designs of exhibition buildings for the 1942 *Esposizione Universale*, Rome, 1938

97. *Tipo D* exhibition building. Plan (BSTA, 024-7444).

98. *Tipo D* exhibition building, dated 18 July 1938. Exterior (BSTA, 024-6044).

99. *Tipo D* exhibition building. Cross-section (BSTA, 024-7298).

100. *Tipo D* exhibition building. Interior (BSTA, 024-7310).

101. *Tipo D* exhibition building. Exterior (BSTA, 024-7311).

inforced concrete to be cast in place into formwork is proved by a composition of pictures for the monumental variant of the *Civiltà Italiana* pavilion that depicts the centring system that Nervi himself developed and patented between 1938 and 1939 against the backdrop of another magnificent work to be erected for the E42.

Nervi's set of designs for the pavilions to be used for E42 also includes two examples of ribbed structures, to be constructed using powerful and visible geometric patterns. In one version, referred to as pavilion C, Nervi demonstrated the extent to which the reticular structure that he had recently built in the Orvieto hangars, and was developing with a view to prefabrication, possessed an irrepressible aesthetic quality justifying its appearance on the roof of a pavilion for E42, featuring a square plan measuring 65 metres on each side. In the other, he performed a creative leap in borrowing the "hyperboloid cross-section" of the design for the 70-metre diameter pavilion and weaving in two arched profiles at its base, transforming the features of that continuous shell into an X-shaped skeleton with curved bars and a graceful profile, like that of the cantilever beams of the *Stadio Berta* (there is no other explanation why he would have classified the pavilion with the X-shaped structure as a variant of type B). The grid evokes an air of majesty, as in the metal structures of furniture by Ludwig Mies van der Rohe. Nervi went on to explain that the structure was planned in steel.[16]

In one single pavilion design, Nervi addressed the call that emerged during the debate over autarky by rethinking traditional construction materials and techniques and renouncing the exclusive use of reinforced concrete in order to save on the steel in the reinforcements (this is described as "*Padiglione Tipo A*," almost as though to underscore its constructive realism and immediate usability by using the first letter of the alphabet). The pavilion would therefore have featured the direction of six thick stone walls, arranged in parallel 10 metres apart and as long as necessary, with spaces communicating through doors and a glazed portico planned at the head. The walls feature an isodomic device and correspond to Nervi's reflections at the time on autarky and the "nobility," "expressive force," and "eternal appearance" of "building out of blocks of good stone"[17]. Only in the beams at the top of the walls and in the roof over the long spaces (and probably also in the place of the portico), did Nervi plan to employ the reinforced concrete that he had used to produce the industrial saw-toothed roofs, but with a sophisticated profile as a result of their waved, asymmetrical cross-section, with an uncluttered visible face, without any ribs, and designed to allow the light to slide over it without creating ripples of shade.

The shell was also to reappear in various studies for the project submitted by Nervi for the invitation to tender launched in 1939 for the *Palazzo dell'Acqua e della Luce*, winner of the second *ex aequo* prize, where it assumes the most levitating and fantastic "nature of form" of the revolving hyperboloid that sprouts huge jets of water. Among the various studies for this latest version of the *Palazzo dell'Acqua e della Luce*, Nervi imagined freezing the jets of a fountain into veritable ribbed parabolic arches.

[16] *The Works of Pier Luigi Nervi*, London, The Architectural Press, 1957, p. 44.
[17] Nervi, *Per l'autarchia* cit.

PADIGLIONE TIPO 'D.

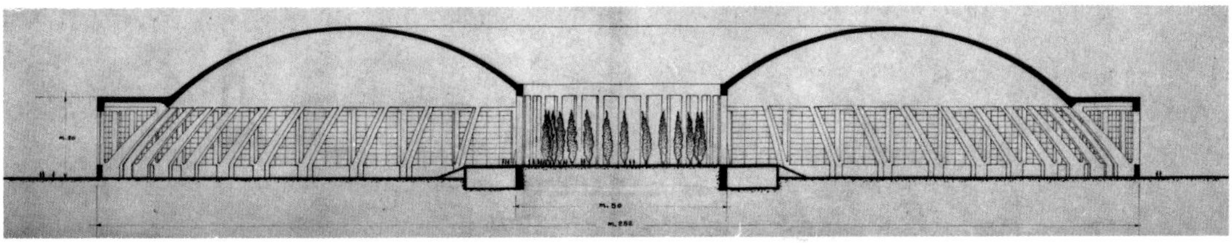

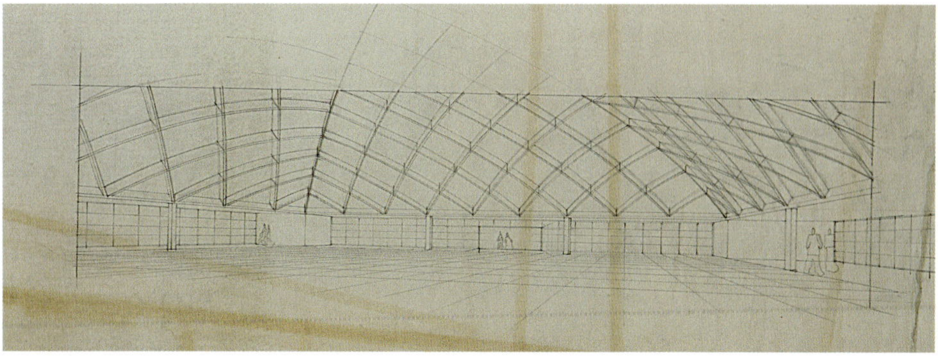

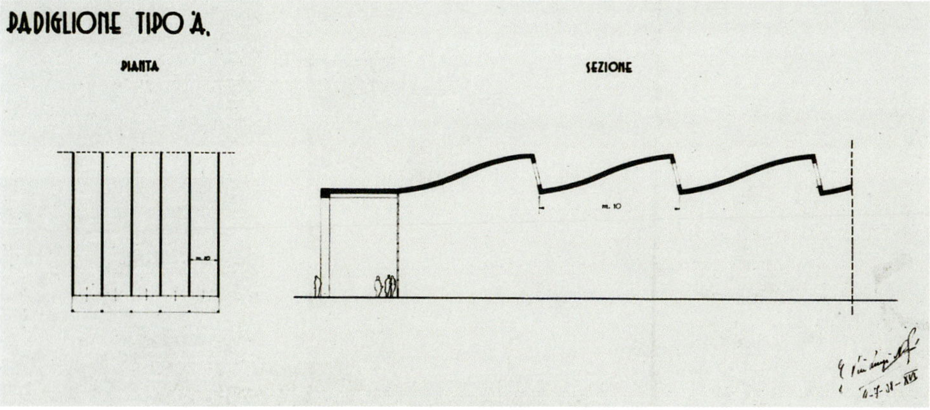

102-105. P. L. Nervi, designs of exhibition buildings for the 1942 *Esposizione Universale*, Rome, 1938

102. *Tipo D* exhibition building. Interior (CSAC, coll. 270/4).

103. *Tipo B* exhibition building. Variant with X-shaped structure (*The Works of Pier Luigi Nervi*, London, The Architectural Press, 1957, p. 44).

104. *Tipo A* exhibition building. Plan and cross-section (Maxxi - PLN, F07343).

105. *Tipo A* exhibition building. Interior (BSTA, 024-2808).

Precast concrete blocks for the *Arco dell'Impero*

Nervi and his firm were involved in what was supposed to be the most prestigious technical and symbolic work of E42 and Mussolini's autarkic regime. To crown the general plan *Esposizione Universale*, Libera designed, between 1937 and 1938 and in conjunction with the engineer Vittorio Di Berardino, his Monumental Arch or *Arco dell'Impero* that was perfectly semicircular in shape with an elliptical cross-section, varying from base to top and with a 200-metre span, to be erected using concrete alone and with a cavity designed to carry visitors to a panoramic terrace. Libera specified that the work was to be "in non-reinforced concrete, left exposed without any cladding, but only hammered on the surface."[18] Nervi & Bartoli was consulted at the suggestion of Cipriano Efisio Oppo, Vice-Commissioner of the E42 Ente Autonomo, to study technical details of its construction and then submitted two proposals in December 1938.

Di Berardino had already shaped the contour of Libera's arch in accordance with autarkic building principles in order to be able to withstand wind loads and do without a metal reinforcement; thereby, as Nervi explained, the "pressure line due to the self-weight corresponds almost entirely to the

[18] *Progetto di atrio di arco simbolico all'E42 a Roma, 1937-1940*, in, *Adalberto Libera. Opera Completa* cit., p. 162.

106. P. L. Nervi, study of a construction system for the *Arco monumentale E42*, designed by A. Libera and V. Di Berardino, for the 1942 *Esposizione Universale*, Via dell'Impero, Rome, 1939 (CSAC, n. inv. pra 1364, coll. 267/3).

107. Società Nervi & Bartoli, patent IT 369975, *Procedimento di costruzione di centine di grandi dimensioni, e centina ottenuta con tale procedimento*, 5 January 1939 (ACS).

108. P. L. Nervi, preparatory model constructed in the *Politecnico di Milano* Testing and Models Laboratory for the construction system of the *Arco monumentale E42*, designed by A. Libera and V. Di Berardino for the 1942 *Esposizione Universale*, Via dell'Impero, Rome, February-March 1939 (Maxxi - PLN, F05430).

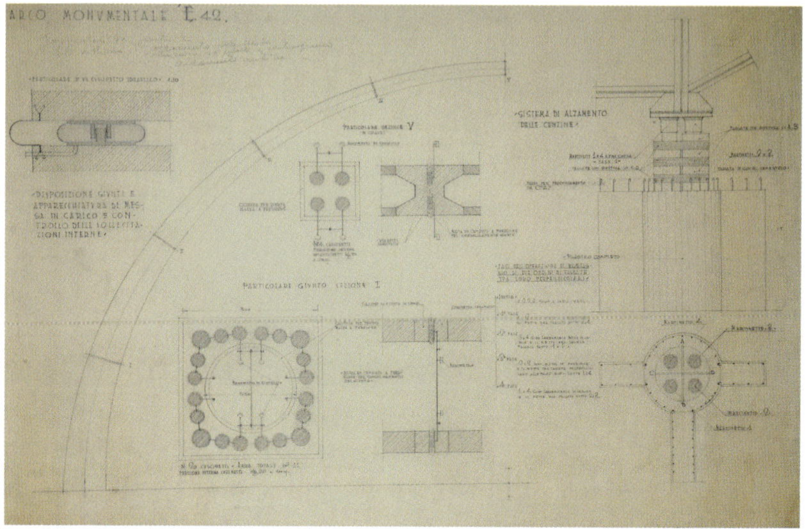

pre-established semi-circular shape," making it possible to eliminate "tensile stress" and the "need for metal reinforcements."[19] The profile that was continuously variable throughout each section of the arch, determined by static laws, contained the germ of Nervi's idea, from the early 1950s, of shaping his reinforced concrete pillars according to geometries that would determine exceptional plastic figures, in a quest for the same static logic perceived by Di Berardino. Nervi's opinion of Libera's choice of a semi-circular shape was still critical and couched in terms that are important for grasping his idea of structure: "The unnaturalness of this shape, due to an abstract conception formed in the mind of the designer, over and above any relationship with the static reality, gave the entire work a specific and exceptional quality. It could even have been an emotive element, but it would certainly have been distinctly unpleasant if the same arch had been an element of clear static functionality."[20]

As soon as he was commissioned to build the *Arco dell'Impero*, Nervi began to think about ways of moulding the non-reinforced concrete within the shapes established by Libera and Di Berardino, and about the system of scaffolding to support its weight during the pouring. His main concerns were still about problems generated by differential thermal expansion between the concrete mixture and the centring, which he had been encountering since the time when he built the *Gioco della Pelota* roof and that, given the size of the *Arco dell'Impero*, would become critical to the stability of the structure during its construction. It is typical of Nervi's intuitive process, now that he had many building projects behind him, that he should have imagined a way of pouring the concrete that would allow for the inevitable movements of the centring and the concrete mixture until the arch was closed. The Arch therefore became a mechanical construction in the ancient Roman manner, made out of colossal long wedges of concrete, or sections, symmetrically arranged along the two sectors of the structure and joined together by special articulated joints and distributed at regular intervals (there were nine in total), to transmit the weight and allow mutual rotations induced by the settling of the centring. Therefore, despite his reply to Piacentini's proposals of an autarktic concrete, Nervi found himself reinventing the contemporary building site in order to build a colossal arch out of concrete alone, almost as though he were an ancient master of the *opus caementicium* and perfectly aware of the "bold, deep significance" of excluding iron due to the particular "symbolic nature of the work," with the aim of arousing in the spectator "thrilling evidence of absolute primacy."[21] For the size, manufacture, and assembly of the elements in a form of concrete reduced to artificial stone, Nervi must have again delved into the hoard of studies carried out on Brunelleschi's dome.

Once he realised the need to break down the arch into long wedges to prevent fractures arising while pouring the concrete, it was natural for Nervi to focus on the method of preparing the concrete mixture that was to become typical of his masterpieces of the 1940s and 1950s: the on-site manufacture of parts. He again displayed his building site expertise by suggesting that the long wedges to be hinged together should be built by assembling several concrete "blocks" or "*tavelloni*" of reduced size (50 x 25 x 12 centimetres) so that they could be easily set up and aged, for about a year, to avoid the effects of shrinkage once they were installed.[22] Regarding the changes to the way he built his hangars, Nervi saw the use of "standard construction" as a means of having absolute control over the quality of the concrete. Nervi only briefly described how the blocks were manufactured, "using very select material, mechanically compacted by vibration or compression."[23] That the

[19] Pier Luigi Nervi, *Un arco monumentale in conglomerato non armato*, in, "Costruzioni-Casabella," XV, August 1942, No. 176, (pp. 23-25), p. 23.
[20] Nervi, *La tecnica ed i suoi orientamenti estetici* cit., pp. 10-11.
[21] Nervi, *Un arco monumentale in conglomerato non armato* cit., pp. 23 and 25.
[22] *Ibid.*, p. 24.

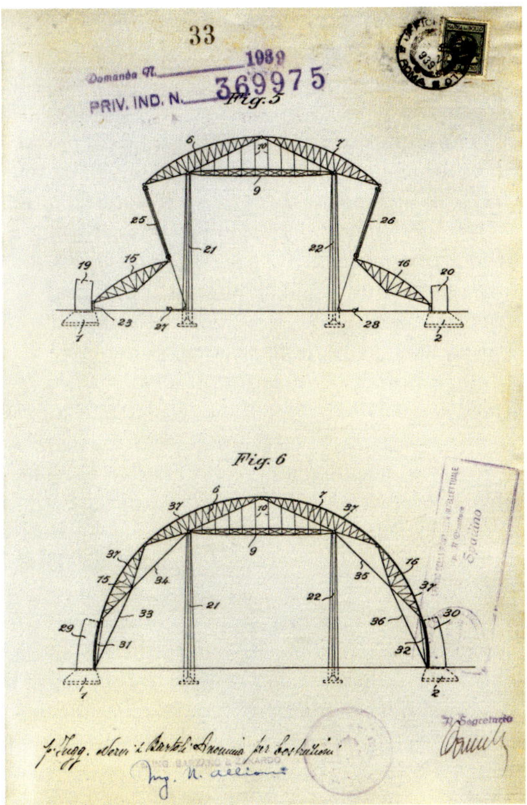

blocks had to be laid in rows meant that the concrete construction had to meet the criteria of bricklaying, possibly similar to the method that allowed Brunelleschi to build his dome. Nervi also stated that he would examine the "laying" of the blocks "in order to achieve the maximum connection between them,"[24] revealing that he had paid special attention to the way they would be interwoven. The entire system was to have been enclosed with a cladding that Nervi believed was worthwhile due to its effects on the thermal control of the concrete; it canceled out diurnal temperature effects.

The other sophisticated device necessary to the construction of the *Arco dell'Impero* was the joint between the long wedges that Nervi planned to construct using hydraulic bearings to allow rotation during assembly of the arch. The joints were to be consolidated by means of cement mortar injected under pressure into bags, in accordance with a device that he had patented and tested for the loading of soil. Therefore, broken down into long wedges, and with the joints transformed into devices for exercising forces, the arch would become a post-compressed structure that could be disarmed once a balance had been struck and the cement mortar had set.

The decision to build the arch with the blocks to be laid in 10 long wedges jointed by means of hydraulic jacks went hand in hand with the creation of a type of centring that did not require major scaffolding (Nervi's thorough knowledge of Brunelleschi's work can again be discerned in the background). The system invented by Nervi is based on building a metal truss centring on the ground, subdivided into four sectors, of which the central ones are hinged together, and raising it into the arch construction position by means of an ingenious device: the construction, section by section, of two reinforced concrete columns on which rest the two central centrings from the time they are constructed on the ground. The progressive construction of the columns was made possible by four hydraulic jacks arranged in the centre of each column; the jacks were raised in pairs in order to position concrete blocks beneath them, repeating the operation until the column stem reached the desired height. The sliding formwork made it possible to build the reinforced concrete columns with cross-shaped reinforcement around the stack of blocks beneath the jacks. Even the method of raising provisional columns using jacks and sliding formwork constituted a significant chapter in the development of concrete pouring technique without significant scaffolding.

Nervi's studies for the construction of the *Arco dell'Impero* were translated into two patents filed in January 1939, both relating to the system of "large centrings"[25] and the "gradual raising of construction elements or parts of constructions."[26] The patented text envisages the use of his system for the construction of a great round arch as well as the structures that were central to the concerns of a regime that was now heading into World War II – hangars.

Even though he was sure that he could build the *Arco dell'Impero* out of blocks of artificial stone, Nervi also had to develop a reinforced concrete version, perhaps in the wake of concerns raised by Danusso about the efficacy of a construction system based on concrete alone.[27] Working in conjunction with Danusso himself and Di Berardino, Nervi began to check his construction assumptions about the *Arco dell'Impero* in reinforced concrete and its resistance to wind loads at the Milan laboratory in February 1939. The tests were carried out on a 1:200 scale model but broken off a few months later, following the decision to build the arch out of aluminium instead of reinforced and non-reinforced concrete. Nervi remained convinced of the value of the prefabricated block solution laid without a metal reinforcement: "Even without stressing the very different aesthetic result of a cement arch – a true blend of the most typical structural element of *romanità* with today's spirit and technical opportunities – you may be sure that every spectator, through the simple utterance, *'there is no iron in it'* would have had a full sense of the act of will that had been made and the victory achieved."[28]

The design for the arch was a prelude to the technical development of Nervi's concrete, from the material cast in place to the assembly of parts manufactured on-site. Studies and experiments conducted by Nervi & Bartoli between the late 1930s and the early 1940s aimed to implement techniques of manufacturing small elements – as light and manageable as possible – that could nevertheless be used to build magnificent structures. The construction of small sheds was merely a starting point for the creation of larger and larger structures.

When he was engaged in developing a centring for the *Arco dell'Impero*, Nervi went back to considering the matter of constructing a viaduct such as that in Viterbo, without centrings, but this time also including magnificent arched portal frames. The viaduct project in 1939 is a demonstration of Nervi's will to avoid resorting to the costly temporary structures that he had to provide for Libera's arch. The shape of the structure itself, whose design stemmed from the lamellar grid used for the hangars, was outlined on the basis of a construction without centrings. Nervi therefore transformed the arches of the viaduct into a pattern of woven bars that gave rise to triangular-meshed spatial geometries at the base, while a right-angled mesh frame grew between the arches in a pyramid formation, as in the Viterbo viaduct. The formless curvature of the arches was calculated so that they could be built without centrings with an unobtrusive overhang, sector after sector, up to the arched top, to be built using a special small, overhead centring. The viaduct would have looked like an almost invisible truss, with unprecedented geometries generated by static and building site rules, silhouetted against the sky as a symbol of beauty arising from pure technical reason, with an ethereal texture as demonstrated by the hangars photographed before construction of the roof covering. It was now clear that after examining the dome, Nervi was faced with the same technical problems as Brunelleschi and, like the Florentine builder, he intended to invent an installation system without formwork and centrings for the fundamental material of his work, reinforced concrete.

109. P. L. Nervi, design for a viaduct, 1939 (BSTA, 024-6919).

[23] *Ibid.*
[24] *Ibid.*
[25] Società Ingg. Nervi & Bartoli, Anonima per Costruzioni, patent IT 369975, *Procedimento di costruzione di centine di grandi dimensioni, e centina ottenuta con tale procedimento*, 5 January 1939, ACS.
[26] Società Ingg. Nervi & Bartoli, Anonima per Costruzioni, patent IT 370205, *Procedimento e dispositivo di sollevamento graduale di elementi costruttivi o di parti di costruzioni, e di realizzazione contemporanea dei sostegni destinati a sopportare tali elementi o parti*, 4 January 1939, ACS.
[27] Nervi, *Scienza o arte del costruire?* cit., p. 117.
[28] Nervi, *Un arco monumentale in conglomerato non armato* cit., p. 25.

chapter five
prefabrication for military and industrial purposes

The Via della Magliana laboratory and experimental models

Following the outbreak of war, the debate over autarky reached a turning point when, in September 1939, a decree imposed an absolute ban on the "use of reinforced concrete and iron" in public and private works, with the exception of those of "special public interest."[1] From then on, and particularly after Italy entered the war in June 1940, Nervi & Bartoli ceased any business activities for private individuals and devoted itself to the design and construction of military works for the Ministry of War. The firm continued to construct buildings only for Società Solvay in Rosignano.

After the Orvieto hangar building sites and those of the various underground tanks, it became clear that Nervi & Bartoli was one of the most prestigious Italian engineering firms engaged in Italian military defence works. It is therefore unsurprising that the Orvieto hangars became prototypes for the study of a rapid manufacturing system that could offer the Italian Royal Air Force the opportunity to build shelters for aeroplanes throughout Italy. In any case Nervi never stopped his research into the construction of a lamellar grid after Danusso and Oberti's tests carried out with a view to the building of the first two hangars in Orvieto. At the end of the 1930s, he therefore began to experiment with a construction system for prefabricated hangars made out of reinforced concrete against the backdrop of other studies of this kind, the last of which concerned the prefabrication of blocks for the E42 Arch. In order to carry out tests not on models, as Danusso and Oberti did, but rather on full-scale prototypes, Nervi built his own laboratory in Rome that was nevertheless based on the Milan Polytechnic Testing and Model Laboratory and perhaps the one that Muggia had once set up at his business headquarters.

[1] R.D.L. [Regio Decreto Legge] No. 1326, 7 September 1939, *Nuove disposizioni che vietano l'impiego del cemento armato e del ferro nelle costruzioni ed in alcuni altri usi.*

110. P. L. Nervi, Nervi & Bartoli laboratory and shed, Via della Magliana 238, Rome. Precast reinforced concrete roof trusses of the shed, 1939 (Maxxi - PLN, F05282).

Thus in 1939, on land owned by his business in Via della Magliana, near EUR, Nervi set up a proper research laboratory to study methods of prefabricating reinforced concrete with the aim of producing roofs of various kinds and particularly those to be used in the construction of military hangars, which belonged to the category of works of "special public interest." The choice of prefabrication was Nervi's technical response to the urgent military need for rapid construction of aviation facilities. In the wake of experiments carried out during the war years, prefabrication was in any case becoming the preferred method for installing reinforced concrete and its by-products.

"Laboratory for construction materials and adjoining rooms" was the title of the design for the shed on the Via della Magliana. His building permit was issued on 11 June 1939 (work on the shed was already underway by the summer).[2] The chronology of events suggested that Nervi and his company were working on behalf of the regime, with a view to building further hangars. Furthermore, Chief of Staff Giuseppe Valle informed Nervi of the regime's imminent need for hangars that could be rapidly assembled.[3] The same shed also provided Nervi & Bartoli with an opportunity to test a system that was to become crucial to Nervi's research during the war years: the on-site prefabrication of reinforced concrete elements for large roofs. As a sign of the urgency of coming up with an economical and rational roofing system, the construction of the shed went hand in hand with the construction of some prefabricated roof prototypes for hangars.

Above and beyond the shapes of the individual parts manufactured in reinforced concrete for the various types of roofs tested in Via della Magliana, for the shed and for the hangars, what enabled Nervi to envisage manufacturing large structures based on the assembly of beam segments, as Henry Lossier already planned,[4] was that he found a method of connecting the parts with a node of great mechanical strength. The novel feature consisted of the special types of that node, produced by welding the reinforcement irons left protruding from the prefabricated parts (a technique known to Nervi since 1923) and sealing the connections by means of concrete made with high strength cement to be poured into special timber formwork that moulded the shape of the node. The tests carried out in Via della Magliana were based on monitoring the strength of the nodes joining the prefabricated parts of the various structures, as well as on the difficulty of assembling individual parts imposed by the different geometries.

Nervi planned to use arched beams to produce a segmental vault and reticular roof trusses for the reinforced concrete roof of the shed, each divided

[2] See Greco, *Pier Luigi Nervi* cit., p. 179.
[3] Alberto Bologna, interview with Renato Valle, 22 February 2012.
[4] See Greco, *Pier Luigi Nervi* cit., p. 178.
[5] Pier Luigi Nervi, patent IT 377969, *Sistema costruttivo per la realizzazione di ossature resistenti di volte, cupole e in genere di sistemi statici, mediante elementi costruiti fuori opera ed uniti con giunti in conglomerato cementizio armato*, 9 November 1939, ACS.

into four sectors that would be joined using the special joint (this was the solution implemented). The development of this type of shed went hand in hand with that of the vaulted roofs of the hangars, again in reinforced concrete and made out of prefabricated ashlars or beam segments. While work was continuing on the shed, Nervi produced two hangar roof prototypes with different geometries alongside the shed itself. The types depended on the aircraft access, which could be through the head or along the side. In both types of structures, Nervi tried out two kinds of truss girders, one reticular and the other Vierendeel.

For the hangars with access through the head, Nervi designed great arches with Vierendeel or reticular girders and transverse stiffening joists, again Vierendeel or reticular. The prefabricated parts were joined together by welding the iron and then sealed using a special cement mortar. Construction of an arch prototype with two beams and joists, both of two types, allowed Nervi to check the difficulties involved in the reinforced concrete prefabrication system used to produce the Vierendeel and reticular geometries. The fact that Nervi subjected a prototype made up of both types to load testing confirms that he was interested in optimising the manufacture of individual elements and choosing the beam geometry that was easiest to achieve using the moulds.

For the hangars with access along the side, Nervi followed the diamond-shaped mesh reticular structure of the two Orvieto hangars to the letter, but now he was able to break down the entire truss into lamellae that could be prefabricated and fitted on site due to his special technique for the junction node. At this point, the reinforced concrete became a perfect transcript of the metalwork for the diamond-shaped mesh lamellar trusses, meaning that the Orvieto prototypes could be used for large-scale construction. Nervi also experimented with Vierendeel and reticular girder geometry in the case of the diamond-shaped mesh truss. He likely also carried out studies – perhaps also tested using relevant prototypes – for other concrete structures, most of which were designed for military purposes, such as cement casings for mines, underwater torpedoes, and floats. In the studies and experiments carried out in the Via della Magliana laboratory, Nervi discovered a special use concrete; where the resistance sections of the structural bars were minimised, the casings became cartilage and the geometries of the structures assumed new plastic forms.

The technical characteristics of the experiment carried out on prototypes of roofs made out of prefabricated elements became the subject of a patent filed in November 1939 that covered, in the words of Nervi, a "construction system for producing the resilient skeletons of vaults, domes, and static systems in general, by means of elements manufactured off-site and joined using joints made out of reinforced concrete mixture."[5]

The fact that the example of a vault shown in the drawings attached to the patent was described as a "type of common lamellar structure" confirms that the main inspiration for his diamond-shaped mesh reinforced concrete skeleton was metal roofs. The "skeletons" were made out of prefabricated "elements" or beams of various configurations, connected by means of right-angled "nodes" to be made out of concrete poured at the convergence of four beams by means of four wooden tablets. In the text of the patent, Nervi also suggested a variant, for which documents are available to determine whether it was tested in the Via della Magliana laboratory, which included truss girders to be produced by means of pairs of pipes connected by slender metal trusses; the pipes were fitted with a metal reinforcement and filled with concrete, and

111-113. P. L. Nervi, Nervi & Bartoli laboratory and shed, Via della Magliana 238, Rome.

111. Prototype of roof made out of prefabricated elements, 1939 (Maxxi - PLN, F05264).

112. Prototype of roof made out of prefabricated truss girders. Photograph of joint detail, 1939 (Maxxi - PLN, F05393).

113. Prototype of roof made out of prefabricated Vierendeel girders, 1939 (Maxxi - PLN, F05265).

114. P. L. Nervi, patent IT 377969, *Sistema costruttivo per la realizzazione di ossature resistenti di volte, cupole e in genere di sistemi statici, mediante elementi costruiti fuori opera ed uniti con giunti in conglomerato cementizio armato*, 9 November 1939 (ACS).

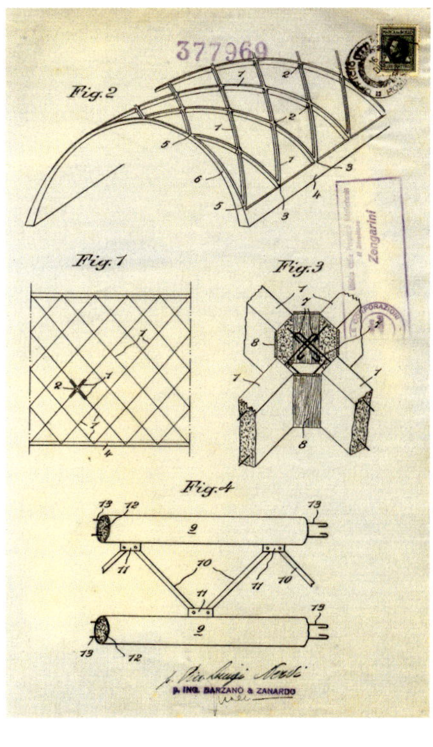
377969
Fig.2
Fig.1
Fig.3
Fig.4
p. ING. BARZANO & ZANARDO

may have been made out of steel or "cement-asbestos." In Nervi's view, which was always focused on concrete, those pipes mainly acted as a permanent form for the creation of slender reinforced concrete bars, even though the text of the patent claimed that the pipes could absorb any "tensile stresses" (it was no coincidence that he also envisaged the more autarkic "cement-asbestos" formwork pipes).

Hangars and prefabrication

In the same period when he was building the shed and the prototypes of prefabricated structures and studying the application of concrete to various types of military ordinance, Nervi was busy with his company designing and building, usually intended for military purposes, on behalf of the Italian Royal Navy, the Italian Royal Air Force, Società Piaggio of Pontedera, Società Chimica dell'Aniene, and Società Solvay.

In addition to some sheds with reinforced concrete trusses built in Trapani or Rosignano, Nervi & Bartoli constructed a series of underground air raid shelters in the military base on the island of Maddalena. These were protected with a reinforced concrete vault or two slabs reinforced with a metal network or "grid."[6] For the ILVA plant, in Portoferraio, they built shelters in a tunnel with a "vault in cement bricks" and one protected by a large vault in reinforced concrete covered by a "penetration slab."[7] Again in 1939, Nervi & Bartoli were commissioned to develop a design, which was subsequently not built, for a hangar with warehouse for Società Piaggio, at Pontedera, which had been commissioned by the War Ministry to produce fighter aircraft and bombers. For the structure of the warehouse saw-tooth roof and the hangar vault, Nervi went back to the diamond-meshed grid pattern he had used for the Orvieto hangars (It is possible that he was still planning to construct the hangar out of cast-in-place reinforced concrete, and not to use the prefabricated system that he was developing in the Via della Magliana laboratory. This is suggested by the presence of expansion joints in the structure and the roof covering in Perret hollow bricks).[8]

Once Italy had entered the war, the Royal Air Force relied on Nervi & Bartoli to build a series of hangars in various mainland locations. These all had to be built in accordance with criteria that had been tested on prototypes in the Via della Magliana laboratory, with two types of truss girder structures prefabricated in reinforced concrete. Thus, between 1940 and 1943, the company produced pavilion hangars, measuring 100 x 36 metres (two at the Orvieto airstrip, in 1940–1941; two at the Torre del Lago Massaciuccoli seaplane base, in 1943; and two at the Orbetello seaplane base, in 1942), and arched hangars measuring 45 x 55 metres (one at the Monte Corvino Rovella airstrip, now Pontecagnano, near Salerno, in 1940; two at the Stagnoni airstrip near Marsala, in approximately 1940; one at the Milo airstrip near Trapani, in 1941).

The fact that the hangars were produced in reinforced concrete led Nervi to return, in 1940, to the question of the "level of autarky" to justify the economic efficacy of his solutions.[9] Having invented a prefabrication system,

[6] Nervi & Bartoli design No. 1814, CSAC, n. inv. pra 208, n. id. 13458, coll. 270/6.

[7] Drawing No. 1856-2, CSAC, n. inv. pra 166, n. id. 13384, coll. 140/1.

[8] Nervi & Bartoli design No. 1821, CSAC, n. inv. pra 164, n. id. 13372, coll. 140/1.

[9] Pier Luigi Nervi, *Per la massima autarchia edilizia,* in, "Costruzioni-Casabella," XII, March 1940, No. 147, p. 3.

115-116. P. L. Nervi, design of plant and hangar for Società Piaggio & C., Pontedera, 1939.

115. "Schizzo prospettico degli shed (Perspective sketch of saw-tooth roofs)" (CSAC, n. inv. pra 164, n. id. 13372, coll. 140/1).

116. "Schizzo prospettico dell'aviorimessa (Perspective sketch of hangar)" (CSAC, n. inv. pra 164, n. id. 13372, coll. 140/1).

117-120. P. L. Nervi, design for hangar, type 100 x 36, 1939-40.

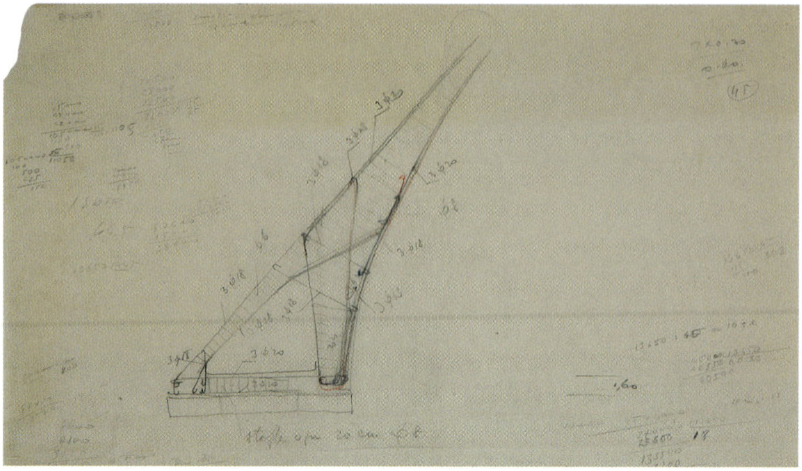

117. Spur reinforcement (CSAC, n. inv. pra 129, coll. 269/5).

together with the use of Innocenti scaffolding pipes for the assembly of parts, allowed him to state that Nervi & Bartoli's structures boasted the "maximum autarkic coefficient," because the entire manufacturing process took place with a view to a drastic reduction in materials for the "cement structure formwork" and for "shoring up temporary reinforcements."[10]

Nervi explored two types of foundation for the pavilion hangars, to be used depending on the type of soil, with or without piles (the piles could be in reinforced concrete or pine wood), and two types of spurs branched at the base. The shape of the spur became the object of careful study, with regard to the size and function of the parts, which revealed concerns similar to those Nervi had experienced when specifying the *Stadio Berta* cantilever beam parts. One of the fundamental details which shows the direction of Nervi's structural lines concerns the shape of the contours at the bifurcation at the base of the spur and the use of concrete in the foundations. In his initial sketches, Nervi wished to give authority to the main support pillar that rose up to the arched vault profile from the foundations. It was thicker than the tilted post, and the reinforcement irons declared its nature and trained the flow of forces into an abrupt change of direction to direct them into the ground along the vertical. The pillar and post were bound to the foot and set on a continuous foundation. In his final solution, Nervi decided to strengthen the post, which lay along the line of descent of the forces, and therefore reduce the size of the pillar to that of the abutment structure strut. This was required above all to hold the inclined post upright during the stage prior to the construction of the vault as well as to stabilise and take a part of the weight of the vault (the pillar foundations and post became separated). The design of the bifurcated spur already encapsulated the essence of the static and symbolic functioning of other similar structures that Nervi would use for future vaulted roofs.

For the construction of the central and corner spurs, Nervi redesigned the formwork to be produced by the carpenters, significantly reducing them until they were no larger than the various parts, without any information about the possibility of installing boards.

[10] *Ibid.*

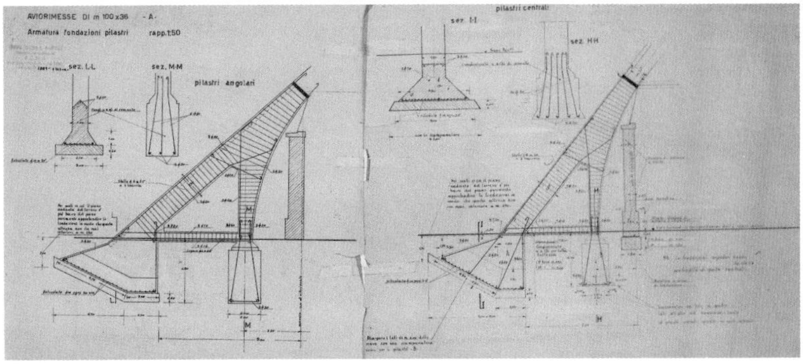

118. "Armatura fondazioni pilastri (Pillar foundation reinforcement)" (BSTA, 024-7289).

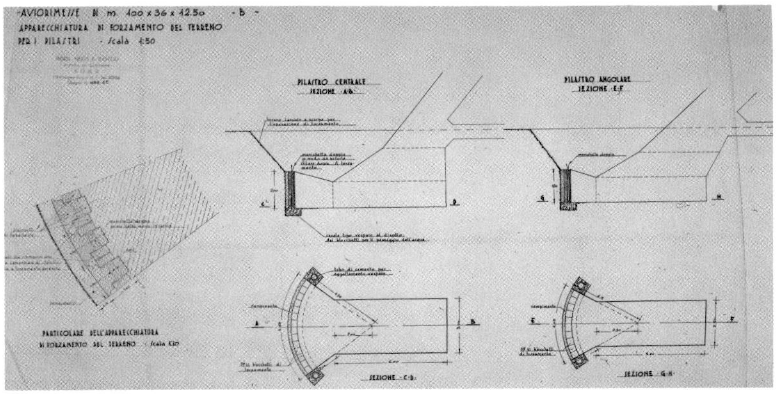

119. "Apparecchiatura di forzamento del terreno per i pilastri (Ground forcing device for pillars)" (BSTA, 024-7288).

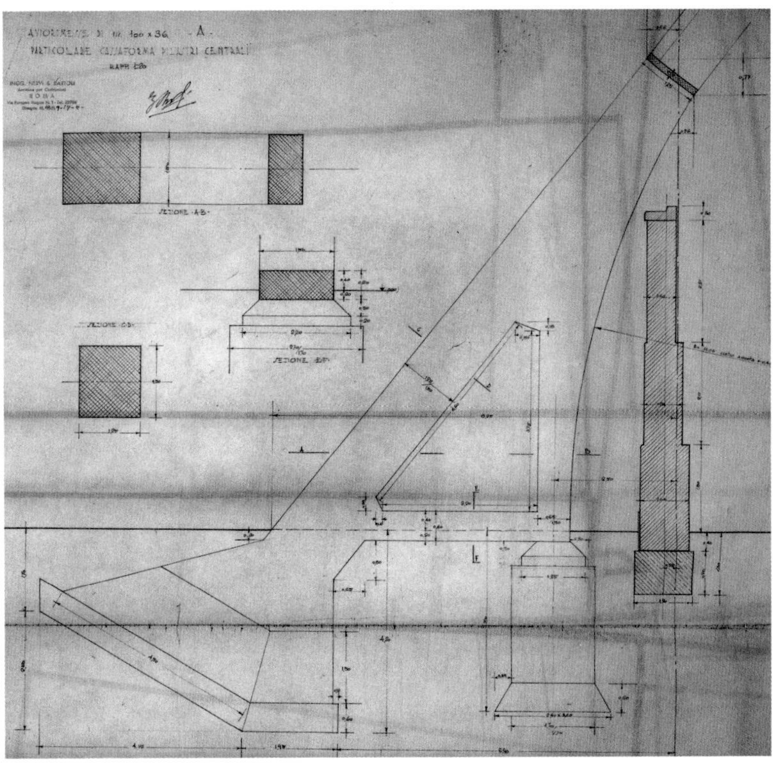

120. "Particolare cassaforma pilastri centrali (Detail of central pillar formwork)" (BSTA, 024-7339).

121-122. P. L. Nervi, hangar, type 100 x 36, Orvieto military airport, 1940-41.

121. Wooden model to explain the prefabricated part assembly geometry, about 1940 (Maxxi - PLN, F05386).

122. View of building site, 1940-41 (Maxxi - PLN, F04236).

Tests carried out on the strength of the nodes in his prefabricated reticular structure cannot have completely persuaded Nervi of the efficacy of the system, because he also used cast-in-place reinforced concrete skeletons for construction of the hangar vault. The final structure was indeed composite, made up of a network of arches consisting of prefabricated truss girders in reinforced concrete and a stiffening skeleton in cast-in-place reinforced concrete, consisting of intertwined beams, a ridge beam, four corner beams and large arches. This stiffening skeleton developed from the point where Nervi imagined particular stress would arise, where the vault rested on the pillars. Apart from the ridge beam, the four corner beams and the interlaced diagonals radiating symmetrically from the two middle pillars, the other cast-in-place reinforced concrete structures-interrupted the diamond-shaped mesh with their geometry in order to stiffen the prefabricated reticular structure.

The truss girders were prefabricated by means of reusable wooden formwork. During the assembly of the pieces of the vault and the construction of the stiffening skeletons, Nervi & Bartoli used Innocenti poles to erect the scaffolding for the wooden centring on which the prefabricated girders of the truss were laid. The truss girders were joined at the nodes by welding the irons and pouring high-strength concrete into temporary formwork, as had been tested in the Via della Magliana laboratory. This operation was accompanied by the production of stiffening beams and arches, using wooden formwork built directly on the centring and reinforced concrete manufactured on-site. In the Orvieto building site, where Nervi & Bartoli built their first two hangars with prefabricated truss girders, Nervi had a small-scale wooden model built of a slice of the vault, at the convergence point of two diamond-shaped grids of the vault webbing, to show the workers how they should install prefabricated elements on curved surfaces. Nervi used the system that he had patented in 1938 to consolidate the ground under the spurs.

The prefabricated truss girders were raised by means of a hoist positioned on the ridge beam. The workers could move the truss girders and lay them by hand. One stage of assembling the first series of prefabricated truss girders, without special construction equipment, is documented by a photograph showing the construction of hangars in Torre del Lago.

123-124. P. L. Nervi, hangar, type 100 x 36, Orbetello military airport, 1942.

123. View from land side (Acm-EPFL, 0172.02.0003.2056).

124. View of the structure (Acm-EPFL, 0172.02.0003.2088).

125-127. P. L. Nervi, hangar, type 100 x 36, Orbetello military airport, 1942.

125. View of the structure. Corner pillar (Acm-EPFL, 0172.02.0003.2057).

126. View of the structure with roof covering (Acm-EPFL, 0172.02.0003.2089).

127. View from the lagoon (Acm-EPFL, 0172.02.0003.2055).

128-129. P. L. Nervi, hangar, type 100 x 36, Torre del Lago Massaciuccoli military airport, 1943.

128. View of the structure of the first completed hangar and assembly of the roof of the second (Maxxi - PLN, F04254).

129. Detail of the spur (Acm-EPFL, 0172.02.0003.2059).

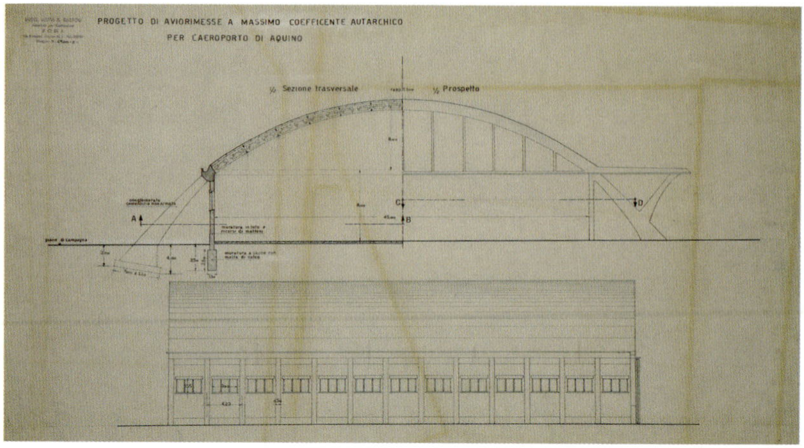

130. P. L. Nervi, "Progetto di aviorimesse a massimo coefficiente autarchico per l'aeroporto di Aquino (Design of hangar with maximum autarkic coefficient for Aquino airport)," 1940 (CSAC, n. inv. pra 25, n. id. 12769, coll. 268/1).

In order to cover the prefabricated lamellar truss, a covering of cement asbestos panels was fitted, secured to a framework of purlins in reinforced concrete, also prefabricated on site. Construction of the roof, including the covering, took about one month of work.

The 45 x 55 metre type of hangar, which the aircraft entered on the short side and lacked any columns, was developed in 1940 for Aquino Airport, near Cassino. The structure of large pillars and edge beams was made out of non-reinforced concrete alone, and Nervi described it as a prototype of a "hangar with maximum autarkic efficiency."[11] The hangar was roofed by installing prefabricated elements alone, which were again joined using the technique used for the other hangars, to form a series of load-bearing arches connected by truss girders or alternating joists. Nervi & Bartoli constructed the hangars based on this model in Marsala, Trapani, and Monte Corvino Rovella.

Construction of the Nervi & Bartoli hangars is documented by photographs demonstrating the different values that Nervi attributed to the two types. A photo shoot by Vasari illustrates the spectacular construction stages of the hangars at Orvieto, Torre del Lago, and Orbetello; with the reticular structures against the sky, before fitting the roof covering. These made a decisive contribution to the international success of Nervi's prefabrication system during the post-war decades. Those photographs became further visual manifestations of the concept of structural truthfulness, and their value was even equal to that of the perspective designed by Hendrik Petrus Berlage for the skeleton of the Municipal Museum of Art in The Hague.

The set of hangars showed Nervi all the economic potential and the degree of control over the finishing of concrete works permitted by prefabrication. Nervi observed the destruction of the hangars at Orvieto, Torre del Lago, and Orbetello by the retreating German army in 1943 in light of the load tests to which he had subjected the prototypes in the Via della Magliana laboratory. The ruins demonstrated to him the absolute monolithic quality achieved by a structure made up of pre-fabricated elements due to the techniques developed for the nodes.[12]

[11] Design Nervi & Bartoli No. 1900-1, CSAC, n. inv. pra 25, n. id. 12769, coll. 268/1.

[12] Nervi, *Scienza o Arte del Costruire?* cit., pp. 120-121.

131-132. P. L. Nervi, hangar, type 45 x 55, Stagnoni military airport, Marsala, 1940.

131. Side view. Installation of roof covering (BSTA, 024-6009).

132. Front view (BSTA, 024-6010).

133. P. L. Nervi, hangar, 45 x 55 type, Milo military airport, Trapani, photograph dated 25 June 1941 (BSTA, 024-6008).

Virtuoso trusses for autarkic silos

In 1940 when Nervi & Bartoli were engaged in building the first prefabricated hangars for Orvieto and the design for the hangar with "maximum autarkic coefficient" for Aquino, and when Nervi was holding forth on "maximum building autarky," the firm designed a construction system for tanks intended for ammoniacal substances (still with autarkic characteristics) in brick and concrete masonry with a reduced metal reinforcement, to be built without the use of wooden formwork. It was planned to pour the concrete into the cavity formed by the brick walls, while pozzolanic cement was to be used to waterproof the structure. The slender paraboloid vault of the "mechanical packaging factory" for Società Solvay in Rosignano, built in 1940, set upon reinforced concrete buttresses, was also designed in accordance with the construction principles of autarkic building. As before, the aim was to do without the formwork and reduce the presence of reinforced concrete, yet still create reinforced brickwork. Nervi therefore laid a thin layer of 5-cm-thick reinforced concrete or a "reinforcement vault" above the vault of brick joists, with a reinforcement that he referred to as "reticulate," made up of 8-mm diameter bars, laid out into a web measuring 20 centimetres on each side, arranged in a lozenge.

It was, however, the designs for the soda silos, dating from 1941 and commissioned for the Società Solvay plant in Rosignano, that demonstrated Nervi's will to seek an autarkic structure and construction system, characterised by a reduced use of formwork, the use of bricks and the application of certain technical innovations tested in the Via della Magliana laboratory and implemented in the first prefabricated hangars in 1940.[13] His desire for "maximum building autarky" and the prefabrication system led him to prefer separating the load-bearing skeleton of the silo, in reinforced concrete, from its wallcovering, in reinforced brick.

In his designs for the Società Solvay silos, Nervi defined two different types of structure, created from the same soda storage system, by precipitation of powder from a gallery at the top: a single huge container measuring 35 metres high, 80 metres long, and 39 metres wide resulting from the juxtaposition of different skeletons, as in the 1939 viaduct design, suited to the operation of the silo, with forked branching buttresses of the same kind used for the

[13] Design Nervi & Bartoli No. 2010, CSAC, n. inv. pra 29, n. id. 12787, coll. 268/3, 268/4.

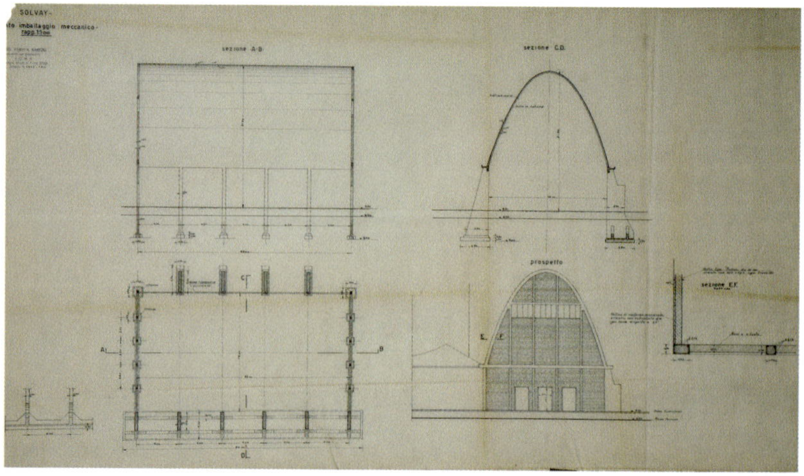

134. P. L. Nervi, design for "Fabbricato imballaggio meccanico (mechanical packing building)," Società Solvay, Via Piave 6, Rosignano Marittimo, 1940 (CSAC, n. inv. pra 29, n. id. 12787, coll. 268/3).

135. P. L. Nervi, designs for soda silos for Società Solvay, Rosignano Marittimo, 1941.
Version with structure made out of precast reinforced concrete elements (CSAC, n. inv. pra 29, n. id. 12787, coll. 268/4).

stadium stands and parabolic arches, as in the salt warehouse at Margherita di Savoia; or three separate silos for which he proposed two variants for the structural form, one a hyperbolic paraboloid (maximum diameter of 17 metres, height of 45 metres) and another with a diamond-shaped truss over the cylindrical surface (diameter of 33 metres, a height of 43 metres).

In the version with the colossal single silo, the reinforced concrete of the skeleton was broken down into two kinds, one manufactured on-site in the forked buttresses and the other prefabricated for the Vierendeel girders, developed in the Via della Magliana laboratory and destined to be welded together to construct the parabolic arches. It was planned to make the infill or "vaulted walls" out of concrete hollow blocks patented in 1937 (the surface in contact with the soda was lined with a layer of reinforced concrete plaster with a grid, or "wired cement").

In the landscape of the Rosignano industrial complex, the colossal silo was to show off the texture of its materials, the reinforced concrete of the unclad skeleton and the intertwining of the joints between the concrete blocks making up the arched infillings. It is difficult to escape the impression that in the structure of the silo, and its apse-shaped floor plan, it had been Nervi's intention to create a monstrous and rugged industrial cathedral of exposed concrete.

With the aim of limiting the quantity of reinforced concrete, Nervi proposed a version of the three silos, each in the shape of a hyperbolic paraboloid and with a shell made out of reinforced masonry. The choice of geometric figure followed the same approach that led him to adopt parabolic arches for the salt warehouse, because the greater width of the hyperbolic paraboloid at its base allowed the soda dropping from above to pile up without excessive pressure against the shell. Due to this characteristic, the shell was planned with their ring walls of 30 centimetres, in two-headed bricks, but surrounded by rings of reinforced concrete with a diameter of 2.10 metres to reinforce it like a barrel against any pressure exerted by an unpredictable and uneven distribution of soda. The rings were integral with a 10-centimetre-thick reinforced concrete strait jacket enfolding the wall between the rings, and they were contoured for the collection of rainwater (the brick shell in contact with the soda was lined with cement mortar). The use of formwork was virtually eliminated for the separate construction of the rings; the same layer of

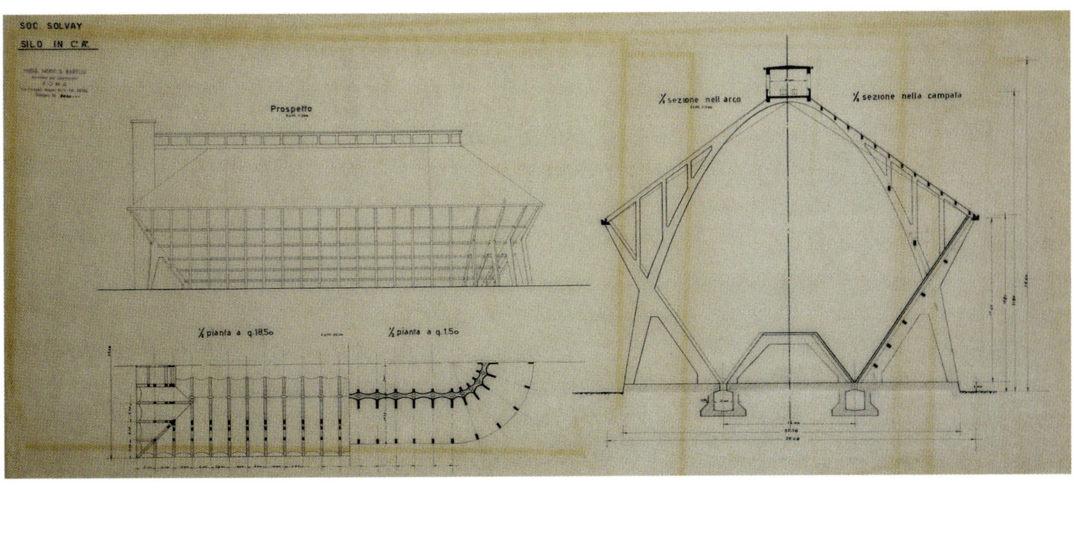

136-137. P. L. Nervi, designs for soda silos for Società Solvay, Rosignano Marittimo, 1941.

136. Version with structure made out of reinforced masonry with a hyperbolic paraboloid profile (BSTA, 024-6916).

137. Version with diamond-mesh truss structure in reinforced concrete (CSAC, n. inv. pra 29, n. id. 12787, coll. 268/4).

concrete applied to the shell between the rings would not have required any formwork (Nervi likely planned to apply it with a cement gun).

The reinforced concrete rings that gave the hyperbolic paraboloid its original pulsing energy, served to reduce the construction materials and revealed the growing construction made up of rows of bricks, round by round; as Brunelleschi would have said, they were the secret cipher of their creator. As in the case of the 1939 viaduct, Nervi was again setting out to make autarkic use of the reinforced concrete, where every technical detail was inspired by the saving of material and the provisional works to be installed on site.

The variant with the three cylindrical silos is the only one of those developed by Nervi for Società Solvay to ignore the mound of soda dropping from above. In this case, the starting point of the creative process could be seen as the desire to apply a certain form of structure to the silo. Nervi imagined he was returning to the diamond-shaped mesh reticular structure of the hangars and using this to create a fantastical interweaving of lines to be set above a base frame whose bars traced segmented arches. In this case, as in the colossal single silo, the reinforced concrete appears in both versions, namely executed on site and prefabricated (the beam segments in the drawings had the geometry of the truss version tested in the Via della Magliana laboratory). The infill turned and detached from the cross grids was planned in masonry or brick, using the same technique as for the single silo. The slenderness of the skeleton lines and their partial detachment from the base ultimately make the reinforced concrete work look like evaporated lace, perforated and transparent, spread out before the mighty vaulted wall onto which they would project slender shadows.

In the cylindrical silos, Nervi sought to achieve a clear distinction of the structure, partly to use the masonry for the infill, but above all in order to use the diamond-shaped mesh truss. After the construction of the hangars, this had become the calling card of Nervi & Bartoli's state-of-the-art, autarkic reinforced concrete. When we consider the variants for the structure of the *Stadio Berta* shelter and stairs proposed in the various designs, and those of the diamond-shaped truss for the hangars, there can be no doubt that Nervi was constantly seeking iconic figures for his structures, which could make his work original and distinctive. He continued to work with the same creative approach during subsequent decades, sometimes threatening to betray his own static intuition.

Above and beyond the limitations inherent in the choice of the cylinder shape, the entire design of the silo's truss of diamond-shaped lines is animated by Nervi's imponderable creativity in guiding the invisible static forces toward a specific aesthetic and a graphic harmony supported by reassuring static figures, namely the markings on the arches, the evident St. Andrew's cross, or the

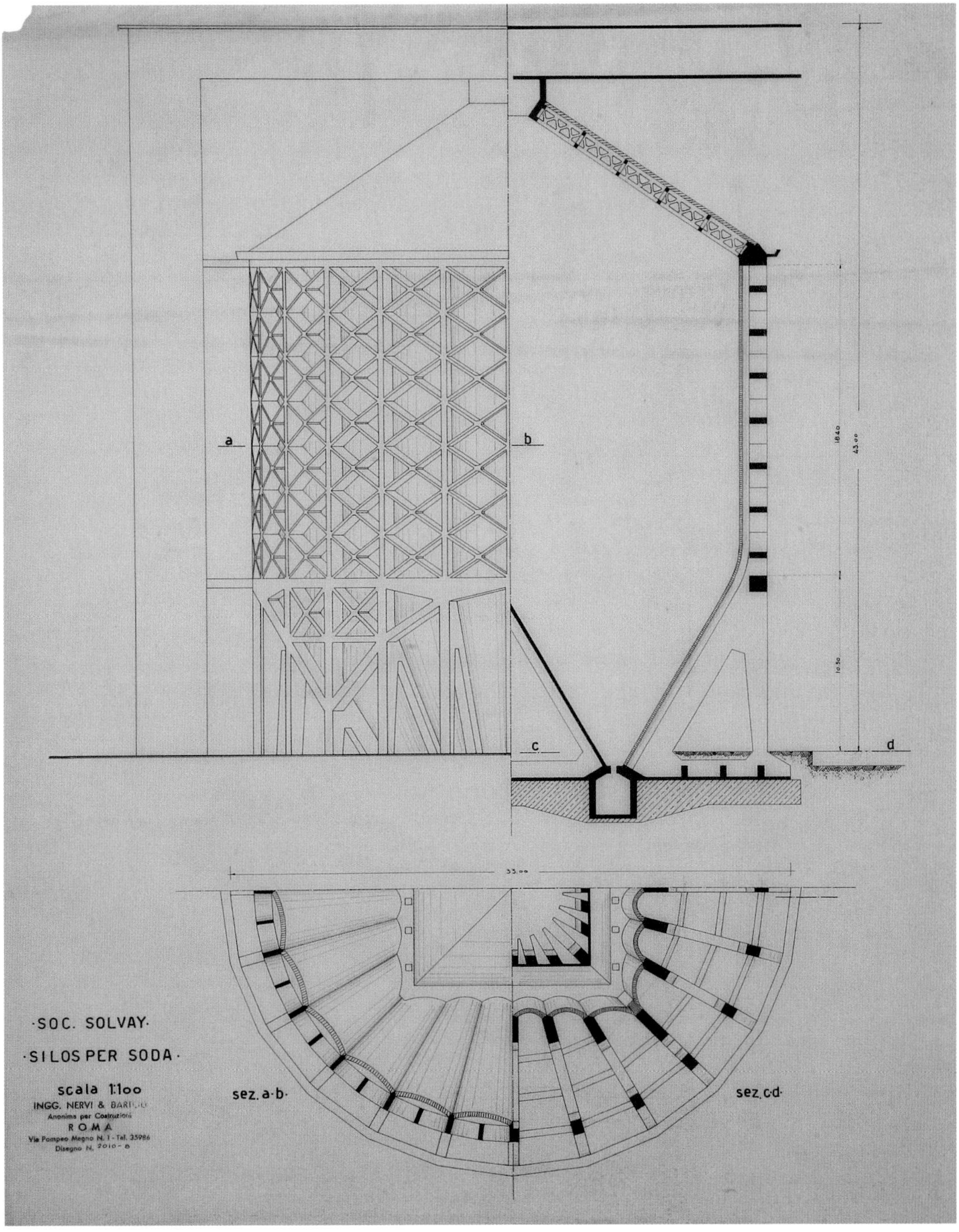
a
b
c
d
·SOC. SOLVAY·
·SILOS PER SODA·
scala 1:1oo
INGG. NERVI & BAR
Anonima per Costruzioni
ROMA
Via Pompeo Magno N. 1 - Tel. 35986
Disegno N. 2010 - B
sez. a·b·
sez. c·d·

mighty pillars. Together with the hangars, this version of the silo demonstrates the emergence of a creative direction toward a specific way of manufacturing structure and space, where, by virtue of the prefabrication and the assembly of the media, the technique and art of building transforms the elemental framework of reinforced concrete into an interweaving that increasingly resembles those of the vaults or rose windows of Gothic cathedrals, the ribs of Guarini's churches, or the veins on leaves observed under the microscope. Nervi was preparing to lay the basic premises for the structural geometries that were to experience a new impulse in the years around 1950 through the works of Richard Buckminster Fuller, Konrad Wachsmann, and Louis Isadore Kahn.

In the clearly-drawn lines of their ribs, both versions of the three silos – hyperbolic paraboloid and cylinder – reveal Nervi's predilection for a structure always characterised by the visibility of the forces at play, differentiated according to the programme yet distinguishable in their originality because of the generating process that always infuses them with the same vitality of form and increasingly turns them into works of art through the mastery of technique. It is no coincidence that in subsequent publications Nervi favoured the solution of hyperbolic paraboloid silos, because only in that form is the structure perfectly moulded to the function of the container.

Hulls in reinforced concrete with a geodesic structure

In addition to the hangars, air-raid shelters, and underground tanks, Nervi & Bartoli was developing and constructing other military works intended for defence at sea, often commissioned by the Italian Royal Navy and including floating outposts, underwater weapon systems, ordnances such as mines and torpedoes, anchorage buoys, floating obstructions, rafts and transport barges, or ships of various kinds. As part of the process of designing these works, between 1939 and 1942, Nervi planned a concrete reinforcement to be produced using a thin shell technique that he had been testing since the days of his partnership with Muggia for the construction of liquid and gas tanks. The nature of the armature, the quality of the mixture composition, and the manner of application were to become, from 1943, the subject of practical experiments that would culminate in the invention of the cementitious product that Nervi would use to build ships as well as magnificent structures in the immediate post-war years.

Everything stemmed from his desire to achieve an ideal cementitious product that was perfectly uniform, not subject to cracking, and always along the same lines as his *sidero-cemento*. The patent for submarine mines and torpedoes, filed in December 1939, contained valuable information about the criteria that Nervi pursued to that point in order to obtain a surface in "cementitious material" or asbestos-cement that was as waterproof as a metal surface. In addition to ensuring the laying of the compound by vibration and "mechanical compaction," Nervi specified that the casing should be kiln dried and then dipped in special substances with the aim of sealing the "tiny internal canals that are found even in the most compact cementitious materials [...]."[14] All these chemical processes ultimately led to experiments conducted with the aim of improving the manufacture of reinforced concrete.

[14] Società Ingg. Nervi & Bartoli, Anonima per Costruzioni, patent IT 379019, *Involucro per mine o torpedini subacquee di materiale cementizio, di cemento-amianto o di altra sostanza non magnetica*, 9 December 1939, ACS, p. 3.

[15] *Ibid.*

Between 1940 and 1941, when Nervi designed rafts, military outposts, or floating obstructions, he broke down the works, as in the case of the hangars, into units that could be prefabricated in the form of caissons. To create the walls of these caissons, he developed a metal weave to reinforce the concrete and prevent cracking phenomena. For this purpose, he used a square mesh grid with sizes that varied according to the floats (15 x 15 or 20 x 20 centimetres), to be produced using round bars measuring 6 millimetres in diameter, which he referred to as "reticulate" and formed the true reinforcement of the concrete, as in the case of the netting used to rig the layer of reinforcement spread over the parabolic vault at Rosignano. He also laid a dense wire mesh over the netting on the walls of the caissons on the surface in contact with the water in order to prevent the concrete from cracking. The wall thicknesses ranged from 2 to 3 and 4 centimetres. The parts in contact with the water were to be coated with a water-repellent layer of asphalt or *idrobitosit*, inserted between two adjacent walls. The wall surrounding the assembly of individual units was to be made out of gunite cement only in the case of floating obstructions.

The concern to obtain a uniform material that would not crack is reflected in the name that Nervi sometimes used to refer to the specific nature of the material used in the casings of his floats: "wired cement walls."[15] The definition is already evidence of his insight that he could improve the cement product by examining the "netting" and "net" – the term "wired" captures the essence of the novel nature of his ferrocement. In some ways, the "metal fibres" embedded in the compound of his *sidero-cemento* already contained the precursors of what had now become his "wired cement walls." A "wired" reinforced concrete made its appearance in approximately the same period in the thin prefabricated *tavelloni* that he used for the roof of the fuel depot in Siracusa, designed in 1941 for the Italian Royal Air Force.

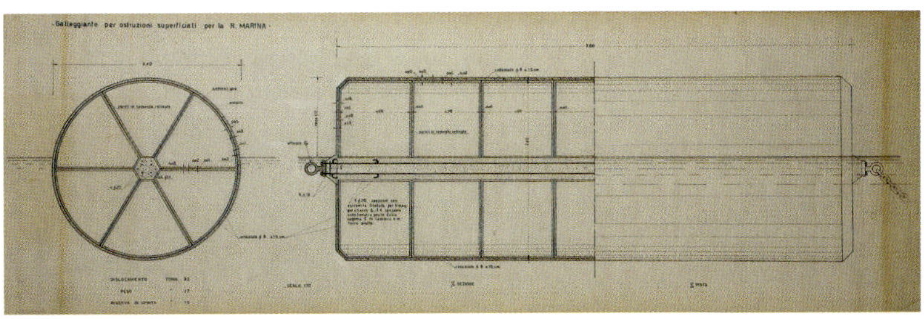

138. P. L. Nervi, design for "galleggiante per ostruzioni superficiali per la R.[egia] Marina (float for surface obstructions for the Italian Royal Navy)," about 1940 (CSAC, n. inv. pra 127, n. id. 13261, coll. 269/4).

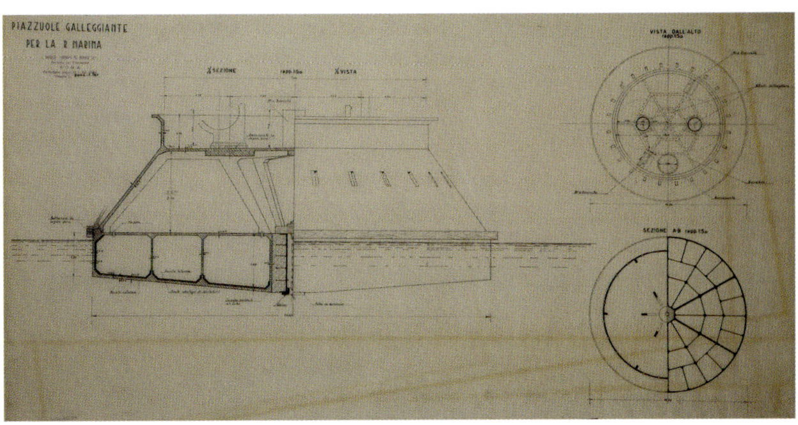

139. P. L. Nervi, design for "piazzuole galleggiante per la R.[egia] Marina (floating emplacements for the Italian Royal Navy)," about 1941 (CSAC, n. inv. pra 87, n. id. 1373, coll. 269/2).

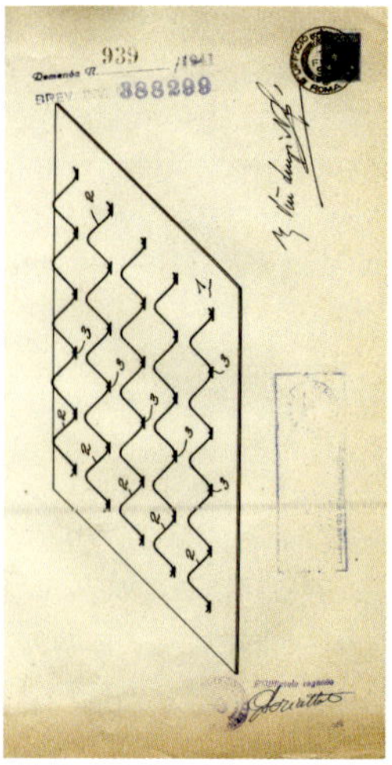

140. P. L. Nervi, patent IT 388299, *Procedimento per la protezione degli scafi di cemento armato dall'azione degli agenti atmosferici e delle acque, o da azioni nocive esterne comunque esercitate, con miglioramento delle caratteristiche statiche degli scafi medesimi*, 13 February 1941 (ACS).

Sometimes Nervi's projects for the Italian Royal Navy were less about practical solutions and more about future possibilities. This applies to the circular military outpost, or "floating emplacements" intended for surveillance and aerial defence of the coastlines, set upon caissons and equipped with living quarters for the crew and deck officer and an emplacement for gunners. The caissons were planned in accordance with the same criteria as the other floating works, divided into segments to be prefabricated and assembled, with a 6-centimetre-thick shell reinforced at the corners and to be made out of reinforced concrete with nets, netting and metal round bars, to be made up using "aluminous cement." Given the warlike outcome of Mussolini's policy, Nervi's design for a floating hotel seems to assume the value of an early demonstration of the utility of reinforced concrete in the construction of military devices such as these floating emplacements.

Nervi did not limit himself to designing various instruments of war. In 1941, after the tests he conducted on the hangars, once again at the laboratory in Via della Magliana, he started experimenting with the manufacture of reinforced concrete structures for the construction of hulls, confirming the crucial role played by that laboratory in the development of military hardware. He therefore designed planking systems, experimented with the prototype of a composite hull structure, and filed several patents. Those early experiments gave rise to a set of ships, built from 1943 to the 1960s with the earlier vessels providing a test bed for developing the technique of reinforced concrete into that of ferrocement.

Nervi's studies and experiments focused on the production of naval structural work in reinforced concrete, following in a tradition that originated, apart from celebrated 19th-century examples, in the early 20th century. He studied the technical nature of planking, with the aim, specifically illustrated by designs for rafts or floating emplacements, of eventually using reinforced concrete to produce this key element of the hull.

It is unknown whether any experimental tests were carried out on the type of planking that was the subject of a patent filed on 13 February 1941, although during the course of the year Nervi produced a structural prototype for hulls in reinforced concrete with wooden planking.[16] The planking described by Nervi served to provide the reinforced concrete hull with a protective layer, in the form of sheet metal or wooden boards, against impact and water corrosion. The originality of the invention lay in having turned what would have been traditional planking into a permanent form for the reinforced concrete (clamps or coiled bars riveted to the wood and welded to the sheet metal to ensure the adherence of the concrete). Nervi stated the limits of the resistance of reinforced concrete against the corrosive action of the water, explaining the reasons behind the specific configuration of all his previous large roofs, from the *Stadio Berta* grandstand to the hangars, where he always used a brick slab, and not a reinforced concrete covering, as a material to be laid on top of the skeletons.

In 1941 he built a prototype of a hull structure at the Via della Magliana laboratory that was still based on the same principle of separation between the parts of the reticular structure and the roof covering that he had developed for the hangars. Reinforced concrete and wood were the two materials

[16] Pier Luigi Nervi, patent IT 388299, *Procedimento per la protezione degli scafi di cemento armato dall'azione degli agenti atmosferici e delle acque, o da azioni nocive esterne comunque esercitate, con miglioramento delle caratteristiche statiche degli scafi medesimi*, 13 February 1941, ACS.

[17] Pier Luigi Nervi, patent IT 395090, *Procedimento di costruzione di scafi con ossatura resistente in cemento armato e fasciame in legno*, 26 January 1942, ACS, p. 3.

141-142. P. L. Nervi, hull prototype, Via della Magliana 238, Rome, 1941.

141. View of the structure (Maxxi - PLN, F07381).

142. Structure load test (Maxxi - PLN, F07380).

used in the prototype, the former for a skeleton and the latter for every type of planking. Construction of the prototype served to check the solidity of the connections between the reinforced concrete beams and between these and the timber board planking. The floor plates of the sides and the ridged hull deck beams were produced by assembling prefabricated truss girders and also by welding the irons. This transverse framework was stiffened by keelsons running along the juxtaposed keel and along the deck, in the form of reinforced concrete beams manufactured on-site. The wooden planks, including the keel, which contributed to supporting the timber planking laid on the hull skeleton, were fastened by means of long bolts passed through the keelsons themselves.

Encouraged by his experiences on the resistance of structures carried out repeatedly in Danusso and Oberti's laboratory, Nervi tested the prototype hull structure to examine the performance of his critical nodes, such as that between the reinforced concrete keelson and the wooden keel. It is possible that the scientific instrumentation and assistance with checking the tests was offered by the laboratory affiliated with the Istituto Nazionale Fascista per gli Studi e la Sperimentazione nell'Industria Edilizia based in Rome and led by Professor Antonio Martinelli, who was to carry out other similar checks for Nervi a few months later. The experiments were carried out in the presence and under the control of the engineer Enrico Mengoni, of the Italian Naval Register.

After building and checking the performance of the prototype, a patent was filed in January 1942 for the "construction of hulls with a strong reinforced concrete skeleton and timber planking," in which Nervi specified ways of manufacturing the beams in "extremely specialised and fully recoverable formwork,"[17] producing nodes between the beams and securing the planking

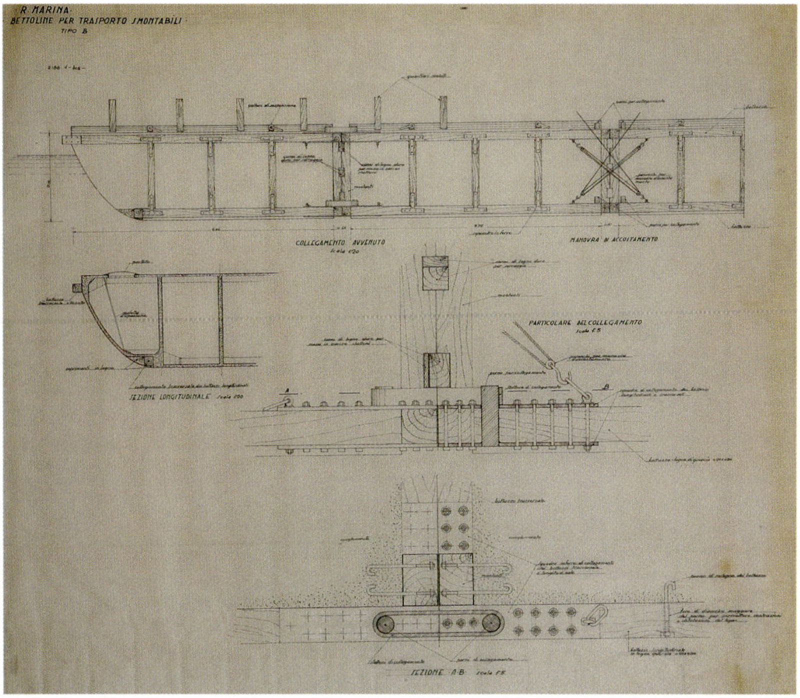

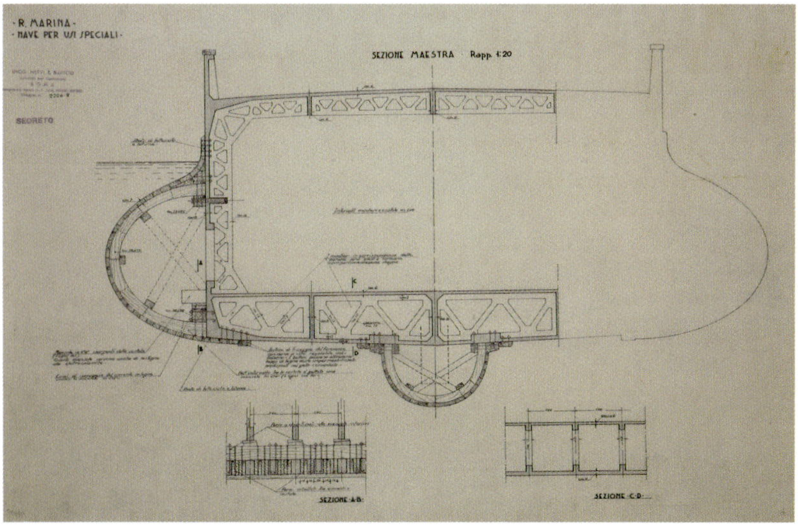

143. P. L. Nervi, design for "bettoline per trasporto smontabili (demountable transport barges)" for the Italian Royal Navy, 1942 (CSAC, n. inv. pra 126, n. id. 13260, coll. 269/4).

144. P. L. Nervi, design for "nave per usi speciali (special-purpose boat)" for the Italian Royal Navy, 1943 (CSAC, n. inv. pra 1032, n. id. 14827, coll. 266/4).

boards – not the same as the type patented in 1941. The removable transport barges developed in 1942 represented an initial partial application of the mixed structural system of reinforced concrete and wood, although they continued to be fitted with a reinforced concrete shell, like the other floating devices designed for the Italian Royal Navy.

A further development of the technique documented in the 1942 patent led Nervi to file another patent on 25 February 1943, for the "construction of hulls with a strong skeleton in reinforced concrete and timber planking."[18] One application of the composite technique of prefabricated beams, cast-in-place beams, and timber planking was the design for an Italian Royal Navy cargo transport boat measuring approximately 10 metres in width, with a concrete slab reinforced with double netting.

The text of the January 1942 patent on the hull structure contains an adjective crucial for describing a type of grid well known to Nervi. While the attached drawing documents a framework of "resistant skeletons" arranged "in accordance with the transverse and longitudinal system," Nervi explains in the text that it is also possible to arrange the same beams "in accordance with two inclined directions," in order to form what he defined as "a 'geodesic' type of structure."[19] It is clear that he intended to use this geometric pattern to transfer the skeleton he had tested in the hangars to that of the hulls.

Nervi now employed the definition of a geodesic structure that had been used since the end of the 1930s to indicate the metal latticework of an aeroplane fuselage[20] to refer to naval skeletons; and this definition revealed, perhaps even to Nervi himself, the universe of geometric figures from the grids he had assembled in the Orvieto hangars. A few months after the patent, Nervi used the same expression to describe the diamond-shaped grid of the Orbetello with the prefabricated truss girders photographed before laying the roof covering, and that of the first Orvieto hangars. That grid was now defined as a "geodesic structure," generator of a "new essentially modern aesthetic derived from the bold wisdom of calculation."[21] In Nervi's terminology, the definition of "geodesic structure" to describe all his hangars finally replaced the term "reticular structure" that he had used for the first Orvieto hangars.[22] There can be no doubt that changing the structural definition from reticular to geodesic was affected by the current debate in Italy around 1940 within the field of military aviation. The change in definition was decisive in emphasising the originality of Nervi's reinforced concrete structure and also played a fundamental role in placing his hangars at a strategic point in the genealogy of those structures – by then referred to as geodesic – that were to emerge in the wake of experimental research carried out around 1950, particularly in the United States.

[18] Pier Luigi Nervi, patent IT 407889, *Procedimento di costruzione di scafi con ossatura resistente in cemento armato e fasciame in legno*, 25 February 1943, quoted in, Greco, *Pier Luigi Nervi* cit., p. 284.

[19] Nervi, patent IT 395090 cit., p. 1.

[20] Aldo Guglielminetti, *Costruzioni Geodetiche*, in, "Rivista Aeronautica," vol. XV, February 1939, No. 2, (pp. 319-327), p. 322, and Gaspare Santangelo, *Sul calcolo delle strutture geodesiche aeronautiche*, in, *VI° Convegno Nazionale di Aerotecnica nel XX anniversario dell'Associazione Italiana di Aerotecnica. Roma, 13 giugno 1940 – XVIII*, Roma, Istituto Poligrafico dello Stato, 1941, pp. 592-611. See also Mario Sassone, Edoardo Piccoli, *The Grid Structures of Nervi's First Hangars in Orvieto: Innovation in Context*, in, Abel, Arun, Chiorino, *op. cit.*, pp. 162-166.

[21] Nervi, *La Tecnica ed i suoi orientamenti estetici* cit., p. 11, captions to illustrations.

[22] Nervi, *Un'aviorimessa di cemento armato* cit., p. 5, captions to illustrations.

A technique based on absolute style

The studies into hulls were not merely one of the many activities conducted by Nervi & Bartoli. For Nervi, their design and testing became an essential exercise for understanding the existence of absolute laws, dictated by speed and resistance factors, and for clarifying his concept of style. When he first defined the roofs of his hangars as "geodesic structures" in an article entitled *La tecnica e i suoi orientamenti estetici* (published in 1942 by a review of Gio Ponti entitled "Lo Stile," prefaced with introduction by Bardi), Nervi used boats, planes, and cars as an example to emphasise the inevitability of forms derived from a "strict adherence to models of physical nature," from "natural laws," from a "'natural' balance of forces," from "construction sincerity," and from "designing *in accordance with nature*."[23] His avowal that the striking increase in the sizes of the roof structures over the spaces meant reducing formal arbitrariness, with the aim of defining a kind of absolute style, was really the prelude to his designs for the magnificent vaults that he began to draw a few months after the article in order to establish the potential of his technical invention of ferrocement. Even though, in that article, he criticised the arbitrary semicircular shape of Libera's arch, there is no doubt that having to explore its construction disclosed to him the existence of those extreme limits of structure, substantially similar to those experienced in the design of the ships. He was thereby preparing to explore in order to discover the "immutability" and the "universality" of the "new style" imposed by technique. Nervi was adamant that his generation had witnessed the "advent of forms that cannot be abandoned and the first steps of a style that will be definitive forever."

Nervi was readying himself to be one of the greatest proponents of the style of the laws of nature and engineering. It was in the context of the assembly checks on prefabricated truss girders in hangars and hulls, and following his statement, in *La tecnica ed i suoi orientamenti estetici*, about the great arches, large bridge girders, and "roofs over large areas" that in 1943, regarding the halt in work in 1942 on Mazzoni's *Stazione Termini* railway

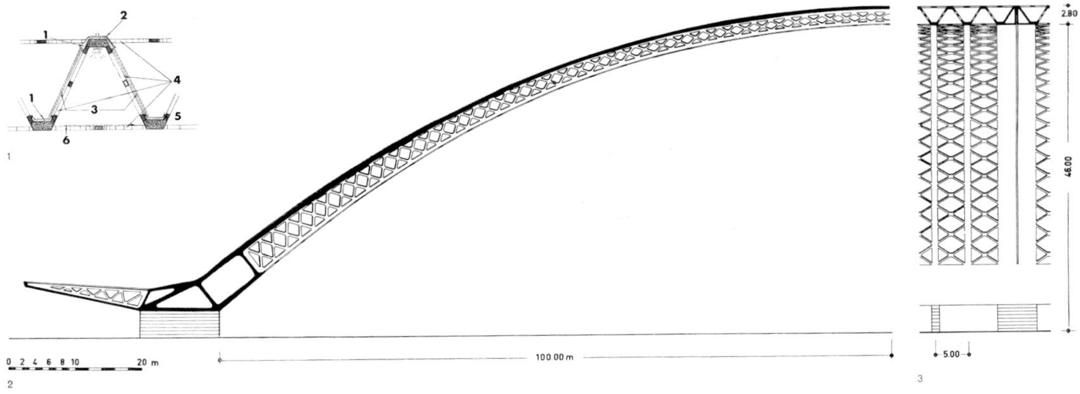

145-146. P. L. Nervi, design for railway station with 200 metre span, 1943.

145. Perspective drawing (P. L. Nervi, *Scienza o arte del costruire?*, Roma, Edizioni della Bussola, 1945, plate XVIII).

146. Detail of roof (*The Works of Pier Luigi Nervi*, London, The Architectural Press, 1957, p. 46).

station in Rome, he prepared a design for a great roof for a railway station, with a segmental vault featuring a span of 200 metres that was 46 metres high at its apex and constructed using prefabricated truss girders that were jointed like those of the hangars and hulls. The study for the construction of Libera's arch would be useful to him because it had reassured him that it was possible to design centrings suitable for this kind of structure. In a departure from the tests carried out in the Via della Magliana laboratory for the arched, vaulted roofs, Nervi envisaged configuring a reticular structure with a corrugated cross-section that was 2.8 metres high in order to achieve the exceptional fixed span established as a goal for Libera's arch. This was to be the same kind as that built by Freyssinet for the Orly hangars, but braced by diamond-shaped trusses, again prefabricated, woven between the corrugated bands. The juxtaposition of the two geometric trusses, which he had already adopted for the 1939 viaduct and for the great silo in Rosignano, gave rise to a kind of three-dimensional vaulted structure that was both reticular and geodesic. The height and the pleating of the structure meant that Nervi was able to pierce the vault and turn it into an overhead *broderie* permeated with diffuse natural light from the windows: a spectacular apotheosis of the evolution of cement vaults toward the filiform weaves of 19th and 20th century metal roofs. The continuous pattern of slender interwoven lines was nevertheless placed in jeopardy by the construction detail that, until then, had always been contested by Nervi, even though it was recommended in reinforced concrete manuals: the expansion joint. His entire roof was therefore divided, at 60-metre intervals, by special solid corrugated beams juxtaposed along the expansion joints to stiffen the various vaulted sectors and make them independent.

When he looked at the perspective drawing of the station, Nervi recognised in the structure the ability to arouse an "authentic architectural emotivity" and he was aware of its "monumental grandeur."[24] The structure shared this quality with Libera's arch, and it accompanied Nervi's subsequent designs for similar great vaulted roofs.

[23] Nervi, *La Tecnica ed i suoi orientamenti estetici* cit.
[24] Nervi, *Scienza o arte del costruire?* cit., p. 122.

chapter six
towards slender construction

The invention of ferro-cementitious felt

The armistice of September 1943 came at a turning point for Nervi's research into cement materials and his business output. All of 1943 was for him – a year marked by the discovery and testing of the particular kind of reinforced concrete that would allow him to set out triumphantly in the direction of works that were even more magnificent than Libera's *Arco dell'Impero*, with a yard of workers and equipment organised in response to the economic austerity of the day. It was during that year that Nervi definitively improved the technical and physical characteristics of ferrocement, as it was used during the 20th century. He also transformed the "wired walls in cement" and the sheets of netting, networks, and cement that he had previously planned for various types of float into a product that was as ductile and foldable as metal foil; it was uncrackable once it had set into its strongest forms, although it was subject to deformation. Any wooden formwork had been permanently removed from Nervi's manufacturing process, and he had achieved his aspiration for an autarkic product as the dissolution of the regime inexorably approached. His discovery of a way of arranging the metal and cement compound to create a form-resistant framework enfolding the space occurred only after perfecting the construction technique of the naval hulls.

The experiments carried out in Via della Magliana for the construction of hulls and floats are reflected in a patent that Nervi filed on 15 April 1943 entitled, "Perfecting the Construction of Slabs, Plates and Other Reinforced Cement Structures."[1] The "perfecting" concerned the genealogy, started by Monier, of inventions and patents that are not explained in Nervi's document and devoted to the kind of concrete reinforced with wire mesh. Even the title

[1] Pier Luigi Nervi, patent IT 406296, *Perfezionamento nella costruzione delle solette, lastre e altre strutture cementizie armate*, 15 April 1943, in, Greco, *Pier Luigi Nervi* cit., pp. 287-288.

clearly reveals the purpose of restoring authority to the kind of "reinforced cement structures" that had been overshadowed by the international success of the Hennebique skeleton system.

Nervi coined the definition of "ferro-cementitious felt" *(feltro ferro-cementizio)*, instead of making use of the existing term of ferro-cementate *(ferrocementato)* or ferrocement *(ferrocemento)*. The term "felt" added to ferro-cementitious is an indication of his dissatisfaction with a definition that did not fully express the characteristics of his product, which was obtained by superimposing multiple layers of mesh and cement mortar, compressed together through the use of operations similar to those used in the production of felt, which is not a conventional fabric but a material obtained by pressing together wool fibres and possibly animal hair. It should not be forgotten, if we are to grasp all the possible nuances of his choice of the word felt, that this material was also used to manufacture hats, which are included on the list of items drawn up by Nervi to illustrate the concept of strength through form. Nervi's ferro-cementitious felt allowed him to construct any possible figure designed to be strong by virtue of its intrinsic form, in the same way as bent or moulded felt.

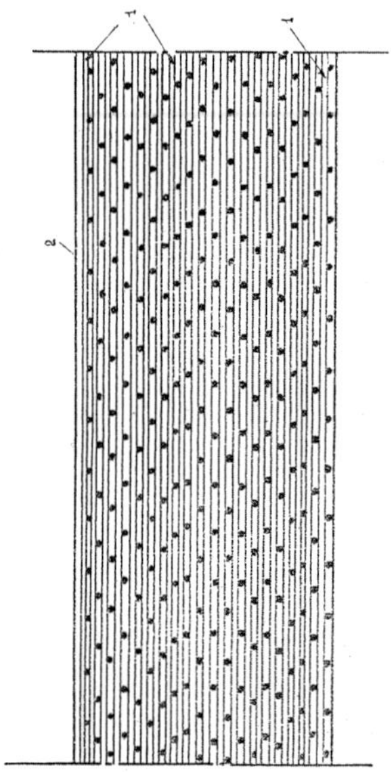

147. P. L. Nervi, patent IT 406296, *Perfezionamento nella costruzione delle solette, lastre e altre strutture cementizie armate*, 15 April 1943 (C. Greco, *Pier Luigi Nervi. Dai primi brevetti al Palazzo delle Esposizioni di Torino. 1917-1948*, Lucerna, Quart Edizioni, 2008, p. 288).

In the patent, the characteristic metal frame, generated by superimposing several meshes, produced what Nervi defined as "equiretinate cement structures." The adjective "equiretinate" summarises, more neatly than the previous definition of "retinate cement walls," the decisive role of the framework of layers of mesh and netting in determining the mechanical characteristics of the product, confirming that Nervi did not manage to find a satisfactory stratification of metal meshes until 1943. The number of layers of mesh and the diameter of the wires had to be chosen in accordance with the structure to be produced, and this also applied to the possibility of including "small diameter" iron reinforcement bars between the meshes. At the time of the invention, Nervi was aware of the spatial potential of his material, although its predominant purpose remained that of producing devices for war and floating ordnance, at least according to a literal reading of the patent, which enumerates the various applications and in which the category of "roofs" seems to have accidentally slipped into a list that includes tanks or hulls.

Nervi emphasised the novelty of his ferro-cementitious felt compared to similar techniques of ferro-cementate or ferrocement, reinforced with only one or two meshes and in which the "cement part was predominant." His ferro-cementitious felt was distinguished precisely due to the greater quantity of iron, the form of the various superimposed layers of metal netting, the reduced thickness, and the role of the cement, which is relegated to a simple waterproof and protective sealing material. The text of the patent also confirms his aspiration to achieve a cement product with the same properties as sheet metal, which had been evident from the 1939 patented on underwater mines and torpedoes.

The reduction in the water content of the compound was another factor that proved decisive in eliminating the phenomena of cracking during shrinkage, as well as being an essential prerequisite for combining cement

[2] Nervi, *Scienza o arte del costruire?* cit., pp. 26-27. See Giuseppe Albenga, Gustavo Colonnetti, Arturo Danusso, Aristide Giannelli, Giulio Krall, Antonio Martinelli, *Tecnica del cemento armato. A cura dell'Istituto Nazionale per gli Studi e la Sperimentazione nell'Industria Edlilizia*, Roma, Edizioni della Bussola, 1946.

[3] Nervi, *Scienza o arte del costruire?* cit., p. 27.

[4] See Pier Luigi Nervi, *Nuove possibilità per le costruzioni navali in cemento armato*, Roma, Ingg. Nervi & Bartoli – Anonima per Costruzioni, undated, Maxxi, p. 14; Pier Luigi Nervi, *Il ferro cementato nelle costruzioni navali*, in, "I° Congresso Italiano dei Trasporti promosso dall'Automobile Club di

mortar and metal without having to resort to formwork. Nervi's special compound was no longer laid by pouring "liquid mud" but by manual application, achieved by pressing a thick mixture against the netting (the use of the term "pouring" for the application of the cement mortar appears inappropriate in the patent text). The process consequently also avoided other defects due to casting a more liquid compound, namely heterogeneity, due to the segregation of aggregates and poor protection of the bars.

Nervi received confirmation of the experiments he had conducted in Rome when he took part, from 3 to 5 May 1943, in a Refresher Course for Engineers, held at the Istituto Nazionale Fascista per gli Studi e la Sperimentazione nell'Industria Edilizia in Rome, where he attended what he described as the "very fine lectures" of Giulio Krall and Gustavo Colonnetti.[2] While Krall showed the auditorium, through calculation, the strength of a reinforced concrete dam with double curvature, Colonnetti undermined the scientific validity of any calculation because the mechanical behaviour of reinforced concrete does not obey Hooke's law in practice. In Nervi's imagination that "unreal world,"[3] which is the basis for the application of mathematical calculation, became the utopian goal of the new material, which should be possible to manufacture isotopically and elastically with the very aim of meeting the ideal conditions laid down by Hooke's law. In this way, someone who had always questioned the effectiveness of mathematical calculation was actually using the unreal axioms of that discipline to find the creative drive to allow himself to radically question all developments in reinforced concrete, as they had been established from the time of the Hennebique system. This was the point at which Nervi became a visionary of the subject and reinvented the entire creative, technical, and theoretical edifice of reinforced concrete with an acrobatic leap backwards that took him to the starting point of the product's invention, when Monier reinforced it with square mesh metal netting. Nervi was now convinced that he could continue his explorations of the universe of structural hyperstaticity, undertaken with scientific criteria during the development of the hangar structures, because he now realised that the product he had tested for the shipyards could allow him to avoid the internal restraint conditions arising in those structures as a result of thermal stresses.

Between spring and summer 1943, he continued with his experiments to develop a technique for replacing the timber planking of naval hulls with his ferro-cementitious felt: at Torre del Lago, in the company of the engineers Mario Codebò and Villa, of the Italian Naval Register and in Rome, under the direction of Martinelli, at the Istituto Nazionale Fascista per gli Studi e la Sperimentazione nell'Industria Edilizia, and the Centro Studi ed Esperienze della Regia Marina.[4] The construction of a hull for a 130-tonne motor sailer provided him with an opportunity for testing his renewed kind of ferrocement. The test results brought recognition by the Italian Naval Register and the Navy Engineering Corps that Nervi's cement product was a material that could possibly be used in shipyards.[5]

Milano con la collaborazione del Collegio degli Ingegneri di Milano, del Touring Club Italiano e della rivista "Interauto." Milano, 11-15 giugno 1947," Sezione Navigazione Marittima, No. 14, (pp. 1-3), p. 1 (attached to Pier Luigi Nervi, letter to Alberto Sartoris, 31 July 1947, Acm-EPFL); Pier Luigi Nervi, *Il ferro cementato e la sue possibilità costruttive*, in, "Atti e Rassegna Tecnica della Società degli Ingegneri e degli Architetti in Torino," II, March-April 1948, No. 3-4, (pp. 66-67), p. 66.

[5] Pier Luigi Nervi, *Il ferrocemento: sue caratteristiche e possibilità*, in, "L'Ingegnere. Rivista tecnica mensile," XXV, January 1951, No. 1, (pp. 17-25), p. 18.

148-149. P. L. Nervi, 130-tonne motor sailer in ferro-cementate, 1943. Under construction, 1943.

148. Skeleton during assembly (Maxxi - PLN, F08597).

149. Skeleton during assembly (P. L. Nervi, *Scienza o arte del costruire?*, Roma, Edizioni della Bussola, 1945, plate XXXII).

150-151. P. L. Nervi, samples of slabs with "equidiffuse" reinforcement built at the Istituto Nazionale Fascista per gli Studi e la Sperimentazione nell'Industria Edilizia, 1943.

150. Samples of slabs (Maxxi – PLN, F11873).

151. Load tests carried out on samples of slabs. "Solettina in ferro cementato. Notare la elasticità e flessibilità del materiale (Ferro-cementate slab. Note the elasticity and flexibility of the material)." (Maxxi - PLN, F05468).

Following the change of material and the application method, Nervi described the planking of the motor sailer as a "planking slab,"[6] a definition that was already in itself significant with regard to the potential application of the product to the building industry (it is likely that Nervi thereby saw the planking as referring to a grid made up no longer of boards but of a reinforcement). The first thing assembled in order to manufacture the planking slab was reinforcement made up of iron bars, what Nervi typically referred to as "reticulate," but was now interwoven in accordance with a double framework of lines, with square and diamond-shaped mesh, almost as if to ensure strength through the superimposition of reticular and geodesic grids (it was an interweaving similar to that used in the great station vault, albeit on a different scale). Six layers of metal netting were then applied to this reinforcement. The next step was "cementing," with the application by hand of a cement mixture with the proportion of components selected to achieve a strong product made using only very fine aggregates that could penetrate the dense metal weave and block each section of the planking slab.

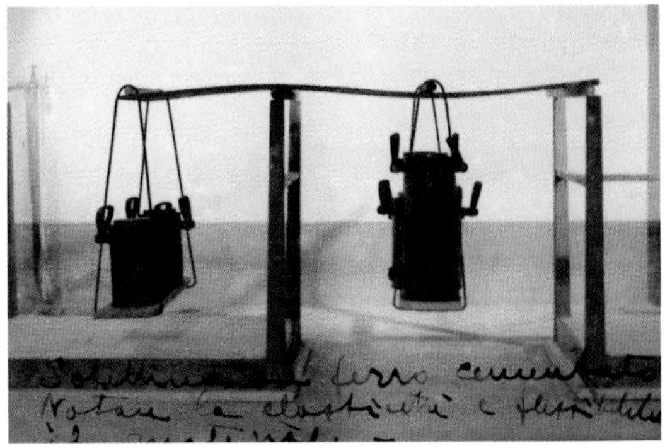

Research into the "structural and executive procedures best suited to the sensitive field of cement naval constructions" led Nervi "towards a type of construction" that, in his opinion, was able to "offer good development prospects and broad applications, including to civil and industrial building."[7] We cannot rule out that the decisive turning point for the whole evolution of his idea of structure, through the extension of his findings for hulls to the building industry, with appropriate modifications, was supported by his reflections on the plasticity of reinforced concrete and Hooke's law.

It is not possible to establish the exact sequence and nature of the tests conducted by Nervi, but it is likely that the experiments carried out in Rome, unlike those conducted in Torre del Lago, were intended to convert the type of ferrocement used for hulls to a kind suitable for "civil and industrial building." The possibility that the insight of applying the improved and modified ferrocement to building was supported by the lectures attended by Nervi seems to be confirmed by the fact that tests on the strength and static behaviour of samples carried out in Rome were conducted under the direction of Martinelli, using scientific instruments from his laboratory, which was affiliated to the same National Institute responsible for organising the Refresher Course for Engineers.

The samples produced by Nervi with Martinelli concerned one plate and two slabs. The thickness of the plate was limited to 1 centimetre, and it was reinforced with 12 layers of very fine metal netting (0.4 kilograms per square metre). The metal weave of the slabs was different from that of the motor sailer planking. The 3-centimetre-thick slab sample was reinforced with no fewer than 35 layers of 0.4 kilograms per square metre netting; the 6-centimetre-thick slab sample was reinforced with 20 layers of netting weighing 1.5 kilograms per square metre. The photographs of the samples show weaves of wire in irregular square meshes, overlaid in multiple layers and held together by nodes, again in wire, in order to achieve a mesh density that Nervi had never envisaged in the designs for the various types of float destined for the Italian Royal Navy and which extended almost through the entire width of the slabs. The samples suggest that Nervi was already searching not so much for a shell of the usual ferrocement kind used for a hull but rather for a new cement product capable of becoming homogeneous and tough enough for civil and

[6] Nervi, *Scienza o arte del costruire?* cit., plate XXXII, caption to fig. 1.
[7] *Ibid.*, p. 129.

152. P. L. Nervi and R. Di Fede, 400-tonne motor sailer, Fratelli Menetto shipyards, Società Motovele d'Italia, Pellestrina di Venezia, 1943. On-site preparation of hull structural elements (*The Works of Pier Luigi Nervi*, London, The Architectural Press, 1957, p. 48).

industrial construction. Due to the uniform distribution of the irons, the product met Nervi's desires for a material that could always be adapted to various strength conditions, irrespective of any calculation errors or predicted stress curves. The extraordinary increase in the number of nets transformed the very nature of reinforced concrete into of a product with a mechanical behaviour more similar to that of Hooke's law, in other words endowed with a deformability and elasticity designed to prevent any cracking. It is worth emphasising that during the creative process that led Nervi to the discovery of his product, he was continually plagued, even with a certain degree of surreal obsession, by his memory of the cracks caused by thermal expansion in the *Gioco della Pelota* roof, in the roof of his first hangars and in Brunelleschi's dome.

The studies and tests conducted by Nervi resulted in the start of work on the construction of an initial large ship designed by Raffaello Di Fede, a 400-tonne motor sailer with a skeleton in precast reinforced concrete and planking in ferro-cementitious felt. The work was started in July 1943 by Nervi & Bartoli in the Fratelli Menetto shipyard at the Società Motovele d'Italia of Pellestrina, in Venice. The company also undertook the construction of three 150-tonne motor transporters for the Italian Royal Navy. Unlike the vessel under construction in Pallestrina, its entire structure was in ferro-cementitious felt. Following the armistice, all the naval works by Nervi & Bartoli were nevertheless destined to remain unfinished.[8]

From that time, Nervi envisaged using ferro-cementitious felt for purposes other than military ones. In anticipation of the reconstruction, on 7 September, he designed and patented a rotary machine for the manufacture of "hollow cylindrical bodies" that could be used as poles for powerlines, pipes for water or gas mains, foundation piles, or ships' masts.[9]

Unsurprisingly, as he worked on the small-scale techniques involving metal mesh frameworks and the hand-application of cement mortar for his felt, Nervi spoke out against the ideas of prefabricated building that were under discussion in the middle of 1943. Instead, always with a view to reconstruction and the growth in demand for housing, he again advocated an approach of "maximum autarky" (he now wrote of "the most comprehensive and rigorous autarky understood in a totalitarian and literal sense") and called for a "continuous improvement" in the "static behaviour of materials" and an "internal balance of structures," through experimental investigations that would limit the "squandering" of resources, just as he was doing with his experiments in Torre del Lago and Rome.[10]

Ferrocement, improved as ferro-cementitious felt, was about to become, for Nervi, not so much a technique designed to produce a special kind of reinforced concrete but rather the very foundation for a vision of the structure of the future, without skeletons and infills, where the formwork had disappeared along with the constraints imposed by its manufacture, where intuition was free to imagine figures that were strong by virtue of their shape, where autarkic aspiration was sublimated into extremely slender thicknesses, where the manual manufacturing method provided employment for local labourers and enhanced their worth, where all the scaffolding intended to support the con-

[8] See Nervi, *Nuove possibilità per le costruzioni navali in cemento armato* cit., pp. 19-25; Nervi, *Il ferro-cemento: sue caratteristiche e possibilità* cit., p. 18; Nervi, *Scienza o arte del costruire?* cit., plate XVII, fig. 2; *The works of Pier Luigi Nervi* cit., pp. 48-49.

[9] Pier Luigi Nervi, patent IT 410202, *1° ATTESTATO COMPLETIVO alla domanda di brevetto n. 3153/43 (verb. 340/200) depositata il 15 aprile 1943, avente*

struction was a triumph of lightness, and where the work would take on an unexpected and almost magical levity. Nervi was ready to go ahead with the invention of this structure, more and surer of his alchemical creative formula made up of practical experience, static insight, and visionary drive.

Waved slabs in ferro-cementitious felt: the shed in Via della Magliana

Had it not been for a decisive and radical transformation in the technical make-up of reinforced concrete, there could have been no further development in the method of involving structure and space that Nervi had experimented with in all his shapes for cast-in-place and precast concrete designed between the 1930s and the early 1940s. This transformation enabled it to finally be freed from the need to distribute the concrete mixture and metal reinforcement in accordance with lines that, however complex, were still forced to follow geometric patterns, as Nervi had found in his hangars, viaducts, and silos.

Yet the juxtaposition of a geodesic grid and protective covering, as in the hangars, was already an early hint of the advent of a structure made up in such a way that the grid and the covering merged into a single entity. On closer inspection, the hangar grid patterns foreshadow, partly through their ability to distribute and disperse the stresses through a multiplicity of nodes, an idea of structure based no longer on columns, beams, or lamellae but on a framework that was even more tightly knit than a geodesic frame – and almost textile in nature. If we consider Nervi's progress through the first variations in the Hennebique skeleton, in the building site of the Società Anonima per Costruzioni Cementizie, to the geodesic structure of Nervi & Bartoli, the goal of a continuous weave that simultaneously performs the role of skeleton and covering, which was made possible by ferro-cementitious felt, practically feels like a natural outcome. Nervi's discovery of the principle of strength through form in some of his designs for E42 played a significant role in this development. This special kind of strength gave a decisive impetus for his quest for structural configurations beyond any rules of framework or pattern. Seen from this perspective, the trusswork of the 1939 viaduct, the colossal silo of 1941, and the majestic railway station roof illustrate Nervi's desire to try every possible combination, even beyond the patterns of regular geometry and also his dissatisfaction at still being forced to limit himself to creative exercises involving the juxtaposition of straight or curved bars. Until the early 1940s all his designs had been inspired by this desire to push the boundaries of the opportunities offered by a grid of bars, thus laying the foundations for further progress.

In 1944, as he extended his research into ferro-cementitious felt after coming up with the planking slabs, Nervi was aware that he had broken through into the universe of "resilient membranes" which could be "waved, folded and curved at will."[11] The terminology used in his writings of this period, which was punctuated by the use of the terms "membrane" and "shell" as synonyms, is significant. It indicates that the intuitive and effective definition of "geodesic structure" had already been superseded just two years earlier by the newly discovered technique.

per titolo "Perfezionamento nella costruzione di solette, lastre e altre strutture cementizie armate," 7 September 1943, ACS.

[10] Pier Luigi Nervi, *Aspetti e problemi della ricostruzione edilizia*, in, "Architettura. Rivista del Sindacato nazionale fascista architetti," XXII, June-July-August 1943, No. 6-8, pp. 171-173.

[11] Nervi, *Scienza o Arte del Costruire?* cit., p. 132.

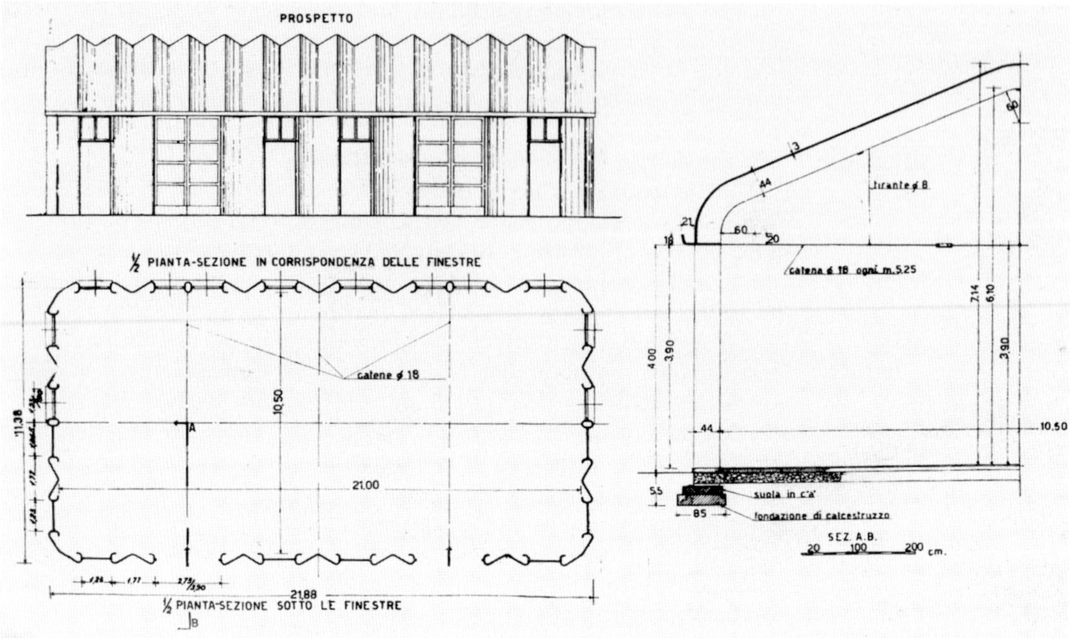

Having discovered the ideal material for a structural membrane now allowed him to invent form-resistant shapes that he could mould according to the direction of stresses. The design of his corrugated reticular structure for the station vault was already a significant progress in this context, because of the profile of the resilient cross-section. At this stage of his work, he went back to consider his pavilion designs for E42 because he was now able to produce those thin shells with his new cement product. He therefore submitted one of those drawings: "Design of a Pavilion for E42 with a Hyperboloid Structure Particularly Suited for Construction in Ferro-cementate."[12]

The construction of a work with the nature of a prototype, the filing of a supplementary patent, and the design for a great roof were the highlights of Nervi's development after the end of the war in the second-half of 1944, and all featured the waved shell shape that was to typify the masterpieces he constructed out of ferro-cementitious felt. The choice of a waved profile should be seen in the context of experiments such as those conducted by Miozzi in the second half of the 1930s on waved vaults. Baroni's roof prototypes also had a significant impact on Nervi's understanding of the potential of cement structures for thin roofs measuring up to 3 centimetres thick and reinforced with metal bars and meshes.

The first use of ferro-cementitious felt for walls and roofs was tested out by the construction of the 21.88 x 11.38-metre shed used to keep tools in the Nervi & Bartoli yard, in the Via della Magliana depot. The shell was 3 centimetres thick and variously folded with wide undulations and pleats to stiffen it and turn it into a building structure. When used in walls, this operation broke the continuity of the shell so much that it generated different forms of static behaviour in the general structure. The presence of doors and windows made it necessary to break down the shell into three modules: one in the form of a simple rectangular panel to be inserted in the position of the windows to close and tighten the system, and two configured as vertical

153-154. P. L. Nervi, Nervi & Bartoli shed, Via della Magliana 238, Rome, 1944.

153. "Capannone ad uso magazzino in ferro cementato (shed for use as a warehouse in ferro-cementate)" (P. L. Nervi, *Scienza o arte del costruire?*, Roma, Edizioni della Bussola, 1945, plate XXX).

154. "Capannone alla Magliana. Particolare ondulazione andante e risvolto portoni e finestre (shed in Via della Magliana. Detail of moving wave and door and window fold)" (CSAC, n. inv. pra 1, coll. 268/1).

structures. The most elaborate of the latter two modules, and also the one that formed the main structure of the walls, was a soft V-shape folded at the ends into truncated scrolls to prevent overturning. This module was defined as a "moving wave"[13] and was 1.77 metres wide, 3.90 metres high, and its undulating structure extended to a depth of 44 centimetres. The profile of this module was destined to undergo significant developments in Nervi's ferro-cementitious felt roofs. The other module for the vertical structure rejected the very nature of the shell because, in this case, Nervi's felt was folded in on itself around a core of wood, left over from the idea of formwork, to generate a single rounded pillar to be inserted at the point where two windows met with their flat panels. The shell of the pitched roof was corrugated in a regular pleated pattern that concluded, in the two heads, in a flat surface, while the pleating was braced by three slender beams, again produced using the same technique (three steel chains combined to buttress the roof stresses). The choice of a pitched roof, as opposed to flat or vaulted roofs, gives the shed the aura of a humble house. The pitch roof continues downward to direct rainwater into the gutter in an unusual way for a traditional roof, but consistent with the plastic nature of Nervi's felt. All corners of the shed shell are more or less rounded, and this curved appearance became a distinctive hallmark of Nervi's technique because it made it easier to bend the irons, which ran everywhere like a diffuse filigree pattern.

To confirm the information provided in the 1943 patent on adjusting the metal netting in the ferro-cementitious felt to the characteristics of individual works, for the equidiffuse reinforcement of the shed, Nervi used a frame that was different from those he had experimented with or assembled for the planking slabs. A considerable number of wooden slats and boards, for permanent formwork, also remained buried in his felt. The main rein-

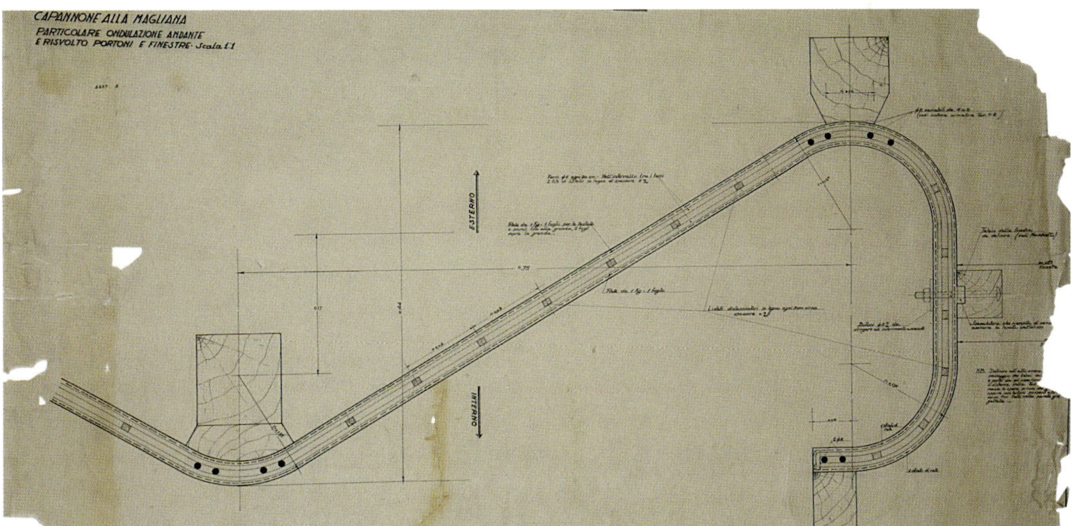

[12] *Ibid.*, caption to plate XXXI.

[13] See drawing *Capannone alla Magliana. Particolare ondulazione andante e risvolto portoni e finestre. Scala 1 :1*, CSAC, n. inv. pra 1, coll. 268/1.

155-157. P. L. Nervi, Nervi & Bartoli shed, Via della Magliana 238, Rome, 1944.

155. Modelling of "equiwired" reinforcements (Maxxi - PLN, F05274).

156. Positioning of "equiwired" reinforcements (Maxxi - PLN, F05277).

157. Cementation stage (Maxxi - PLN, F05271).

forcement consisted of a double grid of 5-milllimetre-diameter bars with a large mesh that varied according to the modules, undulation, and pleating. In the drawing of the "moving wave," the horizontal bars run parallel at a distance of 20 centimetres and are separated by wooden "spacing slats" while the 8-millimetre-diameter vertical bars are concentrated in two pairs at each curve (only one pair was placed at the end of the scrolls). The grid of bars was enfolded by metal netting with a dense mesh, weighing 1 kilogram per square metre – two for the walls and four for the roof.

The method of manufacturing the "equiwired" reinforcement revealed Nervi's structural concept and inventive technique. The number of iron bars was reduced to that considered necessary for the creation of a structure. The vertical rods were concentrated at the curves. These were points of the shell that were form-resistant and aligned with the folds of the roof pleating, where other bars were concentrated almost as if to form an invisible continuous rib. In the parts of the metal netting that were far from the grid of bars, and that could easily have bent under the pressure of applying cement mortar, Nervi used stiffening made up of a framework of wooden slats producing a square mesh measuring 10 x 10 centimetres (in the "moving wave" drawing, these are referred to as "spacing slats"), which nevertheless ultimately contribute to the overall strength of the shell, albeit in a small proportion. The seam between the bars and slats, formed out of knots in the wire, and the assembly of bars, wire netting, and slats; were executed to perfection, almost like an expertly handwoven reed basket.

It is not possible to determine whether the modular design of the walls was only meant to facilitate the construction of doors and windows or whether Nervi had already planned a way to manufacture the shell. In the case of the planking slabs for hulls, the entire iron and netting reinforcement is produced directly in situ, whereas Nervi prepared metal framework sectors for the shed wall shell on the site so that they could easily be shaped. Dividing the shell design into modules allowed this initial method of prefabricating the ferro-cementitious felt to be applied in the "equidiffuse" reinforcement sectors alone. This was the component that engendered the prefabrication, only in 1948, that Nervi extended to the application of cement compound, when he achieved the prefabrication of genuine waved shell segments. In order to prefabricate the "equiwired" reinforcement of the "moving wave," a wooden template was pre-

158-160. P. L. Nervi, Nervi & Bartoli shed, Via della Magliana 238, Rome, 1944.

158. Wooden template for modelling the "equiwired" roof reinforcements (Maxxi - PLN, F05275).

159. View (Maxxi - PLN, F11870).

160. View (Maxxi - PLN, F05279).

pared that rested on the ground, protected in a shed, which was used to shape bars and netting, insert wooden slats, and easily knot the wire.

The workers used a simple wooden scaffolding to install the "equiwired" reinforcements that were joined together during the cementing operations. In order to construct the waved covering of the pitch roof, Nervi used light centrings of wood that were moulded to match the pleated pattern of the shell, onto which the "equiwired" reinforcement could be laid. And the workers covered the equiwired reinforcement with cement mortar, made out of fine sand with a low water content, applied under pressure with a laying trowel and then smoothed with plasterer's trowel to obtain a constant thickness of 3 centimetres and a perfectly polished finish. An observation that Nervi made about the application of the mortar, "easy and quick, particularly on surfaces that are not very inclined or horizontal" but "trickier on vertical faces,"[14] helps us understand the reasons that led him to make the transition to complete prefabrication of the ferro-cementitious felt.

The entire process of manufacturing the ferro-cementitious felt of the shed took place in the context of a craft-scale building site that stressed the manual skill of the workers and precision workmanship at every stage. The final product had the appearance of something made in a workshop, almost like an automotive body shell. The finished shed, left in the natural grey of the compound, with its wide undulations and pleats, looked like an outlandish construction made out of concrete asbestos. Undulations, pleats, and flat sectors created movements, breaks, and chiaroscuro that were alienating and unpleasant because they were devoid of any intent to adapt the moulding of the shell to any known shape. Because of its roughness, the shed became a kind of manifesto of Nervi's courageous quest for the most appropriate figures that his new material could transform into structure. This first result already revealed the main features of the new ferro-cementitious felt aesthetic with primeval, rough, almost cartilaginous, and certainly animalesque shapes.

[14] Nervi, *Scienza o Arte del Costruire?* cit., p. 133.

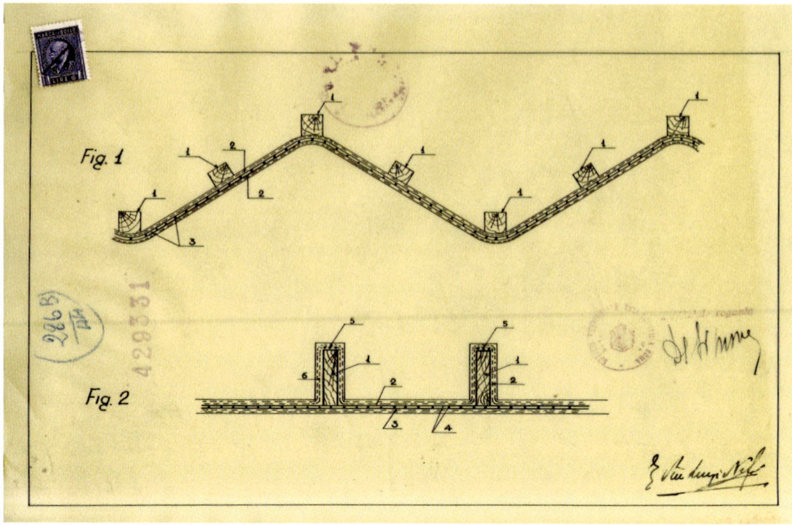

161. P. L. Nervi, patent IT 429331, 2nd supplement to patent IT 406296, 29 September 1944 (ACS).

On 29 September 1944, Nervi filed a second patent supplement on "shell or membrane cement slabs or structures" to his April 1943 patent on ferro-cementitious felt, after his patent for pipe manufacture of September 1943.[15] The formal prerogative of "resilient shell structures" was now encapsulated by the definition of "waved slab," or more generally "appropriately contoured or curved surfaces," sometimes equated with "waved sheet steel." The patent supplement took into account new techniques pioneered in the Via della Magliana shed, from the manufacture of the waved profile, sometimes reinforced by wooden poles, to the application of mortar. The reasons for the combination of metal and wood in the shed's "equiwired" reinforcement are explained in the patent supplement: "When the bars arranged within the netting reinforcement act solely as spacers (this is in the event that the netting reinforcement is statically adequate), they can be replaced by thin wooden slats or any other substance." The fact that the need is indicated for a concentration of "bars at the vertices of the undulations" to serve "a clear static function" is reflected in the shed reinforcement. Nervi did not create any illusions, however, about the possibility of prefabricated reinforcement sectors before their installation, and limited himself to explaining how to construct it out of provisional wooden poles to mould the undulations.

Nervi touched on the matter of the exposed material, particularly the concrete, which was to become crucial in international architecture, especially following the invention of *béton brut*, without any particular theoretical emphasis, when he said that, once smoothed, his ferro-cementitious felt could be left without plaster in its raw state as a "structure without finishing" (*struttura di prima lavorazione*).

While the list of applications newly drafted in the patent supplement continued to include works such as the hulls and tanks for which the material had been invented, now it clearly reflected Nervi's interest in shells used for walls and roofs, partly due to the fact that production of floating military ordnance had ceased. The end of the text foreshadowed the potential of ferro-cementitious felt for the creation of new structural shell shapes, made possible by removing the geometrical constraints of the formwork.

"It is clear," affirmed Nervi, "that this executive procedure can be used with entire ease and economy to make artefacts of any shape because the metal reinforcement in netting, or netting and bars supported by a ribbed skeleton which may or may not be embedded in a cement structure, can be used economically to produce all the shapes that are virtually impossible to execute using ordinary formwork boards."[16]

Nervi's visionary abilities led him to another design drawn up for demonstration purposes, much like the design for the station roof or those for the E42 pavilions. He therefore designed another colossal segmental vault to illustrate the application of ferro-cementitious felt to the roofs he had referred to in his patents on ferro-cementitious felt and in the second patent supplement. While the span of the precast reinforced concrete station roof was 200 metres, that of the ferro-cementitious felt roof for an unspecified space amounted to 300 metres, indicating his intent to measure the superiority of his invented product in terms of size.

Nervi celebrated the final disappearance of a reinforced concrete skeleton in the grand vault. The ferro-cementitious felt slab became a continuous shell corrugated by a series of accentuated folds, a veritable majestic "waved slab" intended to achieve strength through form that would be capable of making up for the absence of main frameworks. Slender 3 x 3-centimetre ribs traverse the undulations every 5 metres to stiffen them, together with slender overhead ties fastened at the top of the waves, some of them visible even inside the shed on Via della Magliana. The uncertainty over the number of ties, which number 7 in the perspective drawing but, according to the drawing showing details of the structure, should be present every 10 metres, when they would amount to no fewer than 40; is a significant indication of one as yet unresolved aspects of the waved slab, which Nervi would later be able to translate into a sophisticated and expressive technical detail. The detailed drawing shows that the slab intrados is traversed every 15 centimetres by ripples that measure 15 centimetres in height and 5 in width, at right angles to the stiffening ribs, in order to prevent any possible inflection of the waves. This detail too, which was destined to disappear in subsequent constructions, indicates that Nervi was somewhat unsure about the stiffness of the waved slab. A perspective view of one sector of the vault shows that these ripples were not represented, which confirms that some significant technical differences were present between the detailed drawing and the perspective views (Nervi increased the number of waved slab stiffening and bracing elements in the detailed drawing). The surface of the slab exposed to the elements was protected by a covering made up of waved concrete asbestos panels, separated by a cavity for air circulation.

Nervi still included expansion joints (spaced at approximately 57 metres) with the same frequency as in the precast reinforced concrete vault. Their positions are marked by a special structural configuration – the undulations are met by a continuous flat band throughout the span of the vault. Those joints also intercept the vault segments where the cement compound must be applied onto the metal grid as one unit in order to produce a uniform waved structure.

Nervi specified thicknesses and reinforcements for the ferro-cementitious felt that amounted to a technical prerogative. The thickness was adjusted to

[15] Pier Luigi Nervi, patent IT 429331, 2nd supplement to patent IT 406296, 29 September 1944, ACS.

[16] *Ibid.*, pp. 7-8.

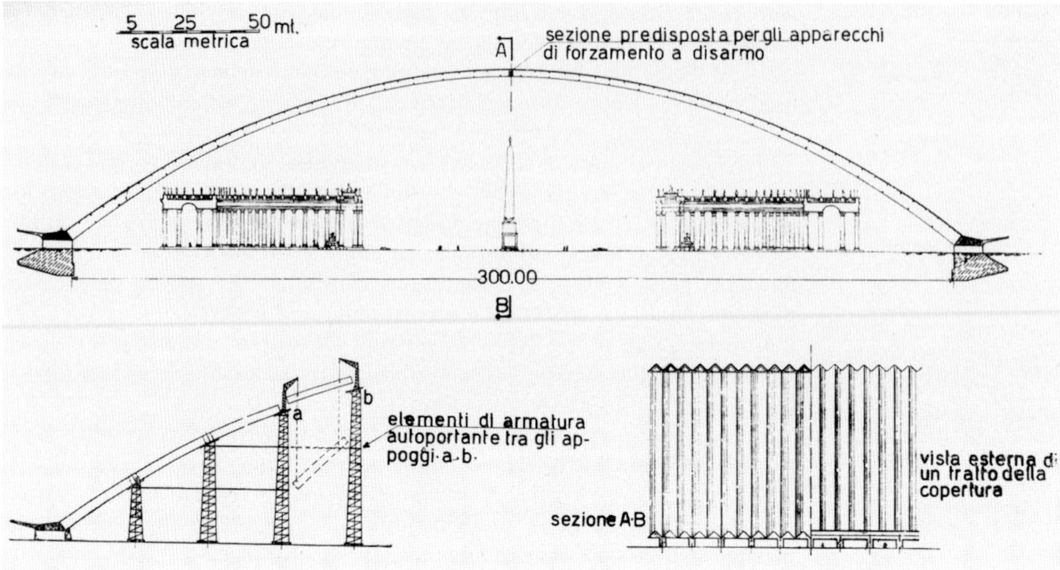

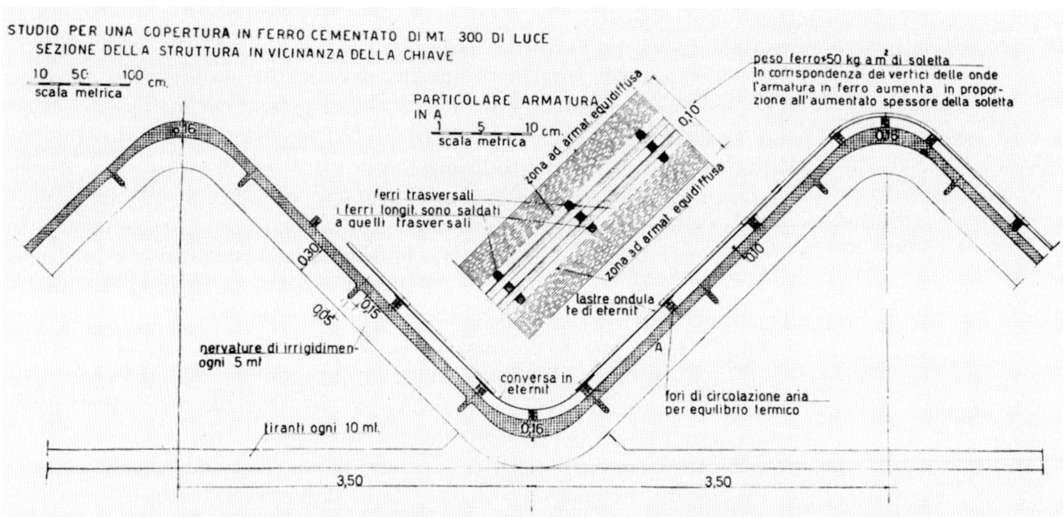

Particolari della copertura di cui alla tav. XXVII. - Da notare il piccolo spessore della struttura e le nervature e tiranti di irrigidimento per evitare fenomeni di svergolamento.

(TAVOLA XXVIII)

162-165. P. L. Nervi, design for a roof in ferro-cementate with a 300 metre span, 1944.

162. Cross-section (P. L. Nervi, *Scienza o arte del costruire?*, Roma, Edizioni della Bussola, 1945, plate XXVII).

163. Detail (P. L. Nervi, *Scienza o arte del costruire?*, Roma, Edizioni della Bussola, 1945, plate XXVIII).

164. Perspective drawing (*The Works of Pier Luigi Nervi*, London, The Architectural Press, 1957, p. 45).

165. Perspective drawing (*The Works of Pier Luigi Nervi*, London, The Architectural Press, 1957, p. 45).

10 centimetres and increased to 16 at the curves from which the reinforcement irons sprouted, as in the waved slab of the shed; interwoven, welded bars ran along the centre of the cross-section, while a dense "'equidiffuse' reinforcement" was present along the edges.

In order to construct the majestic waved vault slab, Nervi envisaged proceeding in accordance with the same criteria tested in the shed, by laying the cement compound directly in situ onto the metal grid of the reinforcement, which was laid over the centrings and already moulded into the waved shape. His criticism directed against prefabrication in 1943 is now reflected in the very concept of a building site where the workers had the job of applying the cement compound directly onto the centrings. At the time when the roof was designed, any prefabrication was limited to the metal grid alone, as with the Via della Magliana shed, which Nervi was sure he would be able to achieve on site. The relevant parts were already shaped into waved sectors in pieces 7 metres wide and 25 metres long, to be lifted by two cranes onto the longitudinal metal scaffolding, where they would slide on rails (the 25 metres corresponded to half of the section of the wave arch between the ties shown in the perspective view).

The "planking slabs" that had been tested for the "cementitious naval constructions" had now been transformed into "waved slabs" traversed by curves like sails billowing in the wind, which enabled them to float in the air and lie over a great void. The exceptional vault envisaged by Nervi is depicted in a cross-section drawing, extending over the Colonnato and the obelisk of Piazza San Pietro in the Vatican City to make it possible to immediately grasp its size and monumentality. Unlike the previous version of the precast reinforced concrete vault, this version made out of ferro-cementitious felt executed in situ is entirely devoid of perforations, and its undulations give rise to shadowy recesses, another very telling indication that Nervi was not yet aware of the structure's strength through form or the extent to which he could make the panel thinner and open it up to the light. The design also serves to illustrate the potential of the structure's "cement with an 'equiwired' or 'equidiffuse' reinforcement," which involved the continuity typical of "planking slabs" for hulls even at this stage of development.

The fact that the list of roofs also includes "domes,"[17] confirms that Nervi intended to use his visionary designs for grand roofs to demonstrate that his ferro-cementitious felt could even shape the myth of *romanità* that he had pursued between the 1920s and 1930s, but was later stripped of any obvious fascist rhetoric and distilled into pure technical aesthetics.

Scienza o Arte del Costruire? An apologia for concrete

All the experiments Nervi had conducted since the end of the 1930s, from the construction of hangars and hulls to visionary projects for roofs made out of reinforced concrete and ferro-cementitious felt lie at the fundamental and innovative core of his technical research, summarised in his first important essay on the creative principles and technical rules of the design and construction of great contemporary structures: *Scienza o Arte del Costruire?*, published in May 1945.[18] The title is a reworking of the 1931 article, *Scienza o Arte dell'Ingegnere?*, which contained some reflections on university teaching that were now discussed systematically and in depth. The subheading makes clear the crucial nature of the technique that Nervi had started to master during his apprenticeship at Muggia and that he was now investigating in all its technical and formal variants: *Caratteristiche e possibilità del cemento armato* (Characteristics and possibilities of reinforced concrete).

The essay was no longer addressed to the training of engineers, as was the case of the 1931 article, although at the end, he advanced a proposal for changing the approach to studies in the engineering faculty. It was written by Nervi for students who were embarking on his course in Construction at the Scuola di Architettura Organica, established in March 1945, and his course in Material Technology and Structural Engineering, delivered at the Faculty of Architecture at Università della Sapienza from 1945–46, when he started a lecturing career that would continue to 1961–62 and included highlights such as the lectures he gave at the Faculty of Architecture in Buenos Aires in 1950, and at Harvard University in 1962.[19] *Scienza o Arte del Costruire?* is directly aimed at instilling the principles of the practical and theoretical art of building to the "student architect,"[20] training the vision of future architects so that they "learn to look for the substantial qualities of materials, know their strengths and their inevitable shortcomings; in particular that they should become accustomed to sensing and feeling, almost as though they were sensitive to the muscle fatigue of a great living body, the play of forces and stresses that develop in the various parts of any resilient structure."[21] Sometimes the essay takes on the cadences of a university textbook. It contains actual lessons

[17] Nervi, *Scienza o Arte del Costruire?* cit., p. 132.
[18] *Ibid.*
[19] See Pier Luigi Nervi, *El lenguaje arquitectónico*, Ministerio de Educación, Universidad de Buenos Aires, Buenos Aires, Facultad de Arquitectura y Urbanismo, Establecimientos Graficos Platt S.A.C.E.I., 1951 and Nervi, *Aesthetics and Technology*

on practical matters such as the role of a drawing or model and even discusses the merits of design criteria. Yet Nervi's main purpose was to educate architecture students to an understanding of the act of building, its theoretical and aesthetic implications, and to introduce them to the secret nature of reinforced concrete.

Scienza o Arte del Costruire? raises the critical issue of the creative processes involved in the built form, investigates the role of calculation and intuition, lays the foundations for an understanding of material behaviour and building site organisation, calls for the divestiture of stone cladding and unnecessary ornaments, and opens up new horizons for the advent of a technology-based aesthetic. Nervi set out to introduce his readers to an understanding of the behaviour of each structure, whether traditional masonry or reinforced concrete, and explained the fact that they are "living bodies" animated by "forces and stresses" that change their existence and the goal of "monolithicity," which involves cracks, as in Brunelleschi's dome – he returns to the scene in the opening pages of the essay – in his early work (the *Gioco della Pelota* roof), in his hangars, which he used as an example to demonstrate how imperceptible changes in the construction process, taking place on the building site, can give rise to different vital forms of behaviour within the same abstract geometric form. Nervi responded to the rhetorical question about the ideal shape for a structure by offering an image of a "natural shape," in other words one that is obtained experimentally by means of models subjected to a range of agents that adapt each profile to come up with a perfect profile, as in a process of natural creation. The aim of *Scienza o Arte del Costruire?* is indeed to guide human invention to a perfect profile by moving the element of chance.

"Being able to create 'melted stones', in any shape, that are superior to the natural version because they have the ability to withstand stress, has in it something magical," he stated with the enthusiasm of an inventor on the threshold of a new era of concrete.[22] He not only believed that structure is "living reality," but also thought the same applied to reinforced concrete, which he defined as "the finest construction system that humanity has so far been able to conceive."[23] In his eyes, it possessed an intrinsic vitality that made it "as variable and changeable as individuals of the same living species can be,"[24] to the extent of demanding control of each stage of its manufacture, from construction of the formwork, positioning of the metal reinforcements, to the pouring of the concrete in situ. and the difficult art of vibration. From this viewpoint, the "on-site" construction of parts, which could then be welded with nodes capable of making the whole thing monolithic, is suggested as a means of achieving the goal of material perfection. Nervi had by then built up a personal aesthetic concept that he had developed on the building site as an engineer and builder and with which he invested every one of his components, even those destined to be embedded in the final result. "[...] the metal reinforcement must always be *aesthetic*; it must give the impression of a bundle of nerves intended to breathe life into the most inert mass of concrete mixture,"[25] wrote the author. He had created and seen the tangle of rods, bars, and brackets; bent and joined in bundles before pouring the concrete

in Building cit.
20 Nervi, *Scienza o Arte del Costruire?* cit., p. 31.
21 *Ibid.*, Avvertenza.

22 *Ibid.*, p. 77.
23 *Ibid.*
24 *Ibid.*, p. 81.
25 *Ibid.*, p. 91.

in a quest for mastery of forces invisible to most people. Reinforced concrete had actually become for him the utopian model of a future civilisation, as he explained in the prophetic words worthy of a latter-day Filarete: "the greater plastic yield of the sections and areas under too much stress means that the part of the task which had been erroneously entrusted to them is transferred to other less loaded parts, which accept it to the exact limits of their possibilities in an admirable spirit of cooperation. If we could achieve something similar in human relationships, poverty and injustice would disappear from our troubled lives."[26] It is impossible to escape the impression that this vision of reinforced concrete is intended to reflect the socialist corporatism that lay at the foundation of the Fascist regime, and an echo of Ojetti's metaphor, described in his opening speech of 1931, of the "fleeting crowd" that tightens "around the more or less yielding iron of authority."

Nervi's discussion of the *properties and potential of reinforced concrete* concludes with a technical discovery, redefined "ferro-cementate," presented as the evolutionary pinnacle of reinforced concrete, able to overcome the "hidden shortcomings," to eliminate or at least reduce cracks caused by thermal expansion: a "practical" and "economic" product for constructing grand roofs or rebuilding the merchant fleet. In *Scienza o Arte del Costruire?* Nervi explained the potential of this discovery in redefining in radical terms the entire production system and the formal world of reinforced concrete. "The formwork," he wrote, "which represents the real weakness of reinforced concrete from a constructive and economic viewpoint, becomes absolutely superfluous. The metal reinforcement made up of netting and bars can adapt with great ease to curved surfaces or any type of skewing. It's intrinsic lightness and deformability mean that it can be supported with light scaffolding, which enormously simplifies the construction of large and very large roofs."[27]

The important aspect of the lesson set out by Nervi in *Scienza o Arte del Costruire?* lies in guiding readers to a thorough knowledge of ways of making reinforced concrete and its by-products, warning them about the technical difficulties in its production based on climate and the source of components, about the uncontrollable phenomena of shrinkage or expansion, about the uncertain results once poured, even with all possible precautions, and about the limitless expressive resources of the structure, without any concession to any aesthetic aspects of exposing the material, and indeed almost wilfully ignoring the appeal of a surface that many were in fact now beginning to turn to in a quest for a constructive foundation to ornamentation. Nervi outlined his vision of reinforced concrete, which was original and radical because it set out to explain the material's creative power in the genesis of the built form, which was something that no textbook had ever covered as comprehensively or with the great certainty that he had achieved through his unique experience of life in building sites, in direct contact with workers, materials and their application. Nervi reached almost lyrical heights in the pages where he confessed his decades-long commitment to achieving mastery of a material that still continued to elude him. "I believe I can sense the general guidelines that could lead to such a result," he wrote on the subject of his vision of a perfect reinforced concrete structure that does not undergo "significant permanent deformation in any part" and stays "as fresh as when it was poured."[28]

There can be no doubt that *Scienza o Arte del Costruire?* resonates in the general debate as an impassioned call issued by the greatest Italian authority on construction material to re-establish an architecture based on technique, seen as cultural vision. Nervi stated that the pathway to regeneration and salvation lay both in the intrinsic values of that technique and in the way it could be applied to the political wreckage from which the new identity of Italian civilisation had to emerge. In his essay, Nervi revealed his desire to finally bury the rhetorical fervour that had sometimes led him to design his structures during the Fascist regime.

Prefabrication tests on ferro-cementate

During the first period of reconstruction, from 1945, Nervi and his firm Nervi & Bartoli resumed their productive activity with works mainly intended for the reconstruction of factories, industrial buildings, or bridges damaged during the war. In many projects and buildings they used, where possible, ferro-cementate and were thus able to test the product extensively. Studies on the "typing" of road bridges, the drive toward "building economy,"[29] and the exhortation to identify repeatable elements in building were reflected in Nervi's decision to continue the approach of manufacturing the ferro-cementate on site, and take with the "equiwired" reinforcement segments of the Via della Magliana shed. The outcome was the first prefabricated parts in ferro-cementate of manageable size, designed to replace the usual brick or concrete asbestos *tavelloni*, and also used with a function that would become essential to the process of building Nervi's future reinforced concrete structures produced on-site: permanent forms. This meant that the ferro-cementitious felt that had partly been created with the stated aim of eliminating the costly manufacture of wooden formwork had now itself become an intrinsic component of a new kind of formwork.

Between 1945 and 1947, Nervi & Bartoli won contracts for rebuilding and expanding the industrial buildings of the *Manifatture Tabacchi* of the Monopoli di Stato (State Tobacco Monopoly Factories) in Chiaravalle, Modena, Benevento, Turin, Naples, and Scafati. The works were constructed with skeletons made out of reinforced concrete cast in place or prefabricated, sometimes even out of metal. According to the building site, Nervi & Bartoli manufactured *tavelloni* on site measuring approximately 2.5 or 3 centimetres in thickness that were flat or have slightly curved ends ("curved cement *tavelloni*"),[30] to be used as permanent formwork for floors or parts of roofs, namely gutter channels, round tiles, and joint covers. When he used ferro-cementate *tavelloni* and joists in precast reinforced concrete, Nervi always specified in his drawings the economic advantages of installing them without additional finishes. "[...] the surfaces of the prefabricated *tavelloni* and secondary joists are perfectly regular and do not require plaster or stucco-works," was the expression that he now generally used.[31]

One of the most sophisticated applications for ferro-cementate was envisaged in a 1945 design for the reconstruction of one arch of the bridge over the River Nera in Narni. Once again, the fundamental question hinged on the centring and the formwork, as in the case of the study for the *Arco dell'Impero*. For the bridge, Nervi considered the mechanics of an arch in wedges and imagined a slender arched box-section structure in ferro-cementate. This would be subdivided into compartments open at the top to allow the concrete to be poured and reconstruct the arch, once two arched sectors had been raised

[26] *Ibid.*, p. 101.
[27] *Ibid.*, p. 131.
[28] Ibid., p. 81.
[29] Pier Luigi Nervi, *Aspetti e problemi della ricostruzione edilizia*, in, "Ricerca scientifica e ricostruzione. Rivista del Consiglio Nazionale delle Ricerche," XV, October-November 1945, No. 4-5, (pp. 360-362), p. 361.
[30] Drawing No. 2376-2, CSAC, n. inv. pra 283, n. id. 13614, coll. 270/6.
[31] Drawing No. 2376-4, *ibid.*

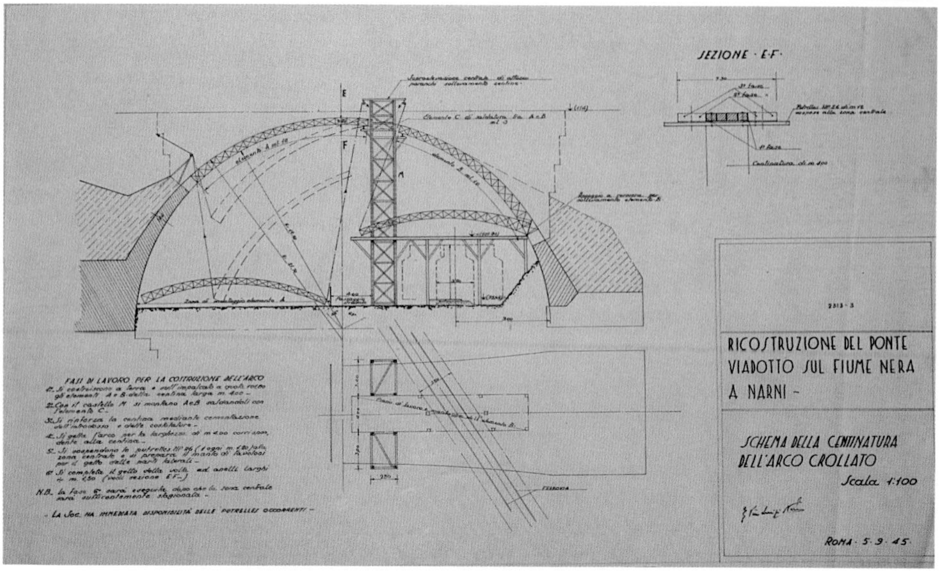

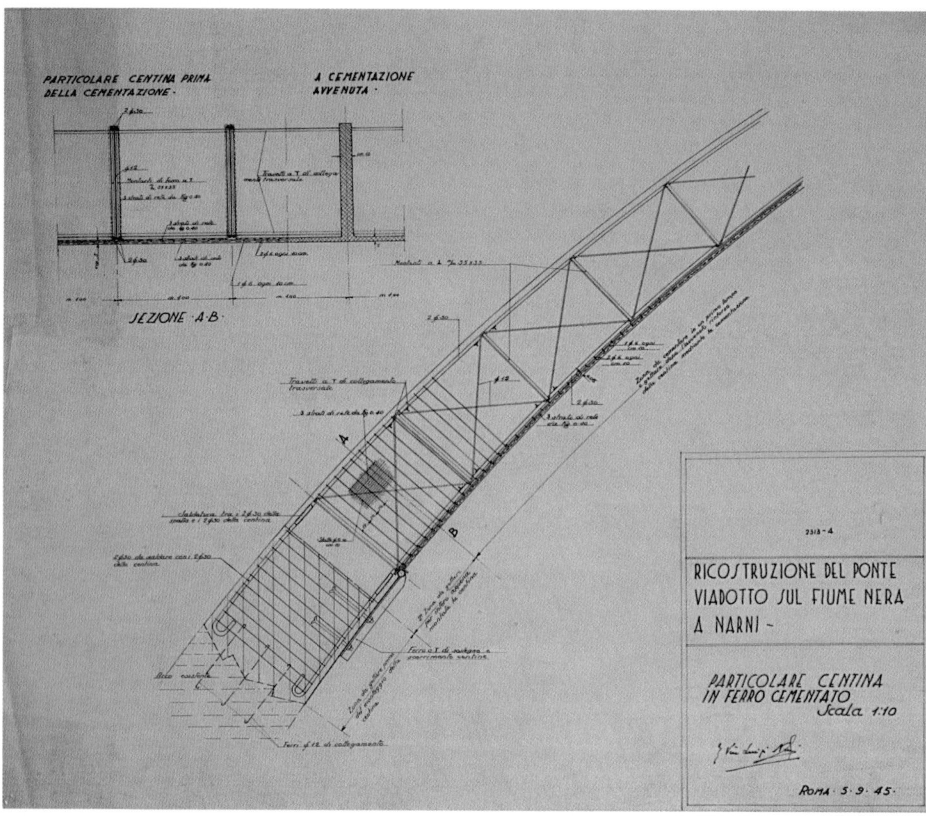

166-167. P. L. Nervi, design for reconstruction of viaduct bridge over the River Nera in Narni, 1945.

166. "Schema della centinatura dell'arco crollato (Diagram showing centring of collapsed arch)" (CSAC, n. inv. pra 286, n. id. 13617, coll. 270/6).

167. "Particolare centina in ferro cementato (Detail of ferro-cementate centring)" (CSAC, n. inv. pra 286, n. id. 13617, coll. 270/6).

in reinforced concrete to support the new structure. The steps involved in assembling the arched box-section structure in ferro-cementate involved raising two sections prepared on the ground using a tower device based on the one planned for the E42 arch; the next step was to cement the reinforcement and, once set, pour concrete into the compartments. What Nervi described as a "centring" was in fact nothing more than a device in ferro-cementate to shape the concrete into a sequence of arches to reproduce a masonry arch system.

The ornamental ribbed shell at Torre in Pietra

Even though all of Nervi & Bartoli's constructions in the years around 1945 deserve mention due to their individual importance in clarifying the construction details of reinforced concrete or ferro-cementate and, more generally, the idea of structure itself, none of them possessed the creative intensity (in terms of the formal and technical results and the issues raised by that creativity) of the design and construction, also in 1945, of the roof of the agricultural machinery warehouse of the company Bonifica Torre in Pietra, at Torre in Pietra near Rome, designed by Michele Busiri Vici. This warehouse allowed Nervi and Nervi & Bartoli to gain hands-on experience that was crucial to the laying of ferro-cementate in the construction of vaulted roofs in reinforced concrete. For this building site, Nervi invented an authentic prefabrication system for the ferro-cementate *tavelloni* used to help build the roof.

The technical nature of the warehouse roof structure was influenced by the type of brick *tavellone* used for floors with joists in reinforced concrete, but Nervi replaced the brick with his ferro-cementate and changed the shape of the *tavelloni* accordingly. The way in which ferro-cementate and reinforced concrete interact in the roof almost appears to be a reflection of an idea that Nervi let drop about the thin vault with a hyperboloid cross-section of one of the E42 pavilions, when he compared the structure to the spokes of an umbrella. Due to the way in which he prefabricated and laid out the ferro-cementate *tavelloni*, the thin vault in Torre in Pietra became a shell traversed by bundles of reinforcement interwoven in accordance with the same diamond-shaped mesh geometry as the reinforced concrete lamellae of the hangars, as though the geodesic structure had become a spider's web of fine ribs.

The warehouse roof is crucial to an understanding of some creative features of Nervi's work. The geometric grid, with its aesthetic qualities determined by the geodesic structures, is sublimated in the way the fine ribs of the barrel vault are interwoven. The vaults of the two semicircular apses are used to complete the geodesic geometry with a perfect spiral grid, almost like vaults, Gothic rose windows, rush baskets, or crocheted mats – even suggesting that Nervi added the apses to complete the shape of his grid. The aesthetics of the grid also guide the arrangement of the chains, which follow the projection of the rib lines. The entire roof design acquires the sophistication of a rippled surface, where the fine rib thicknesses conspire to make structure and ornament look like the same thing.

The "layout diagram," drawn by Nervi in a fine line for the geometric construction of the apse *tavelloni*, is most decidedly written in the language of ornamental iconography. The salient points of the geodesic structure, the "ribs" and the "nodes," are recognisable in that diagram. The layout diagram of the entire roof is designed in the knowledge of the perfect constructability of the geometry in accordance with both technologies available to Nervi & Bartoli, ferro-cementate and reinforced concrete. For this very reason, the design of the grid in the half domes is not a consistent sequence of broken

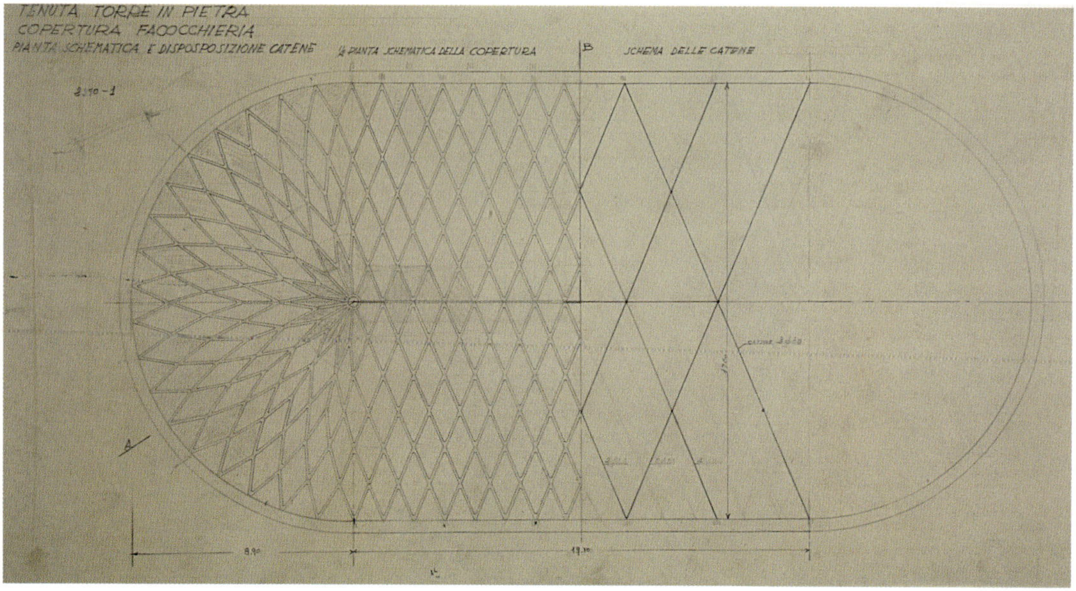

168-170. P. L. Nervi, structure for agricultural machinery warehouse, designed by M. Busiri Vici, Via Aurelia km 28, Torre in Pietra [now Torrimpietra], Fiumicino, 1945.

168. "Copertura facocchiera. Pianta schematica e disposizione catene (Roof of the *facocchiera*. Schematic plan and arrangement of chains)" (CSAC, n. inv. pra 86, n. id. 13161, coll. 267/1).

169 "Copertura locale centrale. Tracciato dei tavelloni delle parti a cupola (Main room roof. Layout of *tavelloni* on the domed parts)" (CSAC, n. inv. pra 86, n. id. 13161, coll. 267/1).

170. "Copertura facocchiera. Sezione trasversale (Roof of the *facocchiera*. Cross-section)" (CSAC, n. inv. pra 86, n. id. 13161, coll. 267/1).

lines in spirals or diamonds of progressively different sizes, but a conjunction of several patterns, in order to avoid creating disproportionate *tavelloni* (excessively large or small). The arrival points of the grids are fixed along the perimeter semi-circumference and are set at certain distances to ensure that the resulting *tavelloni* obtained along the bundle are similar to those of the barrel vault. The starting points, from the centre of the semicircle, are corrected to avoid having to weld sharp figures in a circle. Lastly, the sides of some diamonds are not inclined but follow the lines of the rays in the layout diagram. The peculiar nature of the built form of the ornamental geometry allowed Nervi to pass easily from the layout diagram to the physical thicknesses of ribs and nodes (he gave the latter a consistency that shows they continue to be the focal points of its structure), and list the different parts to be manufactured on site: 410 *tavelloni* were prepared for the segmental barrel vault, in a diamond shape with a slightly curved cross-section and in twelve different sizes; 168 *tavelloni* were made for the apse half-vaults, all the same shape and in four different sizes. In both cases, special triangular *tavelloni* run along the gutter line.

The next step in the creative process leading to the manufacture of the roof *tavelloni* only apparently follows on from the layout diagram, because it is in fact the diagram's very essence and the premise of its figurative and symbolic force. Nervi's power of thought resides in this virtual cycle combining geometry, structure, and construction; he is able to ignore the question of ornament, because the secrets of its genesis are encapsulated in the all-encompassing structural form. Yet we cannot escape the impression that in Nervi's weave, as it was manifested in the Torre in Pietra roof, there lurked a risk of becoming an ornamental play of lines that could not always be perfectly justified by the technical rules of construction and structure, but instead tended toward the gratification of a graphic virtuosity in which the balance between structure and ornament can become uncertain. The transition from a geodesic structure in reinforced concrete lamellae to that obtained through the installation of *tavelloni* and the resulting creative margins opened by freely

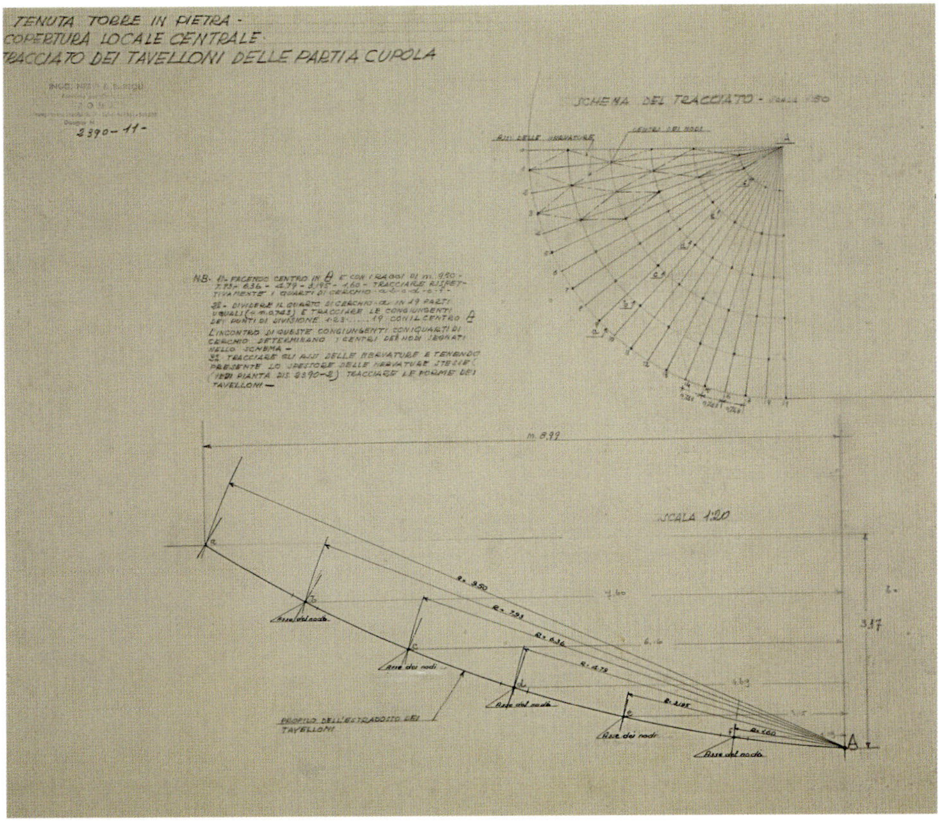
TENUTA TORRE IN PIETRA -
COPERTURA LOCALE CENTRALE -
TRACCIATO DEI TAVELLONI DELLE PARTI A CUPOLA
2390-11-
SCHEMA DEL TRACCIATO
m. 8.99
SCALA 1:20
3.37
PROFILO DELL'ESTRADOSSO DEI TAVELLONI

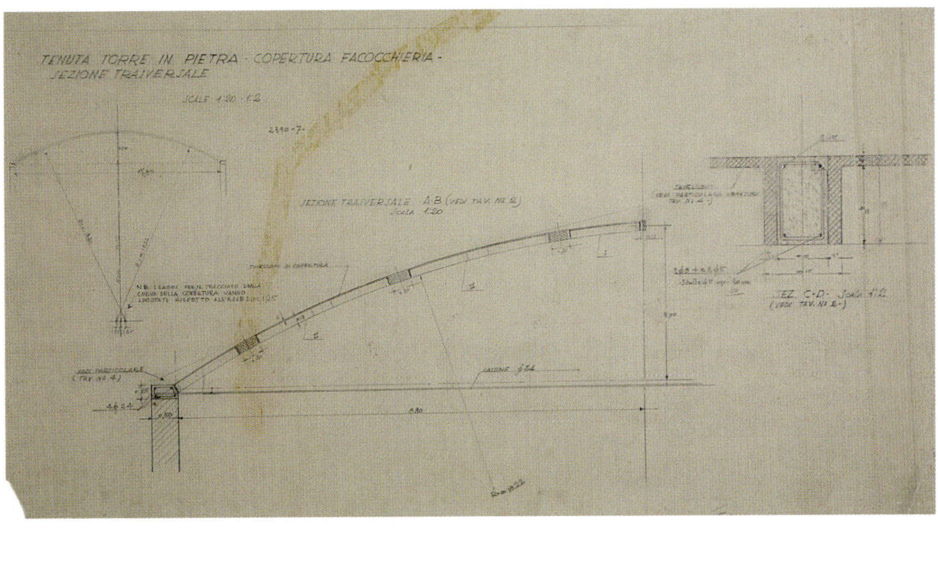
TENUTA TORRE IN PIETRA - COPERTURA FACOCCHIERIA -
SEZIONE TRASVERSALE
SCALE 1:20 - 1:2
2390-7-
SEZIONE TRASVERSALE A-B
SCALA 1:20

171-172. P. L. Nervi, structure for agricultural machinery warehouse, designed by M. Busiri Vici, Via Aurelia km 28, Torre in Pietra [now Torrimpietra], Fiumicino, 1945.

171. Construction of moulds used to produce the *tavelloni* matrices (Maxxi - PLN, F08618).

172. View (Maxxi - PLN, F8621).

choosing the geometry of visible gridlines drove him beyond the boundaries of scientific checks to the secure area of pure form, where he had no guiding instruments except vague formal references to geometries handed down in historical examples. Insofar as the final result may be considered a shell, the fact that it had to be built using *tavelloni* resulted in a visible ribbing. This meant that the structure became monolithic due to the constructive unity of ferro-cementate and reinforced concrete, and hyperstatic due to the nodes. The interweaving left ample room for formal invention that does not exist in the continuous shells in reinforced concrete alone, apart from the impressions left by the way the formwork boards were assembled. Nervi also broke through into an ornamental structure on other occasions.

For the Torre in Pietra warehouse, Nervi the builder invented a process of manufacturing *tavelloni* guided by the desire to establish a system that would allow the production of an increasing number of pieces and to achieve, for each one, a level of surface finish that did not require any coating after installation. At this building site, he started to develop a prefabrication process based on the presence of a mould that allowed him to produce various matrices, each of which he could use to generate several *tavelloni*. This genealogical chain allowed the effective and rapid production of *tavelloni*, achieving as perfect a surface finish as possible so that they could be left exposed. This meant that the entire work looked like the consistent result of a constructive process where each link was rigorously joined to the others, just as the moulds led to the matrices and then the *tavelloni*, obscuring the uncertainties of the grid geometry through the technical precision of the building site system.

When constructing the moulds that would be used to produce the *tavellone* matrices, Nervi had a segment of the curve of each vault modelled on the ground, in the building site, probably using soil or sand. This was then coated with a layer of concrete that was smoothed with a straightedge. The wooden sides forming the diamond-shaped moulds were fitted to this surface (one corner of each diamond shape was secured with a clamp; the sides were kept in place with the aid of a slat and bricks). The reinforcement and metal mesh was laid between the sides of the moulds, and the concrete was applied by hand. Once the concrete had set, the moulds were opened by means of the clamp to remove the ferro-cementate parts, each of which formed a matrix with three kinds of surfaces: the surface in contact with the cement layer at the bottom of the mould and the wooded side timbers was smooth; the internal hand-worked surfaces were rough. Once they had been turned out, the workers repeated the same operations on the matrix – installation of the reinforcements (two layers of netting and 5-millimetre bars) and concrete – ultimately obtaining the *tavelloni* to be installed in the vault, shaped against the smooth surfaces of the matrix and finished sufficiently to remain exposed (thicknesses between 2 and 2.5 centimetres).

Breaking down the structure into individual parts, prefabricated on site, allowed Nervi to avoid having to use the sturdy temporary structures that would have been necessary to lift the reinforced concrete shells during the vault construction stage. The *tavelloni* were therefore laid on a light scaffolding, probably similar to those used for the hangars, and spaced 8 centimetres apart to create empty diamond mesh bands into which the metal reinforcements could be positioned and the concrete poured. At this stage of the vault construction process, the ferro-cementate *tavelloni* revealed their different roles as roofing and stiffening elements and permanent formwork for a filiform geodesic structure in reinforced concrete. Nervi still had to resort

to closing the base of the empty band between the *tavelloni* with temporary wooden boards in order to construct the ribs on site. A detail of the wooden base revealed that the construction system had not yet reached the level of refinement that would characterise later works by Nervi & Bartoli.

Farmers and labourers entered the warehouse in Torre in Pietra to obtain agricultural tools under the shelter of an unexpected and surprising marvel of building technology and of the *properties and potential of reinforced concrete*, in the words of the subtitle to Nervi's book. For Nervi, there was nothing more to celebrate in Italian reconstruction, apart from technology elevated to an absolute value, enclosed in itself like that grid, which he now embodied as an aesthetic form in itself.

The composite system of *tavelloni* in ferro-cementate, with ribs and nodes in reinforced concrete tested in the Torre in Pietra warehouse, spawned the first variants for other roofs and water tanks. The geometric grid of this composite system borrowed the traits of the hangar grid in designs for the pavilion vaults of the fish market at San Benedetto del Tronto, and in the canteen of the company ILVA in Bagnoli, both dating from 1947 – glazed panels replaced the *tavelloni* in ferro-cementate to produce bands of natural lighting. For the water tanks, Nervi & Bartoli proposed two different applications of ferro-cementate, as the material of a continuous shell to be built on site as in the case of the one for the shed in Via della Magliana, or in *tavelloni* prefabricated and put together to produce reinforced concrete skeletons on site, as in the Torre in Pietra warehouse. In other cases, the firm planned to make a truss that would be left exposed. Nervi used the *tavelloni* system for the water tower of the fish market built in San Benedetto del Tronto approximately in 1947.

The on-site manufacture of ferro-cementate *tavelloni*, as tested in the Torre in Pietra warehouse, found a corollary in a 1946 design for a prefabricated house with a circular plan that could be extended over time. The house walls and roof were broken down into a schedule of several "concrete elements" that Nervi specified as being "reinforced with wire mesh or perforated sheet metal left over from machining," revealing the extent to which his ferro-cementate was essentially a derivative of his *sidero-cemento*.

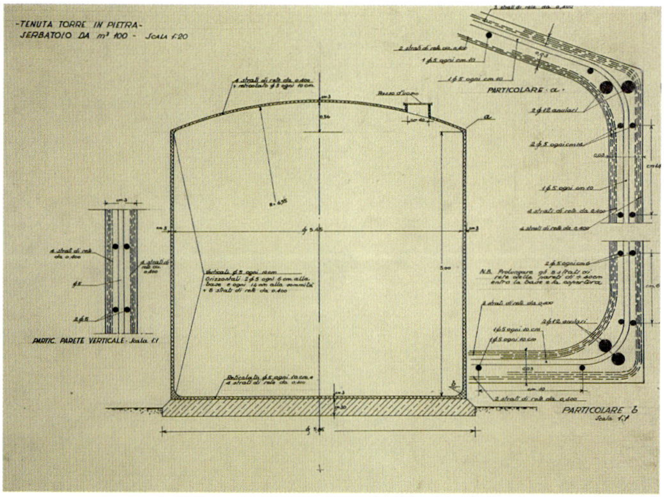

173. P. L. Nervi, "Tenuta Torre in Pietra. Serbatoio da m³ 100 (Torre in Pietra estate. 100 m³ tank)," 1945 (CSAC, n. inv. pra 86, n. id. 13161, coll. 267/1).

174-175. P. L. Nervi, design for fish market, [Via C. Colombo], San Benedetto del Tronto, 1947.

174. View (BSTA, 024-2802).

175. View (BSTA, 024-2799).

176. P. L. Nervi, design of "tipi di serbatoi da 50 m³ di capacità in ferro cementato (types of 50 m capacity tanks made out of ferro-cementate)," about 1946 (CSAC, n. inv. pra 13361, n. id. 13361, coll. 270/1).

177. P. L. Nervi, design for "serbatoio da 50 mc (50 m³ tank)," about 1946 (CSAC, n. inv. pra 881, n. id. 14575, coll. 155/5).

178. P. L. Nervi, water tower for fish market, [Via C. Colombo], San Benedetto del Tronto, 1947. Demolished (Maxxi - PLN, F8619).

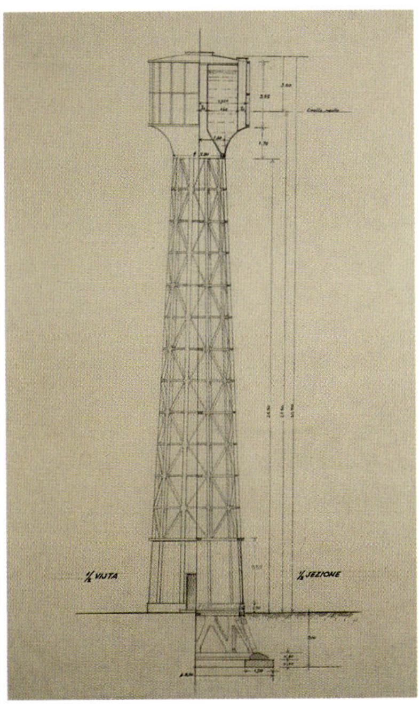

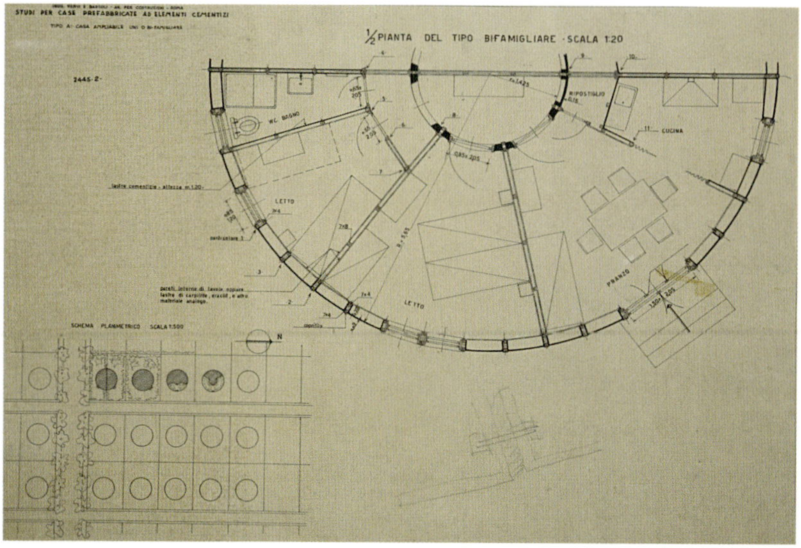

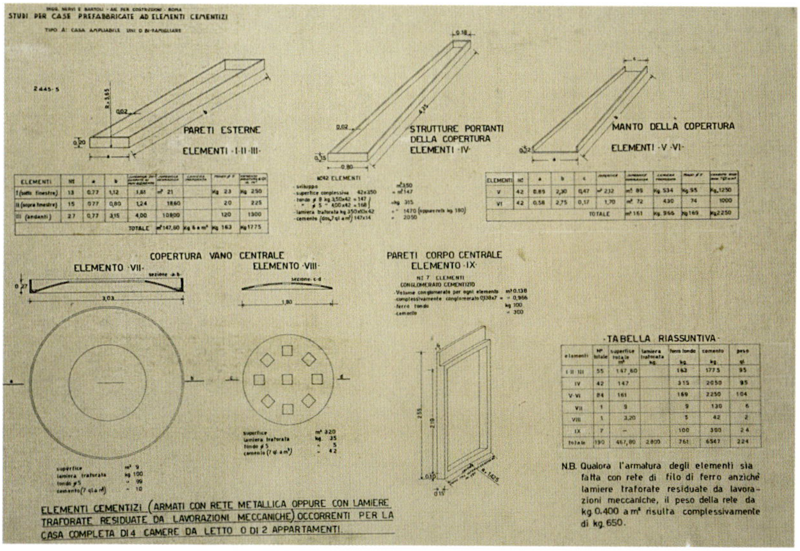

179-180. P. L. Nervi, "Studi per case prefabbricate ad elementi cementizi (Studies for prefabricated houses made out of cement elements)," 1946 (CSAC, n. inv. pra 992, coll. 157/6).

A perforated waved shell at the *Fiera di Milano*

After the shed in Via della Magliana, the hulls and the warehouse in Torre in Pietra, and after various designs for the application of his product, Nervi became increasingly convinced of the innovative potentials of ferro-cementate and talked to his friends and at conferences of the "possibility of forming curved or waved membranes, shell structures of any type," "shapes" hitherto "absolutely unthinkable," that "despite their complexity," "statically resolve a given problem more effectively."[32] He even spoke of "freedom of form"[33] and of "imaginative design" breaking through into "fields from which until recently it had been absolutely precluded." "Spatial resistant structures, simple waved or shell membranes [...] find simple and inexpensive construction possibilities absolutely unattainable with normal structures in cement and even metal."

The vision that drove him to imagine and draw majestic roofs with waved shells in ferro-cementate was put to the test after the first empirical experience with the shed in Via della Magliana, in a decisive scientific trial from a commission, given together with Giuseppe De Finetti, to design *Padiglione 25*, otherwise known as the *Emiciclo*, which was built between 1946 and 1947 in the *Fiera di Milano* complex.

Nervi examined various solutions for the roof of the hemicyclic gallery with its span of 16 metres. These included a geodesic truss, metal structural work, and a sequence, subsequently constructed, of asymmetric sinusoid sawtoothed roofs in ferro-cementate, with horizontal openings across the reinforcement bars. This was the kind of shell that Nervi had already proposed for the magnificent vault intended to demonstrate the possibilities of his ferrocementitious felt and that he had already constructed, in a pleated form, in

[32] Pier Luigi Nervi, letter to Alberto Sartoris, 20 March 1946, Acm-EPFL.

[33] Nervi, *Il ferro cementato e le sue possibilità costruttive* cit., p. 67.

181-182. P. L. Nervi, structure of *Padiglione 25* at the *Fiera di Milano*, designed by G. De Finetti, Quartiere Fieristico, Milan, 1946-47. Demolished.

181. "Fabbricato della piazza centrale. Schizzo prospettico (Building in the central square. Perspective sketch)" (CSAC, n. inv. pra 1400, coll. 266/3).

182. Perspective drawing (CSAC, n. inv. pra 1401, coll. 266/3).

183-186. P. L. Nervi, structure of *Padiglione 25* at the *Fiera di Milano*, designed by G. De Finetti, Quartiere Fieristico, Milan, 1946-47. Demolished.

183. "Armatura della copertura a sheds (saw-tooth roof reinforcement)" (CSAC, n. inv. pra 1402, coll. 266/3).

184. Models produced at the *Politecnico di Milano* Models Laboratory, 1946 (C. Greco, *Pier Luigi Nervi. Dai primi brevetti al Palazzo delle Esposizioni di Torino. 1917-1948,* Lucerna, Quart Edizioni, 2008, p. 223).

185. Models produced at the *Politecnico di Milano* Models Laboratory, 1946 (C. Greco, *Pier Luigi Nervi. Dai primi brevetti al Palazzo delle Esposizioni di Torino. 1917-1948,* Lucerna, Quart Edizioni, 2008, p. 223).

186. Building site, 1947 (C. Greco, *Pier Luigi Nervi. Dai primi brevetti al Palazzo delle Esposizioni di Torino. 1917-1948,* Lucerna, Quart Edizioni, 2008, p. 223).

the Via della Magliana shed. In the *Emiciclo* gallery brought the fundamental innovation that one of the sides of each undulation was cut to provide the gallery with natural lighting. This action of cutting the undulation gave rise to an asymmetrical progression of undulations, transforming the regular shell designed for the huge station vault into a series of saw-toothed roofs.

The waved shell in the shape of a saw-toothed roof became the first element of a roof by Nervi in ferro-cementate to be checked scientifically at the Milan Polytechnic Testing and Model Laboratory. The tests were decisive in establishing the resistance of a waved shell cut by horizontal openings, and the possible kinds of roofs to which Nervi's invention could be applied were also dependent on them. Oberti produced a 1:5 scale model of three saw-tooth roofs to subject to load testing for Nervi.[34] Unlike the previous models for hangars and the E42 Arch that had been made in celluloid, these were built in accordance with criteria that Nervi had adopted to test slab samples, thus using bars and metal mesh and a special cement known as "ultracement." The very nature of the material used and also the scale of the model, which were both out of the ordinary for Oberti and Danusso, indicate the extent to which those tests concerned not only the performance of the structure, as was usual in the laboratory, but also that of the product invented by Nervi, which now had to be tested in its application to a waved shell. The constructed model was also used to verify the stages of the process of constructing the shell in ferro-cementate, as is revealed by a report by Oberti: "Owing to the difficulty of stretching meshes on the model with its double curvature, the plastering of meshes could not be performed, and therefore a form was used for pouring. This procedure proved the impossibility of keeping minimum planned thicknesses because of the rising of meshes (anchored to the form below) with the interposing of the pouring mortar between meshes and the surface of scaffolding. Therefore, thicknesses of about 12 mm were reached, rather than the 6 mm planned, whereas for higher thickness areas, the devised sizes could be kept."[35]

Tests carried out on the model led to some significant considerations about the shape of Nervi's waved shell. Oberti's report also confirmed that Nervi was concerned about the size of the shell and initially considered turning the metal fixtures into a stiffening structure. Nervi's studies for waved shells, carried out immediately after the Milan experiment, concerned the method of perforating the undulations.

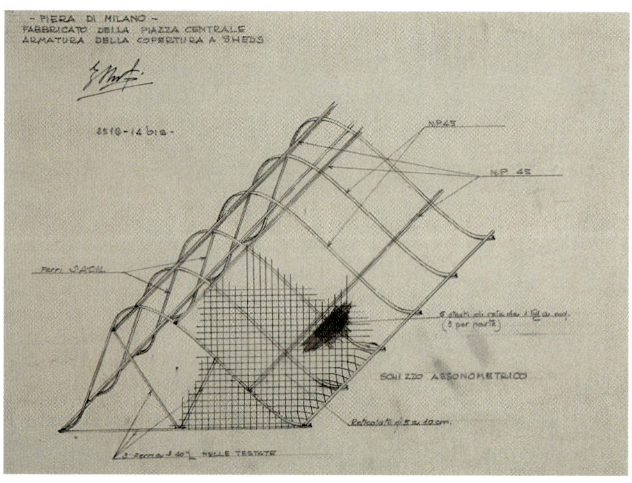

In a study drawing, Nervi depicted a grid of lines forming his now distinctive lozenge motif before the curved window of the *Emiciclo* facade. This design represents a celebration of his ferro-cementate system, perhaps used here to create a protective shield against the sun, thus a kind of original version of the *brise soleil* that both architectural journals and architects, particularly Brazilian, were beginning to popularise in the early 1940s.

Impresa Quaglino of Milan was commissioned to build the waved shells, and they followed the system that Nervi & Bartoli had used for the roof of the shed in Via della Magliana in order to manufacture the ferro-cementate. Each saw-tooth roof was built without formwork by bolting steel piping on-site, curved into lines that defined a resistant waved surface (the surface of each saw-tooth roof was reinforced by metal ribs). Six layers of metal mesh were laid on these; cement mortar was then applied. It could not have been easy for the workers to manufacture the saw-tooth roof on site, fit the waved irons of the main reinforcement, spread and secure the netting on both faces, then apply the cement mortar and smooth it along the waved intrados and extrados. In Nervi's opinion, Impresa Quaglino was responsible for carrying out the work "so hastily and so badly."[36] From then on, he ensured that his own company constructed his future products, but above all, he radically changed the way ferro-cementate was manufactured.

[34] Guido Oberti, *Structural design and testing, by means of models, of some special construction (using ferro-cement)*, in, Guido Oberti, Surendra P. Shah, eds., *International Symposium on Ferrocement*, Rilem, 1981, pp. 2/18-2/23; Neri, *op. cit.*, pp. 46-52.

[35] Oberti, *op. cit.*, p. 2/18.

[36] Pier Luigi Nervi, letter to Alberto Sartoris, 30 July 1947, Acm-EPFL.

chapter seven
ferrocement and large shell roofs

Structures of *Salone B* in Turin, or of the maximum moment of inertia

The stages following the discovery of ferro-cementitious felt, the ways in which it could be applied, the visionary projects, and the experimental building sites, up to the scientific checks with a view to building the *Fiera di Milano Emiciclo* gallery – all this demonstrated that Nervi went about studying his product with the tenacity of an entrepreneur. His aim was to improve each step and every detail to hone the system until it was good enough to bear his own name and that of his company within the production scenario of reinforced concrete and its by-products, just as Hennebique did in his day. While Oberti's tests had confirmed the creative potential of a waved shell in ferro-cementate, even with skylights cut into it, the product's practical applications by Impresa Quaglino had revealed the limits of its manufacture on-site. The methods of implementing variously waved shells that Nervi & Bartoli had already tested in Via della Magliana were nevertheless complemented by the innovative building site in Torre in Pietra, where Nervi had broken down the shell into ferro-cementate *tavelloni* manufactured on site, and the experience – pioneering for the Italian construction culture – of prefabricating the reinforced concrete of his hangars. Therefore it was not surprising that Nervi was heading steadfastly toward the option of manufacturing his "equiwired" reinforcement as well as all of his ferro-cementate on-site. In order to complete this last and crucial metamorphosis in ferro-cementitious felt, built like a ship's hull out of precast ferrocement, Nervi had to reconsider the design of the waved shell. Its continuity as a material weave was so essential to the success of its strength through form, and he had to imagine it fragmented into parts, as he had done to the geodesic structure of the first hangars in reinforced concrete manufactured on-site. Nervi was aware that in order to break down the waved shell, he had to identify the size of the parts based on their production and movement in accordance with a given building site organisation and set-up. He had to invent a joint between pieces with the same characteristics as the node in the geodesic structure, or a replacement that

was still able to confer the same unity as a waved shell manufactured on-site, once assembled.

The thing that gave Nervi the crucial impetus to undertake the final step and reach the goal of what would become "structural prefabrication" in ferrocement was an order to design and execute *Salone B* of the *Palazzo di Torino Esposizioni*, opened on 15 September 1948 to mark the 31st Motor Show. He would have to work with a very tight schedule, from June 1947 to spring 1948, under conditions marked by financial cuts and material shortages, where the only thing available in abundance was low-cost labour. Nervi & Bartoli won the commission to carry out all the extension work on the *Palazzo di Torino Esposizioni*, designed by Fiat engineer Roberto Biscaretti di Ruffia. In addition to *Salone B*, this included re-roofing *Salone A* and building *Salone C* (the latter task was undertaken in 1949).

Nervi built the vaults and a half-dome for *Salone B* based on two shell shapes, waved and ribbed, which were constructed using the product he had invented – and both were set up on reinforced concrete structures. For the roof of *Salone A*, he developed a symmetrical or saw-tooth progression in the undulations of the shell to be fitted between appropriate beams.

It is no coincidence that the definition of ferrocement, the last term used by Nervi to describe his product, emerged during the construction work on the *Palazzo di Torino Esposizioni*, when he introduced the last significant variant in his product. Three decisive evolutionary steps can be identified in the changes in terminology adopted for his product, from its advent as a material for hulls (ferro-cementitious felt), to its application to waved shells manufactured on-site (ferro-cementate), to its radical transformation into a crucial

187-188. P. L. Nervi, design for *Salone A, Torino Esposizioni* complex, Corso M. d'Azeglio, Turin, 1946.

187. Option with saw-tooth roof (Maxxi - PLN, F07406).

188. Option with waved beams (Maxxi - PLN, F07407).

material for structural prefabrication that came about in 1948 when the term became "ferrocement"[1] (even though Nervi still continued to use the definition ferro-cementate during the same year).[2]

The stages that led Nervi to this discovery of the prefabrication of the waved shell in ferrocement for *Salone B* were anything but linear. When he suggested the first solutions for *Salone B*, Nervi was thinking of a structure significantly different from the one he subsequently produced. He still envisaged making the building out of reinforced concrete and ferrocement produced on site, using the construction methods he had employed at the Milan *Fiera Campionaria* building site.

The model of a vault with a waved shell, which emerged from the earliest studies for Turin, his "waved vault,"[3] is the one that he had envisaged in a design to demonstrate the potential of ferro-cementitious felt (then freshly patented) for very large roofs. In order to cover a hall designed for the exhibition of cars, Nervi extensively perforated his shell to the point of threatening its very structural identity in a quest for a diffuse, bright lighting of the space, which was comparable to his idea for the magnificent reinforced concrete railway station vault. The fact that he was not only able to raise the graceful waved shell in Turin along the general lines that he had envisaged a few years before but also to make it even more ethereal by opening the space up to daylight and thus weaving an unprecedented structural form made out of a cement structure with natural light was decisive confirmation that the visionary intuition breathing life into his creativity was rooted in a realistic awareness, as an engineer-entrepreneur, of the practical, technical, and constructive dimension of every shape he imagined. From the outset, Nervi sought to demonstrate, through that work, that he was capable of raising a great concrete structure without expansion joints.

In the early studies, which literally reflected the 71-metre vault span and the sequence of vertical pillars supporting the roof designed by Biscaretti di Ruffia, into which he sunk the sturdy, formless sloping spurs of his structure, Nervi divided the length of the vault shell undulations into 10 sectors. They were identified by thin ribs, also waved and variable in height, referred to as "reinforcement diaphragms," which ran transversely to the waves in order to

[1] See Pier Luigi Nervi, letter to Alberto Sartoris, 1 December 1948, Acm-EPFL.

[2] Nervi, *Il ferro cementato e la sue possibilità costruttive* cit.

[3] Nervi, *Le strutture portanti del Palazzo per le Esposizioni al Valentino* cit., p. 120.

189-192. P. L. Nervi, *Salone B, Torino Esposizioni* complex, Corso M. d'Azeglio, Turin, 1947-48.

189. Option with ferro-cementate vault and steel truss girders, 1947. Perspective drawing (C. Greco, *Pier Luigi Nervi. Dai primi brevetti al Palazzo delle Esposizioni di Torino. 1917-1948*, Lucerna, Quart Edizioni, 2008, p. 229).

190. Option with ferro-cementate vault and steel truss girders, 1947. Cross-section (CSAC, n. inv. pra 966, n. id 14705, coll. 157/2).

191. Option with ferro-cementate vault and steel truss girders, 1947. Detail (CSAC, n. inv. pra 966, n. id 14705, coll. 157/2).

192. Option with ferro-cementate vault, 1947. Perspective drawing (Acm-EPFL, 0172.02.0003.2023).

stiffen them (the distance between the wave centre lines was 2 metres). The profile of the cross-section was unusual within the genealogy of vaulted structures in the way it continually expanded towards the key; yet it was consistent with the opening in the wave sectors, to allow in light, and within each, to increase the resilient cross-section. This progression persisted until the built cross-section, albeit in corrected form.

The waves were shells in ferro-cementate alone in the first two sectors at the ends of the vault. In the other six sectors, which were extensively pierced along the sides to form continuous skylights, the structure became so composite that it could no longer even properly be described as a shell. The curved segments and reinforcement diaphragms were planned in ferro-cementate while true metal truss girders made out of diagonal pipes welded to the L-shaped edging irons ran through the skylight spaces. The detailed drawings for the undulations show that the ferro-cementate reinforcement was made out of four layers of meshing applied to each of the faces of the grid without any break being shown in the hollow under the wave, where there would have been joints if the shell had been prefabricated out of different parts. The connection between the reinforced concrete pillars and the waved vault took the shape of a box-section structure that created a continuous horizontal band. This was to be made in the form of a double-slab floor in ferrocement, through the use of prefabricated ferrocement *tavelloni*, which took the form of caissons with one side open. These were to be laid side-by-side to create the standard space to accommodate the reinforcement and pour the concrete for the joists. The design of the connection between spurs and undulations, the links of the waved segments, and the technical configurations of their reinforcement indicated that Nervi planned to build on site the entire vault structural system, not only the abutment spurs. Details of the shell waves reveal that Nervi was confident that he could rely on the ferrocement and the truss girders along the skylights alone to guarantee the stability of the entire vault, without the need for additional reinforcement running over the undulations.

At the next stage of the project, Nervi applied his own design to Biscaretti di Ruffia's structure and did away with the vertical pillar shape in order to expose his inclined spurs, shaped into a contour distinctive of his work. This expressed the flow of forces generating a space under tension, in accordance with dynamic lines typical of his Fascist era masterpieces. Beams and cantilever beams in cast-in-place reinforced concrete branch off the spurs,

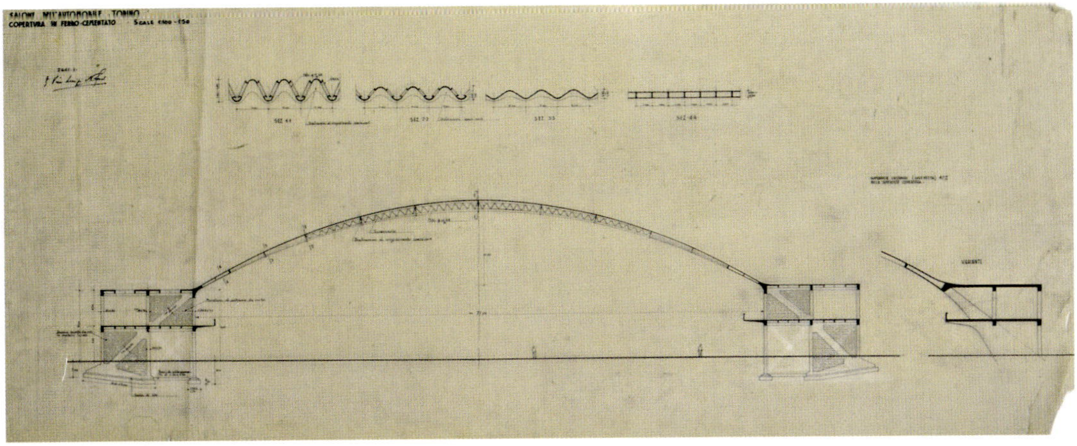
SALONE DELL'AUTOMOBILE - TORINO -
COPERTURA IN FERRO-CEMENTATO

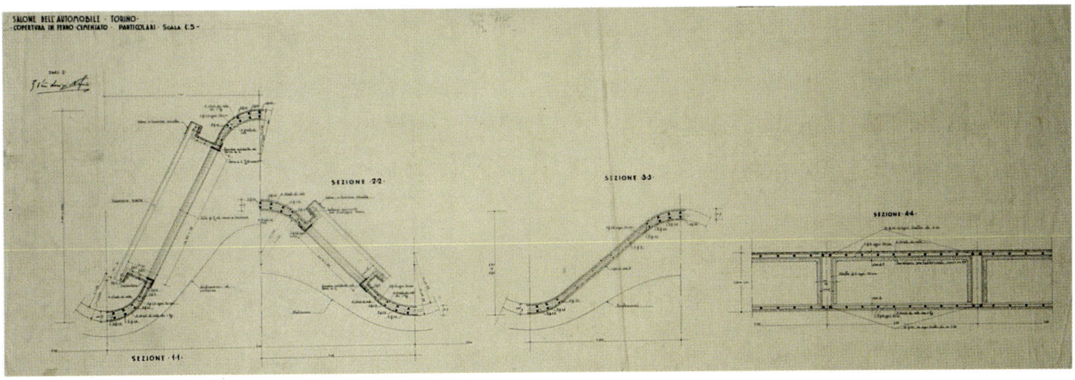
SALONE DELL'AUTOMOBILE - TORINO -
- COPERTURA IN FERRO-CEMENTATO - PARTICOLARI - Scala 1:5 -
SEZIONE -22-
SEZIONE -33-
SEZIONE -44-
SEZIONE -11-

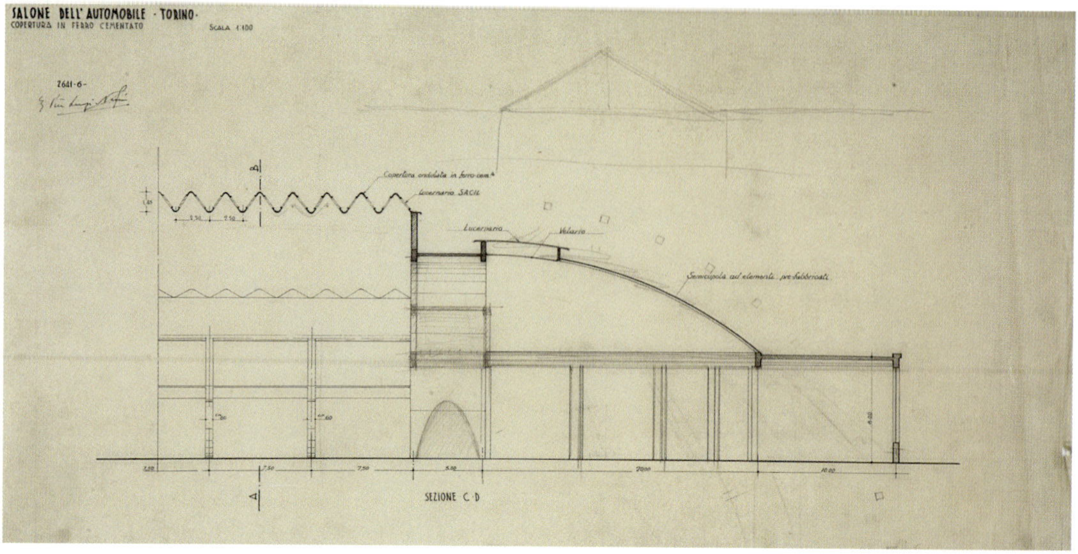

always contoured to express the mechanics of the forces, as was the case with the ones used for the *Stadio Berta* roof. In the case of the spurs, as with the stadium roof cantilever beams, Nervi often published the drawing of his reinforcements to demonstrate the underlying reasons for his special profiles.

Due to the decisive change in the abutment structure lines, the vault assumed the connotations of a velarium with more plastic conviction. This was demonstrated by the fact that it leaned toward the supports along a line of curvature that made it seem as though it were slipping and extending along the bars of the spurs, to cover the galleries and then open up, even at that point, to the effects of daylight. The perspective drawing for the *Salone* indicates the extent to which the leaning of the roof toward the abutments exposed the windows of the galleries and transformed the solid support of the first version into an intangible flow of reflected light punctuated only by fleeting glimpses of the spurs.

In this design, Nervi increased the segments into which the length of the vault waves was divided, and the overall number increased to 13, while the number of perforated parts increased to 9. This significantly altered the

193-195. P. L. Nervi, *Salone B, Torino Esposizioni* complex, Corso M. d'Azeglio, Turin, 1947-48.

193. "Salone dell'automobile. Torino. Copertura in ferro cementato (Motor show. Turin. Ferro-cementate roof, 1947)" (CSAC, n. inv. pra 966, n. id 14705, coll. 157/2).

194. Option with vault in ferro-cementate and inclined spurs, 1947. Perspective drawing (BSTA, 024-6877).

195. "Sezione di un elemento della volta ondulata (Cross-section through a waved vault element)," 1947 (BSTA, 024-7637).

structure, which returned to being a waved shell in ferro-cementate alone. The skylights were no longer continuous but became slashes in a shell firmly anchored to the stiffening diaphragms. Fixtures with vertical brackets of the same type used in the *Fiera di Milano* saw-tooth roof now appeared in place of the metal truss girders. During this stage of the *Salone B* design, the waved shell could already have been planned as a prefabricated structure, as it had to reflect that in the cross-section drawings, the description "prefabricated elements" appears only for the hemicycle half-dome *tavelloni*, now similar to those used in Torre in Pietra. The wording "waved roof in ferro-cementate," meanwhile, appears for the waves, and their profile continues to be similar to that of the previous version, in other words without reinforcements laid over the shell. It is nevertheless likely that in this version of the design, Nervi was contemplating extending the application of the prefabrication to the waved shell, given that it was subdivided into shorter segments that would not be difficult to handle on the building site, and by virtue of the presence of those original "reinforcement diaphragms," which are sufficiently large to act as contact surfaces between prefabricated parts. Nervi was, however, well aware that the transition from a cast-in-place waved shell to a prefabricated shell required special measures to restore the necessary unity to the assembled parts, namely the type of measures that he had implemented in the nodes when he prefabricated the geodesic structure of the hangars. In the waved shell system of *Salone B*, which lacked ribs woven into nodes, Nervi was unable to identify any connection points where he could weld the parts together; the joint surfaces offered by the diaphragms did not lend themselves to this kind of device either. The solution Nervi came up with appears rushed and, in any case, undermined the static behaviour of the shell and the effective symbolic separation between the spurs and the vault, which stretched to cover the galleries. Nervi decided to stretch the wave segments with a reinforcement of metal bars onto which the concrete could be laid – the "ribs" shown in the drawings.

Prefabrication of the shell undulations, now each broken down into pieces as if they were part of a magnificent arch and that Nervi significantly referred to as "ashlars," also required a consequent change in the wave cross-section at the point of the hollow, where bars were arranged that could be bent in situ in order to join two consecutive waves. Nervi planned seven dif-

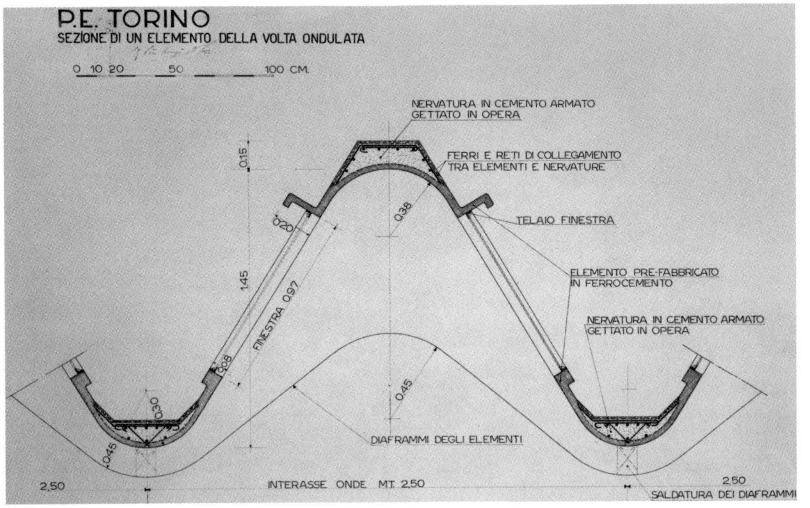

ferent wave ashlars to be prefabricated for each undulation in the shell (the distance between the wave centre lines was now 2.50 metres, and they were approximately 4.5 metres in length).

It was only when Nervi decided to prefabricate the waved shell that he organised the continuous bands curved toward the vault abutments into a series of fans, each consisting of three sectors formed using great *tavelloni* prefabricated in ferrocement as permanent formwork, with the tip of the fan at the top of the spur and its ramifications stretched to form a connection to the shell undulations. These fans are not at all an expression of a formal continuity between parts of the structure but actual essential organs of the prefabricated sector stiffening device, made out of channels that direct the reinforcement bars towards the waves. Nervi built a big triangular connection slab between the pairs of fans, with formwork consisting of wooden boards adjacent to the ferrocement permanent formwork of the fans in which the reinforced concrete could be cast in place. All the cast-in-place reinforced concrete ribs above the undulations are protected against the elements by a layer of ferrocement that Nervi put in place here instead of layers of water-repellent material, due to the properties of impermeability and crack-resistance for which it had been invented and used for hulls, floats, and tanks.

In this way, the bundles of reinforcement bars start together in the pillars, separate out in the fans and then run along the hollows and ridges of the waves with the aim of stitching together the prefabricated parts that are attached to them by pouring concrete in situ to produce ribs (special wooden edges were used for the wave ridges). The rib reinforcement bundles would have given an observer in the building site, or anyone looking at photos taken to document the construction, the "impression" of "nerve bundles" that must bring the structure to life, in the words of Nervi in *Scienza o arte del costru-*

196. P. L. Nervi, *Salone B, Torino Esposizioni* complex, Corso M. d'Azeglio, Turin, 1947-48.
Study of the connection between the fans and the vault, 1947 (BSTA, 024-2795).

197. Photographic collection of P. L. Nervi, photograph of a vault of the church of St Peter and St Paul, Soběslav, Czech Republic, about 16th century (Maxxi - PLN, F04624).

ire? writing on the "aesthetics" of the metal reinforcement.[4] The reinforcement bundles were nothing more or less than climbing tentacles of reinforced concrete of the foundations and pillars that surrounded the prefabricated ferrocementate shell with invisible reinforced concrete "arches," almost reducing the shell to an infill and permanent form for the ribs.

The presence of continuous steel bars between pillars, fans, and waves nevertheless determined Nervi's decision to review the initial separation between spurs and vault. His desire to represent that "nerve bundle" led him on a quest for a design that was even more expressive than the invisible reinforcement constructed. Once the work ended, the engineer's account dwelt on the details of the structure, because contradictory force streams and different technical approaches flowed and coalesced in such an imponderable manner that the design could barely translate them into a built form.

"One particularly sensitive point," confessed Nervi, "was the area of union between the isolated structural elements of the pillars and that continuity of the vault. After much uncertainty, I tried to resolve this transition as simply as I could by grouping three waves of the corresponding vault area on the top of each pillar in the form of a fan."[5]

Visible traces of these uncertainties were left in the study drawings when, after introducing the fan, Nervi also considered continuing the lines in the form as spokes to the first set of blind ashlars. This additional correction clearly reveals his desire to prevent the reinforcement bars from abruptly changing direction as they branched out from the fan to the waves, as actually happens in the constructed option (Nervi actually decided not to extend the lines of the fan, even to the first row of waves). If he had also made that final change to the fan, and had therefore caused the lines of the fan to cross the threshold represented by the continuous horizontal stiffening beam that ran to the abutment of the undulated shell, Nervi would have achieved the pleasing effect of an almost Gothic rib and, what's more, would have highlighted the role of the invisible "nerve bundle." However, in doing so he would have changed the

[4] Nervi, *Scienza o arte del costrire?* cit., p. 91.
[5] Nervi, *Le strutture portanti del Palazzo per le Esposizioni al Valentino* cit., p. 120.

perception of the original significance of his waves, which would no longer have appeared as graceful manifestations of a thin shell; in fact, the substantial changes made to the structure ensured that they no longer did. The "uncertainties" of which he wrote concerned the sophisticated subtleties of the "aesthetic expressiveness" between every line of his work.

Nervi transformed the floors of the two galleries in *Salone B* into structures that were equally expressive of the forces at play, where the moment diagrams are able to manifest their creative power due to the method of prefabrication, which now made it possible to replicate cheaply the sophisticated curved profiles of the joints with the spurs. Prefabricated ferrocement *tavelloni* were fitted between the joists in order to strengthen the casting through its form and create a decorative diamond-shaped hollow.

In *Salone B*, during the design stages, Nervi removed the hemicyclic flat roof planned by Biscaretti di Ruffia, in order to replicate the kind of shell half-dome made out of precast ferrocement *tavelloni* constructed in the Torre in Pietra warehouse. The drawing of the grid was now, however, divided into three sequences by the different radial geometries, sometimes connected artificially but necessary to avoid creating excessively small compartments that would be difficult to transpose into the manufacture of *tavelloni*. This juxtaposition of grids, which is not pleasing when evaluated according to the 19th century ornamental lexicon from which its geometric construction nevertheless depends, was not described by any of his contemporaries. This demonstrates the outmoded nature of its form as well as the strangeness of the matter and the ornamental parameters when applied to an evaluation of Nervi's work, as though his formal expressions merely involved problems of structure. Those awkwardly interwoven ribs therefore appear only as an expression of the load-bearing system, and can also be justified as such by Nervi's calculations. He devoted several pages to calcu-

198-200. P. L. Nervi, *Salone B, Torino Esposizioni* complex, Corso M. d'Azeglio, Turin, 1947-48.

198. Intrados of gallery roof, 1947 (BSTA, 024-2796).

199. Reinforcement of spurs (*The Works of Pier Luigi Nervi*, London, The Architectural Press, 1957, p. 64).

200. "Schema della semicupola per tracciamento dei tavelloni (Diagram of half-dome for *tavelloni* layout)," 1947 (CSAC, n. inv. pra 966, n. id 14705, coll. 157/2).

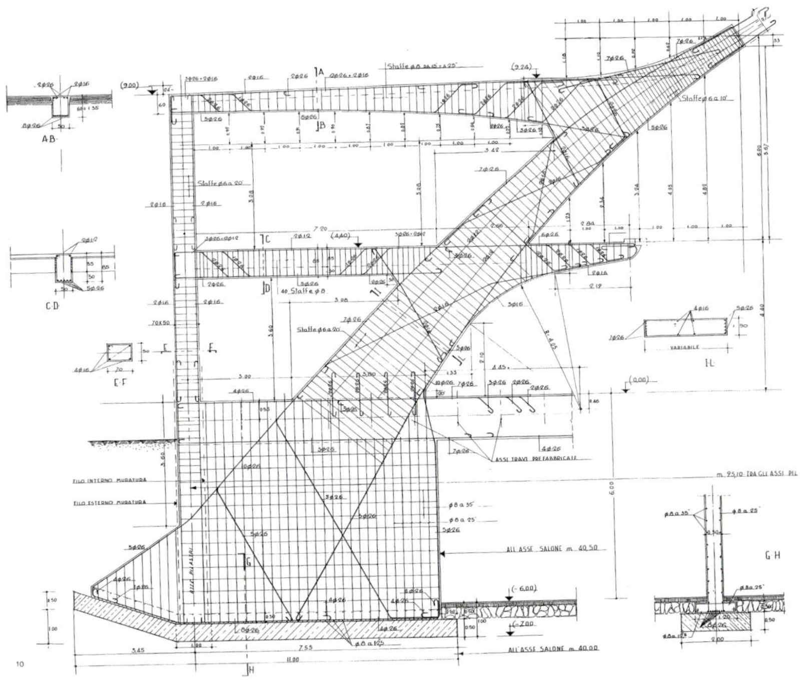

lating the undulating shell, even applying Dischinger's equations to obtain the isostatic lines, due to the difficulty generated by their dissymmetry.[6] Nervi himself warned his contemporaries of the existence of imponderable creative margins in the creative design of that ribbed shell, where he had lost the consistency that normally guided him when coming up with other structural geometries. Those margins of creativity in the choice of geometry for the interweaving of *tavelloni*, which was apparent in the Torre in Pietra vault, therefore returned with a vengeance in the half-dome of *Salone B*. Now Nervi confessed to a lack of sure-fire rational criteria for delineating the lines along which the shell was broken down into *tavelloni* and the extent to which the prefabrication of the ribbed shell in ferrocement was also an aesthetic factor: "It is clear that the constructive system allows great freedom in the geometric determination of the elements and the clearance

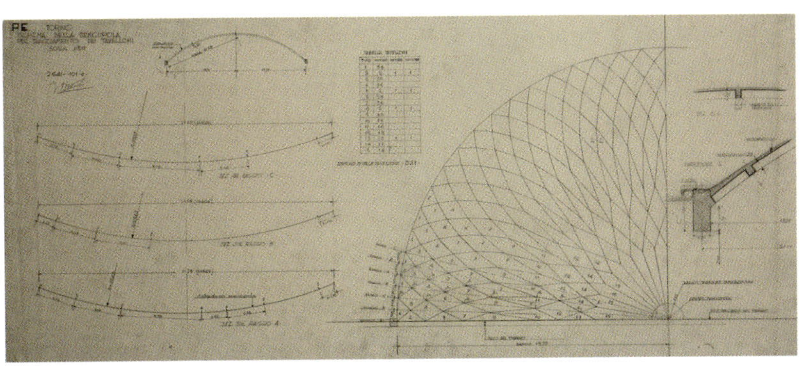

[6] *Ibid.*, p. 121.

201-203. P. L. Nervi, *Salone B, Torino Esposizioni* complex, Corso M. d'Azeglio, Turin, 1947-48.

201. View of flooring intrados, 1947-48 (Maxxi - PLN, F08711).

202. On-site construction of flooring beams, 1947 (Maxxi - PLN, F08707).

203. Building site area, 1947 (CSF).

of ribs that can assume arrangements that are believed to be best to achieve static and aesthetic effects."[7]

The half-dome is buttressed by a gallery covered by an annular floor formed out of tall, narrow joists that Nervi planned to cast in place by means of "wooden formwork that is properly finished and planed in order to avoid having to apply plaster,"[8] and prefabricated arched ferrocement *tavelloni*, to be fitted above the joists as permanent forms in order to build the slab. In order to build the annular floor, Nervi designed a "sliding scaffolding that could be lifted and lowered."[9] This was to become one of the characteristics of the installation of a new type of floor typical of Nervi & Bartoli's future works. Nervi introduced yet another structure between the roof of the hall and that of the hemicycle, an arch with a floor of plastered hollow-brick blocks and concrete.

The construction of *Salone B* began in August 1947.[10] The construction manager, appointed by the client, was Vittorio Bonadè Bottino, while the Nervi & Bartoli foreman was Olivier Olivo. The large number of different parts to be prefabricated, in addition to the parts to be cast in place, meant that the building site had to be efficiently divided into different areas according to the parts, sometimes linked with rails for the transport of materials, and each equipped with sectors for storing cement and for the manufacture and stacking of *tavelloni*, wave ashlars, and joists before installation (all the iron and timber was stored in one underfloor sector of *Salone B*, flanked by rooms where the work could be carried out, which were obtained from the adjacent *Salone A*).

The building site became Nervi & Bartoli's first great enterprise for the structural prefabrication of ferrocement. It was an extraordinary workshop where labour was sublimated into industrial production along Taylorist principles and an assembly line like that used to produce the cars to be exhibited in the *Salone B*. Here, the process of creating the shape of a cement structure reached – in the manner that the construction came together – an unsurpassed pinnacle in terms of the monolithic nature of the assembly and the degree to which the parts were broken down.

The following were produced for the wave ashlars of variable sizes that formed the large shell vault: templates on which to model the "equiwired" reinforcement in bars and metal mesh, as in the Via della Magliana shed building site; moulds for manufacturing the heads making up the ashlars' diaphragms separately; and wooden trestles with a wave-shaped block, known as "removable forms," that could be used to support the reinforcement once it had already been shaped, apply the two heads, spread the cement mortar and break up into a wave-shaped ashlar. The head elements were crucial to stiffening the entire wave-shaped ashlar. Due to this quality, which required careful separate manufacture, they were equipped with rings for lifting the ashlars (a horizontal bar was fastened to these rings so that it could be attached to the hoists). In order to install the wave ashlars, a scaffolding of Innocenti poles was raised, divided into two large opposing sectors, each one corresponding to one quarter of the vault and sliding on rails. The metal guides of a "Decauville rail" were fitted above the scaffolding, along the centre lines of each wave. Each wave ashlar was raised by hoists fixed to the top of the scaffolding and then made to slide along the rail until it was spaced approximately 4 centimetres from the previous one, then secured with cement mortar (approximately 30 were fitted each day). Reinforced concrete ribs were made to strengthen the wave ashlars and ensure the whole shell remained in one piece. The use

[7] *Ibid.*, p. 122.
[8] Drawing No. 2641-104e, CSAC, n. inv. pra 966, n. id. 14705, coll. 157/2.
[9] Drawing *Schema per ponteggio Innocenti, ibid.*
[10] Pier Luigi Nervi, *Le Hall du Palais des Expositions du Valentino, à Turin*, in, "La Technique des Travaux," XXV, September-October 1949, No. 9-10, (pp. 272-278), p. 274.

204-206. P. L. Nervi, *Salone B*, *Torino Esposizioni* complex, Corso M. d'Azeglio, Turin, 1947-48.

204. Installation of prefabricated *tavelloni* for fans, 1947 (photograph by R. Moncalvo; BSTA, 024-7580).

205. Installation of a steel reinforcement between the prefabricated *tavelloni* making up the fans, 1947 (BSTA, 024-7581).

206. Detail of fans (photograph by A. Moisio; Acm-EPFL, 0172.02.0003.2028).

207-210. P. L. Nervi, *Salone B, Torino Esposizioni* complex, Corso M. d'Azeglio, Turin, 1947-48.

207. Putting together the waved ashlars of the vault, 1947 (BSTA, 024-2777).

208. Assembling the reinforcement and formwork to produce the pouring connecting the vault waved ashlars, 1947 (BSTA, 024-2782).

209. Assembling the vault, 1947 (Maxxi - PLN, F01032).

210. View (photograph by A. Moisio; Acm-EPFL, 0172.02.0003.2025).

of ferrocement was extended to other components of the structure of *Salone B*, ranging from the preparation of moulds for the prefabrication of beams to permanent formwork.

Nervi repeated the approaches he had tried out in the Torre in Pietra building site in order to prefabricate the half-dome *tavelloni*. He therefore ordered one segment of the counter-form to be recreated on the ground and applied a system of matrices to generate the ferrocement moulds he would use to produce a series of *tavelloni* of the same size, using the same product. In the technical detail of the *tavelloni*, he introduced horizontal facings along the edge that would directly produce cavities when installed on the ribs. He then used these to lay the reinforcement and pour the concrete for the ribs, thereby doing away with the wooden boards he had used under the *tavelloni* to construct the ribbed shell in Torre in Pietra. Other ferrocement moulds were also used to prefabricate the variously contoured, flat, curved, or caisson-*tavelloni* – still the same product – that were used as permanent elements for floors and roofs, or the reinforced concrete beams of the galleries, "obtaining pourings with very regular surfaces that did not require additional plastering."[11] The moulds were sprinkled with oily substances so that the *tavelloni* could easily be extracted. In some parts, this treatment resulted in surface defects.

The waved ashlars and the various types of *tavelloni* bore the marks of their prefabrication: impressions left by the timber boards for the former and

[11] Nervi, *Le strutture portanti del Palazzo per le Esposizioni al Valentino* cit., p. 122.

211-213. P. L. Nervi, *Salone B, Torino Esposizioni* complex, Corso M. d'Azeglio, Turin, 1947-48.

211. Constructing the half-dome, 1947-48 (BSTA, 024-7585).

212. Intrados of the half-dome (photograph by R. Moncalvo; ARM).

213. Annular floor and half-dome (photograph by A. Moisio; Acm-EPFL, 0172.02.0003.2108).

stains left by oily substances for the latter. Once the reinforced concrete had been manufactured and the ferrocement pieces had been assembled, Nervi set out to make invisible the defect and traces of the construction process, almost as though he were dissatisfied with the degree of surface finish. This marked the stage of correcting the quality of the Nervi & Bartoli concrete and cement, touching up both the work and the photographic representation, as Nervi had done previously in the case of the *Stadio Berta*.

The complex transition between exposed reinforced concrete spurs and the waved shell that had absorbed him during the steps of the design continued to appear defective even when implemented, shedding doubt on the finish quality. The intrusion of the prefabricated parts of the fan in ferrocement, next to the triangular slab in reinforced concrete, cast in place on formwork made up of timber boards, led Nervi to resort to the expedient of smoothing the surfaces with a cement coating, which he also spread on the spurs and on the first two sets of waved ashlars without openings. All these surfaces were then painted white, contributing to defining the polished appearance of all the cement directly visible to visitors to the *Salone*. Only the waved ashlars with the skylights were left untreated, in part relying on the optical effects of backlighting and distance. The application of neon strips along the intrados of the lower vertices of the undulations helped to mask the joints between the ashlars that ran exactly along those lines. Nervi had a coat of thick white paint applied to conceal the oil stains that were visible on the half-dome *tavelloni* (perhaps the same lime paint that he had previously used for a water tank in the 1930s).

Despite the imperfections, Nervi's goal of achieving a perfectly polished finish through his system of ferrocement parts prefabricated in smooth moulds

made out of the same material ensured that the construction project of *Salone B* in Turin established the traits of a concrete aesthetic that was the complete antithesis of the one that was emerging in the still booming building sites of *Unité d'Habitation* in Marseille in 1948, whose concrete was scored by the imperfect impressions left by boards. Through these two works, ferrocement and cast-in-place concrete appeared to set themselves up as methods opposed to a future world of *béton brut*.

Riccardo Moncalvo's photographs of the building stages and Aldo Moisio's photographs of the finished work contributed to the international success of *Salone B* as a crucial work of Nervi's structural prefabrication in ferrocement, sometimes touched up at key points to make all the surfaces of the work look more perfectly polished than they really were. Sometimes, through the artistic testimony of certain shots, those photographs revealed the formal essence of the structure, its very being generated by rational and inexpensive processes of manufacture and assembly. Through the action of contemplative perusal, from particular angles, fixed and disseminated through photography, Nervi himself seems to have discovered the unexpected expressive force of figures that he pursued with the static sensibility developed through his many building site experiences. Detailed photographs of the waved shell, shot by Moncalvo from a viewpoint that took advantage of the scaffolding and the building site when the work was still under construction and published in essays by Argan or Nervi himself, used close-up shots to show the undulations traversed by the stiffening diaphragms. They reveal the still imperfect manufacturing process of the ferrocement, which actually appeared to be tattooed with the impressions left by the counter-form boards. None of the surfaces visible in the photographic print published after considerable retouching reflects what the work actually looked like.[12] Meticulous touching-up with a brush

[12] Argan, *Pier Luigi Nervi* cit., p. 67, and Nervi, *Costruire Correttamente* cit., plate XXV.

[13] Nervi, *Le strutture portanti del Palazzo per le Esposizioni al Valentino* cit., p. 120.

[14] Argan, *Pier Luigi Nervi* cit., p. 19.

[15] Quoted in, Roberto Einaudi,

214-215. P. L. Nervi, *Salone B, Torino Esposizioni* complex, Corso M. d'Azeglio, Turin, 1947-48.

214. Unretouched photograph of the vault intrados, 1948 (photograph by R. Moncalvo; ARM).

215. Retouched photograph of vault intrados (P. L. Nervi, *Costruire Correttamente,* Milano, Hoepli, 1955, plate XXV).

transformed the photograph into no less than a perspective drawing of an unreal skeleton with perfectly polished surfaces, which Nervi now held up as an example of how to "construct correctly." The asymmetrical shot of a wave in the foreground completely distorts the perception of the repetition of identical pieces that technically underpins the vault, and it was taken from so close that it transfigures the wave into a backbone with slenderly-traced ribs sprouting from its sides. There is no doubt that the waved, ribbed shell, if observed from that angle, can be held up as the foundational shape of a structure that was to characterise certain masterpieces, particularly in Turin, ranging from the works of Carlo Mollino to those of Giorgio and Giuseppe Raineri. That photograph reveals the animal strength of Nervi's structure, the way it took on a natural appearance through a process of conceiving the built form and monitoring its technical feasibility at all times during the design stages.

Everywhere the lines and surfaces lean, bend, flex, or arch with evident plastic sensibility to reduce the amount of material used and, as much as possible, make it adhere to the vital drive of the forces flowing through the structure. Sometimes that is even visible in the different densities established by the moment diagrams that were the creative tool Nervi used to progressively define the "aesthetic expressiveness" of the work.[13] In *Salone B*, the structural prefabrication showed itself capable of rendering the forces at play visible and celebrating them in the obsessive repetition of the elements. This marks the point that it became possible to "develop from the 'mesh' to the 'fabric' of construction." The geodesic weave of the lamellae is sublimated in the "extraordinary '*tricoter*' of space" described by Argan.[14] All the elements contribute to a powerful characterisation of the aesthetic expression of structure, without any of them having been designed with a specific formal function – apart from the half-dome grid. It is significant that Nervi, when the work was complete, was attracted by the plastic effect produced in the sequence of undulations by the stiffening plates and discovered their aesthetic role in the perception of the work. "These diaphragms produced patterns that I had never imagined," he admitted.[15]

Each part of the structure ultimately possessed its own principal configuration that appeared natural in the flow of sinuous lines and waved surfaces. They participated in a quest for the maximum inertia, as the structure was no longer subject to the constraints of formwork construction, except in rare cas-

Pier Luigi Nervi: lezioni romane. Lecture Notes, Roma 1959-1960, in, Annalisa Trentin, Tomaso Trombetti, eds., *La lezione di Pier Luigi Nervi,* Milano, Bruno Mondadori, 2010, (pp. 63-159), p. 96.

es. The skeleton of Matte Trucco's Lingotto, which was still timber, therefore relinquished its position to the now bone-like housing of Nervi's *Salone B*.

Because it was an undiluted expression of ferrocement techniques, the shape of *Salone B* in Turin became a symbol of the style Nervi was aiming for. It was absolute and natural, because it was determined by implacable and unchangeable laws. There were no longer any totalitarian ideologies to be celebrated, as with the structural inventions of the 1930s. Now Nervi was striving toward a goal of large dimensions, where statics and construction drastically reduced the room for creative freedom, and technology apparently produced a structure determined at every point along lines consistent with the material and the forces at play – a veritable technological style. The monument erected by Nervi in the shape of *Salone B* was not dedicated to Fiat's industrial products; instead it was a self-erected monument to his own production system. Nevertheless, his work was having a new impact on the ideals of the nascent Italian democratic state. This was now taking place in the name of the laws of economy and the gods of technology, to be enacted without any more rhetorical emphasis – a noiseless, chaste rhetoric of efficiency and precision.

Patent and designs for waved or curved cement structures

The ferrocement prefabrication system implemented in the *Salone B* building site served as tangible evidence of the possibility of raising a grand waved shell inspired by the magnificence of domes by Brunelleschi and Michelangelo. It was immediately protected by Nervi with a patent covering "waved or curved cement structures," filed in August 1948,[16] upon completion of the work in Turin and on the hull of the vessel *Nennele*, which represented for him the "furthest limit of ferrocement achievement"[17] due to the reduced shell thickness (12 millimetres). The circumstances under which the text was prepared ensured that the patent contained the essence of the inventions implemented on-site as well as corrections made to that system in order to avoid defects that had already become apparent to Nervi by the time the documents were delivered to the patent office.

Compared to previous patents on the same cement compound, the very tone of the writing of the 1948 patent revealed the level of mastery achieved by Nervi in the manufacture of ferrocement. It also communicated his desire, after discovering and testing the constructional and static potentials, to improve the aesthetics of the product finish. It could even be said that questions of the aesthetics of profiles and surfaces formed the innovative core of the patent.

Nervi even stated that the profile of the undulations and that of the stiffenings could be chosen at will to achieve a certain "aesthetic effect," given that any cross-section shape with an organised progression appears to be naturally suitable for "waved or corrugated structures" because it stretches toward the maximum moment of inertia. The graphic diagrams also include a corrugated shell with a triangular cross-section, which reproduces the corrugated profile used in the ideal design for a grand vault in precast reinforced concrete for a railway station. "The shape of the undulation is unimportant for static and construction purposes, and therefore broken profiles may be adopted in place of the curved undulation," he stated in peremptory fashion, shedding light on the significance of the waved shell of *Salone B* as the sensitive manifesta-

[16] Pier Luigi Nervi, patent IT 445781, *Procedimento costruttivo per la realizzazione di strutture cementizie ondulate o curve con o senza tensione preventiva*, 26 August 1948, ACS.
[17] Nervi, letter to Sartoris, 1 December 1948 cit.

tion of a static progression implemented in accordance with certain aesthetic choices that he may have intended, in that case, to interpret the aerodynamic curved outlines of car bodies. Nervi himself had postulated the extrapolation of shapes determined by aerodynamic laws in planes or cars to other objects, through contamination and appeal, in the genesis of style. If Nervi's studies on the curvatures of the diaphragms used to stiffen the undulations of the shell of *Salone B* are examined in light of the patent text, where he stressed the "aesthetic reasons" dictating the ways in which those diaphragms "could be shaped," we find confirmation that there is plenty of room for creation in the aesthetics of the technology, as is masterfully revealed by the close-up photos taken at the end of the prefabrication process showing the parts installed in the finished work.

The other aesthetic matters addressed by Nervi in the patent concern the surface finish. When he described the ferrocement formwork for prefabrication of the parts, he emphasised their benefits for the "perfect regularity of the intrados of the elements and therefore the good aesthetic effect of the structure without the need to resort to additional finishing work or plastering." It is worth noting that Nervi only used ferrocement formwork when producing the various kinds of *tavelloni* for *Salone B*. On building sites started after the one in Turin, he changed the type of removable forms made up of boards that he used to prefabricate the vault wave ashlars, which were all marked with impressions of the wood. On the subject of the prefabrication of reinforced concrete parts, he described in the patent "removable formwork" in iron as well as in wood or "other material," evidently with the aim of achieving polished surfaces in this case as well.

The description of the various "waved or curved cement structures" covered the case of flat roofs, which Nervi put to the test with the galleries of the Milan *Fiera Campionaria Emiciclo*, considering variants for the roof of *Salone A* in the *Torino Esposizioni* complex. For that kind of waved structure, always made out of prefabricated parts, he also considered the pre-compression of wave-shaped wedges ("with or without pre-stressing"), and specified the use

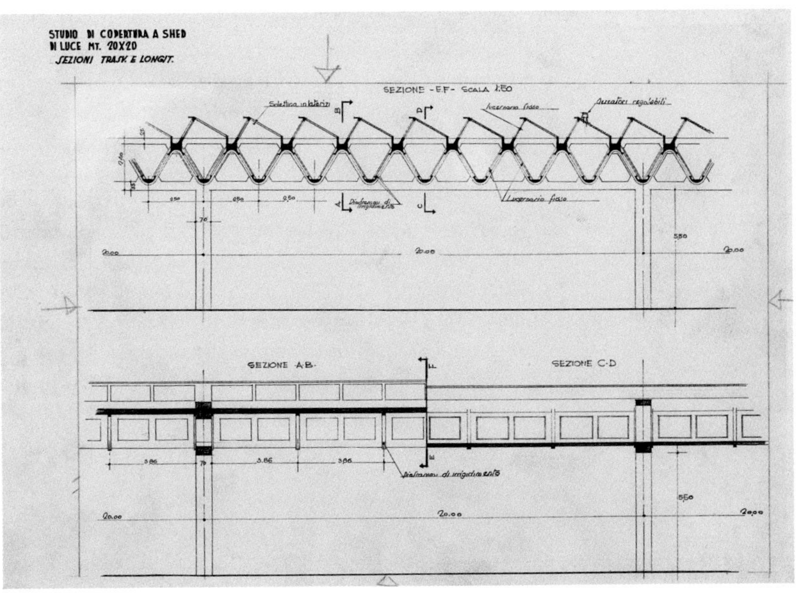

216. P. L. Nervi, study for saw-toothed roof, 1948. "Studio di copertura a shed di luce mt. 20 x 20. Sezioni trasv.[ersale] e longit.[udinale] (Study for saw-tooth roof with 20 x 20 metre span. Cross-section and longitudinal section)" (Acm-EPFL, 0172.02.0003.2091).

217-218. P. L. Nervi, study for saw-toothed roof, 1948.

217. Perspective drawing (Acm-EPFL, 0172.02.0003.2090).

218. Perspective drawing, dated 10 July 1948 (BSTA, 024-2778).

of high-strength steels for reinforcing the "resilient cores" of individual parts that "have been pre-stressed."

When he was engaged in drafting the patent and following what appears to have become his custom for illustrating his inventions in the 1940s, Nervi developed a project to demonstrate a flat roof to be made out of a waved shell made up of prefabricated wave ashlars in ferrocement, exactly as described in the patent. His drawings for a "saw-tooth roof," dated July 1948, depict a structure made out of prefabricated elements in ferro-cementate, with wave ashlars 2.50 metres wide and 3.86 metres long. Those were fitted in horizontal beams approximately 20 metres long that were perforated for bright, diffuse lighting and equipped with saw-tooth roofs using a rainwater collection system to be produced with skylights and brick slabs, because he was no longer willing to generate vaults. The application of pre-compression gave rise to the ferrocement rod stiffening the opening of the skylights in the wave ashlars (the rod does not appear in the same wave ashlars of *Salone B*). To demonstrate the aesthetic effect on the space of arranging the undulations either horizontally or lengthwise, Nervi designed two perspective views of the same hall measuring 100 metres in width and 5.50 metres in height with the same reinforced concrete skeleton organised into a 20 x 20-metre pattern.

[18] Nervi, patent IT 445781 cit., pp. 8-9.
[19] Nervi & Bartoli design No. 2881, CSAC, n. inv. pra 713, n. id. 14348, coll. 290/4.
[20] Nervi & Bartoli design No. 2914, CSAC, n. inv. pra 906, coll. 157/3.
[21] *Bando di Concorso per il Progetto di completamento del fabbricato viaggiatori della nuova Stazione di Roma Termini*, in, "Strutture. Rivista di scienza e arte del costruire," April 1947, No. 1, pp. 41-43;

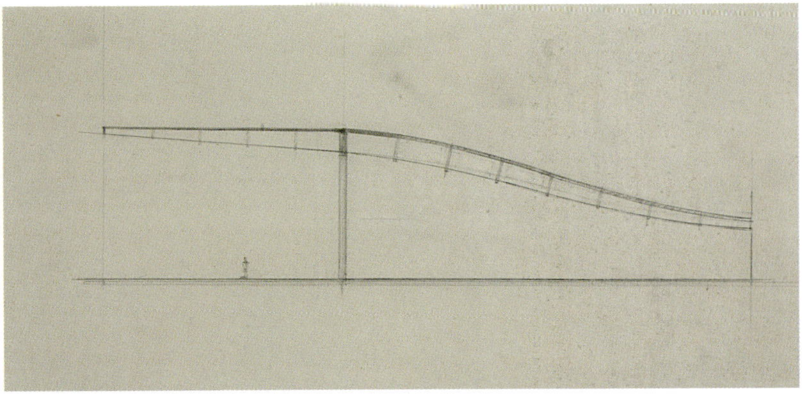

219-220. P. L. Nervi, study for roof of Termini railway station, Rome, 1948.

219. Version with arched profile. Cross-section (CSAC, n. inv. pra 904, n. id. 14627, coll. 157/3).

220. Version with horizontal roof. Perspective drawing (CSAC, n. inv. pra 904, n. id. 14627, coll. 157/3).

The detail of the explanation written at the end of the August patent confirms the level of detail that Nervi had gone into when designing the "saw-tooth roof" in July: "When flat roofs must be made, the construction system remains essentially the same, and it also becomes possible to use high-strength steel, pre-stressed to produce pre-compressed structures, for the resilient cores of the reinforcements arranged on the wave ridges and hollows, with all the specific technical and economic benefits of such structures [...]."[18]

The drawings and patent of July and August 1948 containing descriptions of the various kinds of "waved or curved cement structures" were not intended to remain empty words in the projects proposed by Nervi & Bartoli at the end of the 1940s. The flat roof with waved shell was used in projects constructed and drawn up around 1948 and 1949, for the entrance hall of the *Roma Termini* railway station, for the roof of the Società Sicurezza Trasporti Autolinee (SITA) bus garage in Bari,[19] and for the Fiat branch office in Bari.[20]

Nervi had already considered the *Roma Termini* station when he drew up the station design in 1943. Now, for the design for the same station, he studied some options after publishing the call for tenders and two articles on the competition and its results in 1947 in his magazine "Strutture".[21] At some unspecified time between the design and construction of *Stazione Termini*,

Concorso per il completamento del fabbricato viaggiatori della nuova Stazione di Roma-Termini – Motto: UR (Relazione dei progettisti architetti e ingegneri: Ridolfi, Quaroni, Cardelli, Fiorentino. Ceradini, Caré), in, "Strutture. Rivista di scienza e arte del costruire," December 1947, No. 3-4, pp. 56-57; Saverio Muratori, *Un progetto della Stazione di Roma, ibid.*, pp. 58-61.

221-222. P. L. Nervi, design of roof for SITA bus garage, Bari, 1948.

221. "*Tipo C*" (BSTA, 024-2789).

222. "*Tipo D*" (BSTA, 024-2792).

Nervi proposed a shelter for the passenger building, using the same precast ferrocement elements he had used as a construction material in Turin.

The drawings documented the study of three different cross-section profiles. One was arched in a way that Nervi had never used previously in his thin, waved structures so as to adapt his system to contain a meaningful place for passengers to flow in and out, as in the design by Mario Ridolfi and Ludovico Quaroni as well as that by Annibale Vitellozzi and Eugenio Montuori. Another also had a waved outline, but the line was simpler and similar to that drawn by Vitellozzi and Montuori. The third and final design reflected Nervi's building site approach and took the form of the horizontal assembly of waved ferro-cementate elements, as used in the flat saw-tooth roof of 1948. In the various versions, Nervi always caused the shelter to extend beyond the row of pillars defining the gallery line of closure. The fact that Nervi had a perspective view drawn up only for the horizontally laid-out waved structure could indicate a predilection for a solution that was easier to construct, such as that of the 1948 drawings for the flat saw-tooth roof. Nevertheless, the versions with a waved cross-section undoubtedly resembled virtuoso forms of the kind produced by Torroja.

223-224. P. L. Nervi, design of roof for Fiat branch office, Bari, 1949.

223. "*Tipo B*" (BSTA, 024-2805).

224. "*Tipo C*" (CSAC, coll. 157/3).

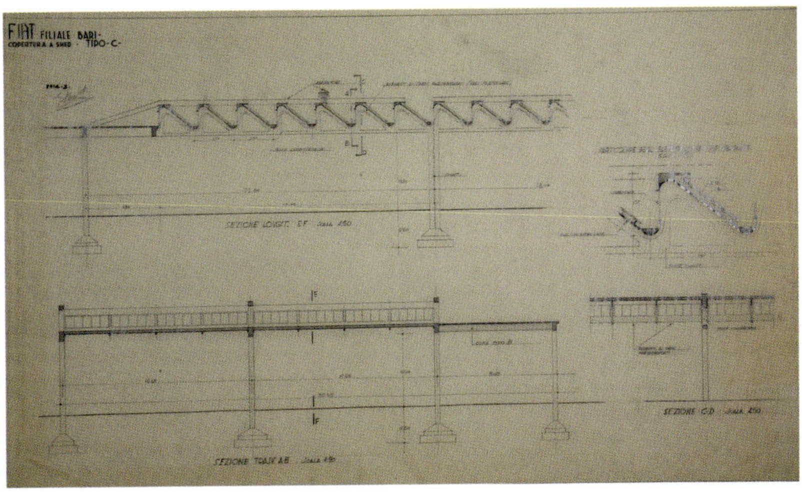

The approach used in the various roofing systems, made out of prefabricated elements in all cases, as they developed from the hangars to *Salone B* in Turin, informed two of Nervi's four proposals for the SITA bus garage roof in 1948 – one made out of truss girders and joists prefabricated in reinforced concrete and one made out of prefabricated waved beams in ferrocement. The other two proposals were based on a kind of box girder that Nervi had already envisaged. Until then, he had never configured it as the main structure of a prestressed concrete roof in various forms (one alternative includes openings for natural lighting).

The variants proposed in 1949 for the roof of the Fiat branch office shed were also exercises in diversified applications of the ferrocement and reinforced concrete prefabrication systems, according to Nervi's fundamental types. In one of the proposals, wave ashlars prefabricated and assembled into inclined sequences served to create large saw-tooth roofs arranged on precast reinforced concrete beams. In another alternative, Nervi envisaged a waved shell to be built out of prefabricated ashlars in ferrocement. This was designed as a variant of the saw-tooth roof designed in July 1948, where the asymmetrical undulation, through its profile, revealed that the purpose was

225. P. L. Nervi, design of roof for concourse of the Porta Nuova railway station, Turin, 1948 (BSTA, 024-2793).

226. P. L. Nervi, design of roof for concourse of the Palermo railway station, 1948 (BSTA, 024-2790).

industrial rather than monumental (Nervi removed the system of protection with skylights and brick *tavelloni*). Nervi also designed the details of his "pre-stressing" system in a series of studies for the Bari shed.

The monumental nature of the structural system implemented for *Salone B* informed the 1948 designs for the roofs of the Turin Porta Nuova and Palermo railway stations. Reasons of spatial quality led Nervi to plan a barrel vault for the Turin roof, to be installed on Roman arches built to a design by Alessandro Mazzucchetti and Carlo Ceppi. In these projects, Nervi proposed extending his system of waved ashlars in ferro-cementate to the creation of a vault head wall, as an alternative to the grid he had planned for the *Fiera di Milano Emiciclo*. Detailed studies of the roof of the Palermo station confirmed his concern with protecting the ferrocement against the elements by means of a covering of *tavelloni* to be rested on special fins created in the waved ashlars. In variants of the undulations – developed for Palermo in all cases – that were symmetrical and non-symmetrical but both with a single side of the wave open for a skylight, Nervi pushed the boundaries of plastic creativity and the formal effects on the space of shells in precast ferrocement.

Nervi proposed one version of the "waved or corrugated structures" with a triangular cross-section profile, as indicated in the 1948 patent, for a 1949 competition design for the construction of a hangar in the Buenos Aires national airport. He envisaged weaving a single parabolic vault over the grand 180-metre span (the notice called for two roofs, each with a 90-metre span),

again using the system of structural prefabrication of ferrocement parts.[22] Now the ribs of stitching between the ashlars, invisible in waved shell roofs such as that of *Salone B*, became proper arches in the pattern of structural lines to the point of revealing, through the evidence of the overall design, the extent to which the corrugated structure is a hybrid made out of reinforced concrete and ferro-cementate, in the latter case reduced to a formwork, wall, and stiffening of a sequence of dispersed arches in a quest for the maximum moment of inertia. "Hangares de hormigón armado," was the title of the drawings. The structure was protected by a covering of brick *tavelloni*, while natural lighting was ensured by sidewalls, to be built out of wave ashlars woven vertically and perforated – now arranged to create "paredes onduladas" with stiffening facing the inside of the hangar. Luigi Moretti captured the Gothic window lighting effects by describing the structure as a "cement cathedral."[23] The corrugated geometry of the roof was echoed by that of the colossal pillars which, unlike Nervi's various spurs, were arranged along the curve of the vault as though it were a cut-off curved wall.

The monumental nature of structural prefabrication in ferrocement was interpreted in a manner that could even be described as historicist when, in

[22] L. M. [Luigi Moretti], *Un progetto di Pier Luigi Nervi per un'aviorimessa a Buenos Aires*, in, "Spazio," I, July 1950, No. 1, (pp. 50-51), p. 50.
[23] *Ibid.*, p. 50.

227. P. L. Nervi, design of hangar for Buenos Aires international airport, 1949 (CSAC, n. inv. pra 913, n. id. 14633, coll. 157/3).

1949, Nervi designed the church of the Sacred Heart in Terni, in conjunction with Sergio Musmeci, who was employed in the Nervi & Bartoli engineering department. Square prefabricated *tavelloni*, with a curved cross-section measuring 15 millimetres in thickness, were shaped to act as the permanent forms of transverse and longitudinal ribs in reinforced concrete. They made it possible to plan a series of barrel vaults with 11-metre spans to be woven between the reinforced concrete arches of the central nave and to form the apse half-dome. The note "cladding excluded," with reference to the octagonal pillars in the drawings, alludes to an estimate for the unfinished work.[24]

Nervi was now experimenting with all possible kinds of roofs using his structural systems and his assembly processes. He even envisaged shells with complex geometry, always arising out of his ideas of roofing large spaces with slender continuous waved structures in ferrocement. This approach was confirmed by a design for a shed to be used for machining operations, where he started from a saw-tooth roof opening in the form of a depressed arch and came up with a type of folded shell or "thin-shell conoids."[25] The waved and perforated roof rested on an overhead trussed backbone structure in reinforced concrete. The comparison with the roof alternative chosen for the same shed, to be produced using prefabricated elements in reinforced concrete, seems to have been proposed to demonstrate the expressive potential of the ferrocement system.

When Nervi emphasised the quality of the surfaces of his ferrocement elements in his 1948 patent, he was aware that at the *Salone B* building site he had been unable to give the wave ashlars of the great vault a polished appearance because he had manufactured them on trestles made out of wooden boards. The "perfect regularity of the element intrados," claimed in the patent, became the goal that led him to review the imperfect prefabrication device he had created at his Turin building site. The result was a type of mould designed to complete the Nervi & Bartoli structural prefabrication system for the waved shells in a manner that would remain canonical until the 1950s. This final refinement of ferrocement prefabrication was accomplished in a building project for a boathouse at San Michele di Pagana, in Liguria, known as *La Tebaide*, which Nervi & Bartoli built to a design drawn up by Daneri in 1948. Nervi & Bartoli was consulted about a contract for the work in November 1948; in December Nervi sent in the preliminary drawings and the first work began in January 1949, with the prefabricated beams being poured at the end of the month; in June Nervi sent an "updated design," while three reinforced concrete arches were completed by the beginning of August. This meant that by 6 August, it was possible to start "laying the first group of pre-

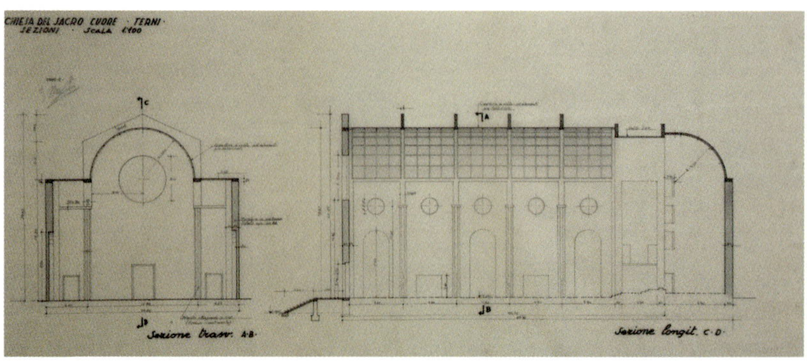

228. P. L. Nervi (with the cooperation of S. Musmeci), design for *Chiesa del Sacro Cuore*, Terni, 1949. Cross-sections (CSAC, n. inv. pra 957, n. id. 14696, coll. 157/4).

229-230. P. L. Nervi, design of a building for machining work, about 1949.

229. Option with thin-shell conoid vaults (BSTA, 024-6012).

230. Option with precast reinforced concrete elements (BSTA, 024-2806).

fabricated beams," and two days later, Nervi and a young engineer employed by his company, Aldo Arcangeli, who had already worked on the site a few months previously, were able to visit and inspect the result of the work. By 25 November, all the work in reinforced concrete was complete, and Nervi & Bartoli "carried away the material."[26]

The five cast-in-place reinforced concrete arches had taut lines, reminiscent of Maillart's bridges, the forks of the *Stadio Berta* roof, the beams in the basement of the church of San Marcellino, or the spurs of *Salone B*. The cross-sections were shaped to be variable in order to consolidate the supports and prevent the structure from tipping over, in a clear desire to put a particular aesthetic stamp on the entire structure. Nervi suggested various solutions for the floors, two with "prefabricated waved beams" in ferrocement including, in one case, the solution of prestressing and one in cast-in-place reinforced concrete. All had a distinctive design that gave rise to a ceiling with an evident plastic relief (in two cases, the arches were even replaced by a skeleton made up of pillars and beams). Nervi specified that the planned expansion at the end of one type of waved beam "was not necessary for

[24] Drawing No. 2885-2, CSAC, n. inv. pra 957, n. id. 14696, coll. 157/4.
[25] Nervi, *Costruire Correttamente* cit., plate. XLVII, and Pier Luigi Nervi, *Structures* (translated by Giuseppina and Mario Salvadori), New York, F.W. Dodge Corporation, 1956, p. 105.
[26] See Luigi Carlo Daneri, *Conte Carlo Felice Trossi – "Il San Michele." San Michele di Pagana*, AD, hand-written record.

231-235. P. L. Nervi, structure of the "La Tebaide" boathouse, designed by L. C. Daneri, Via Brigida Morello 67, San Michele di Pagana, Rapallo, 1934-36.

231. "Particolari (Details)," 1948 (CSAC, n. inv. pra 895, n. id. 14605, coll. 157/3).

232. Modelling of prefabricated beam steel reinforcement, 1949 (AD).

233. Prefabrication of the beams, 1949 (Maxxi - PLN, F08646).

234. Detail of two waved beams placed side by side, 1949 (AD).

235. View, 1949 (G. E. Kidder Smith, *Italy Builds. L'Italia costruisce*, London, The Architectural Press, 1955, p. 254).

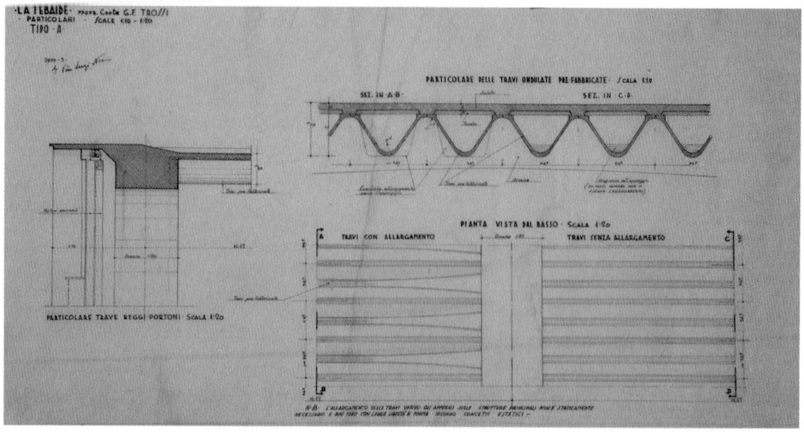

static purposes";[27] the expansion had probably been introduced to render the presence of shear stresses more expressive and contribute to the structural aesthetics informing the variable arch cross-section. "It can be done with great freedom of form according to aesthetic concepts," Nervi noted, revealing a significant aspect of his way of envisaging structure.[28]

The floor for the *Tebaide* was made out of waved beams prefabricated in ferrocement of constant cross-section that follow on from one another to contain a ceiling with a continuous rippling effect (height 0.51 centimetres, length 9.58 metres). Unlike the wave ashlars Nervi usually designed for vaults and flat roofs, the waved beams had stiffening diaphragms in the cavities. Brick *tavelloni* were laid over the beams onto which the slab could be cast in place.

Nervi & Bartoli prepared special moulds on the building site that were used to produce waved beams with a polished surface. The moulds were made out of two long, low, parallel dry-stone walls in the Ligurian building tradition; the cavity was shaped in accordance with the beam profile by means of a mortar coating with a polished surface (gypsum was probably used). Special wooden templates, like the one used in the Via della Magliana and Turin building sites, were used to mould the reinforcement bars and layers of netting that were then laid in the mould and buried by applying cement mortar to a thickness of approximately 2.5 centimetres with a trowel by hand (a special device anchored near the transverse stiffeners made it possible to remove the beams from the mould). A photograph showing the detail of the two waved beams placed side by side bears witness to the formal outcome sought by Nervi when he invented the kind of mould placed on the ground: a smooth surface with no signs of the board impressions that are still visible in the wave ashlars used in Turin.

The contrast of surfaces reveals the different manufacturing processes used for the reinforced concrete and ferrocement. The arches are marked by impressions left by the boards, which were still laid without any particular refinement. The waved floor is perfectly smooth, although visibly stained by the oil used to help extract the pieces.

On the small San Michele di Pagana site, as previously on the Torre in Pietra site, Nervi was experimenting with something that he thought he would

[27] Drawing No. 2899-3, CSAC, n. inv. pra 895, n. id. 14605, coll. 157/3.

[28] *Ibid.*

be able to develop into a structural system for prefabricated brick-cement floors with waved beams, and other practical applications, by placing original aesthetic emphasis on his ferrocement. The example of the *Tebaide*, Nervi & Bartoli's first floor with a waved pattern, demonstrated the formal inventions possible with prefabricated cement structures applied to the flat roofs that Nervi continued to investigate with Arcangeli during 1949, paving the way for the discovery of more new geometries.

The prefabricated house prototype and the Tortona building site

Research into building prefabrication began during the war years, and a study for a prefabricated "single-family" house dating back to 1946 led Nervi to start experimenting with the use of ferrocement with a prototype for affordable housing. The prototype was built in the Via della Magliana laboratory on the basis of a design dating from January 1949 for two types of single-storey "prefabricated houses" with pillars in reinforced concrete and wooden cross-beams laid into a square mesh (1.10 x 1.10 metres). These were enclosed in square sheets of 15-millimetre-thick ferrocement, and the assembly was sealed by joint covers, also in ferrocement.

Using products inspired by economic efficiency and entrepreneurial pragmatism, Nervi & Bartoli continued to erect reinforced concrete skeletons to rebuild buildings damaged by the war. Examples included a monumental livestock warehouse in the *Mole Vanivitelliana*, Ancona, and a refectory at the *Manifattura Tabacchi* in Chiaravalle. The company also designed exceptional vaults with an elliptical profile that were only 4 centimetres thick and ribbed on the extrados for the Cantieri Navali Riuniti workshop, again in Ancona. This was an original technical interpretation of a type made popular by Dischinger and Finsterwalder. It also designed a structural reinforcement grid of a tank for the Azienda Comunale Energia Ambiente in Rome, with beams in reinforced concrete of variable cross-section.

The work realized in one of Nervi & Bartoli's biggest building projects after the Turin project, a complex built between 1948 and 1950 for the Monopoli di Stato in Tortona, confirmed their great variety of techniques and structural figures. These included cast-in-place reinforced concrete, the prefabrication of ferrocement *tavelloni*, and the use of prefabricated elements

236. P. L. Nervi, design for a prefabricated house, 1949. View (BSTA, 024-2794).

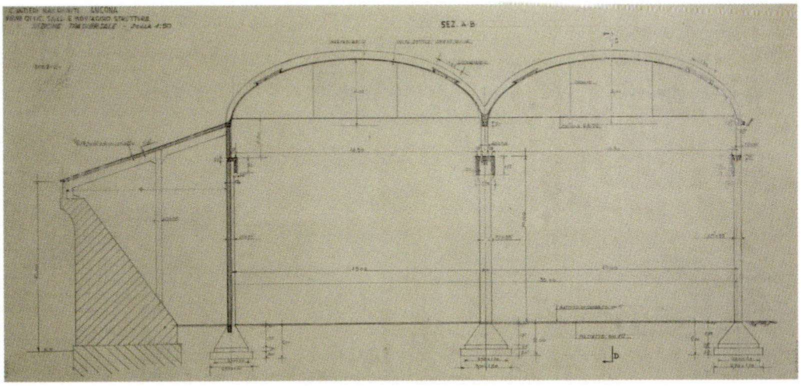

237. P. L. Nervi, "Cantieri Nav.[ali] Riuniti Ancona. Nuova offic.[ina] sald.[ature] e montaggio strutture. Sezione trasversale (*Cantieri Navali Riuniti* Ancona. New welding and structural assembly workshop. Cross-section)," 1948 (CSAC, n. inv. pra 768, coll. 290/5).

in centrifuged reinforced concrete produced by Società Cementi Armati Centrifugati of Trento.

For the salt depot buildings, Nervi erected a skeleton made out of reinforced concrete composed of longitudinal beams and parabolic arches with vault webbing that was laid out of *tavelloni* prefabricated in ferrocement arranged in such a way as to give rise to a diamond-shaped mesh geometry with a geodesic structure. A "wired slab" was constructed over the *tavelloni*. Due to the pattern created by the multiple lines of the structure, the depots took on the nature of a monumental industrial space descending in a direct line from the hangars. However, now Nervi had simplified and exposed the stiffening skeleton that he had constructed in situ for his prefabricated series and woven mainly along geodesic lines.

Before the construction of the two buildings, Nervi carried out tests in Via della Magliana on a square *tavellone* measuring 113 centimetres per side with a thickness of 3 centimetres. In order to assemble the *tavelloni*, Nervi & Bartoli used a mobile crane on rails and a scaffold made out of Innocenti poles with wooden slats at the end to which the "large boards" were nailed.

For buildings that would be used to process the salt, again in the Tortona complex, Nervi designed a reinforced concrete structure to be cast in place, characterised by mushroom-shaped floors. The fact that in March 1949, Ar-

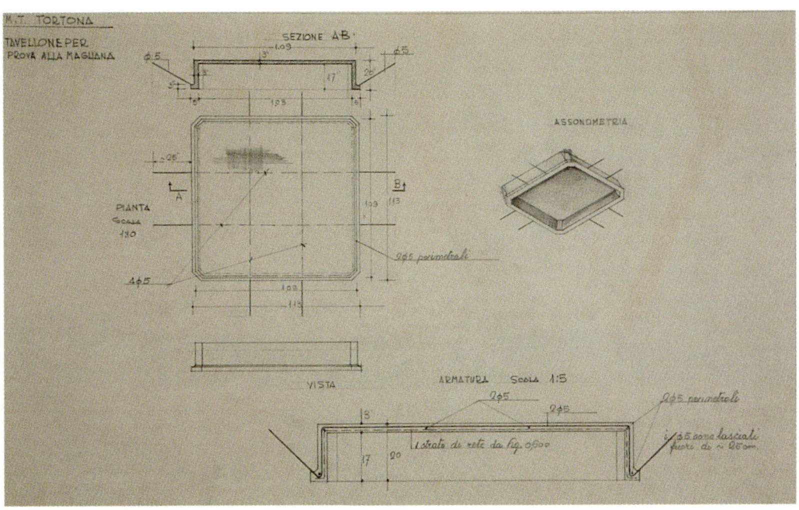

238. P. L. Nervi, buildings for salt store, *Amministrazione Monopoli di Stato* complex, Via Bengasi 5, Tortona, 1948-49.
"Tavellone per prova alla Magliana (*Tavellone* for testing at the Via della Magliana laboratory)" (CSAC, n. inv. pra 245, n. id. 13572, coll. 42/2).

239-241. P. L. Nervi, buildings for salt store, *Amministrazione Monopoli di Stato* complex, Via Bengasi 5, Tortona, 1948-49.

239. Building site, 1949 (BSTA, 024-7161).

240. View, 1949 (BSTA, 024-7172).

241. View, 1949 (photograph by A. Moisio; Acm-EPFL 0172.02.0059.0358).

cangeli calculated the bending moments according to criteria taken from German manuals confirms, together with the waved beams that were being manufactured for the *Tebaide*, that the Nervi & Bartoli engineering department was developing techniques and structures for new kinds of floors.[29]

Salone C for *Torino Esposizioni* and the patent for *"elementi-forma"*

Nervi's creative process now addressed a certain building methodology. In defining an idea for a structure broken down into individual components, he identified the fundamental figures of a vocabulary that was continually subject to expansion and refinement. He was secure in the certainties gained from his building site experiences, in the invention of manufacturing techniques for improving the formal quality of individual parts, and in the refinement of lifting, assembly, and stitching systems. A combination of these factors generated a body that was animated by the correlated juxtaposition of different parts, always with significant variants that reflected further developments in accordance with a sequence based on continuous research.

A meaningful pinnacle of Nervi's creative process in the establishment of ferrocement was achieved by erecting the waved shell of *Salone B*, from which all the variants examined in subsequent designs and building projects stemmed. Between 1949 and 1950, the *Torino Esposizioni* complex became the site of the creation of another grand vaulted roof, that of *Salone C*. This had already been planned when work started on *Salone A* and *Salone B*, but the design and construction had been postponed. When work was already under way on *Salone C*, Nervi & Bartoli was also commissioned to extend *Salone B* by six spans, using the same waved structural elements that they had already built.

When designing the vault of the new hall, Nervi hesitated between replicating the essential structural features of *Salone B*, to maintain the unity of the complex unity, including certain changes in the light of the combination of arches and waved beams adopted for the *Tebaide*, or inventing a structure that was better adapted to the rectangular configuration of the floor plan (50 x 65 metres).

His studies focused on a structure set on corner support points from which the various vault configurations arose. These were all to be constructed in precast ferrocement elements, resulting either in a pavilion vault without ribs, like the hangars or the San Benedetto del Tronto market, or a cross-vaulted configuration with four vault webbings and ribs that would have given the hall an ancient aura. The two versions of the vault were distinguished by the configuration and the importance of the cast-in-place reinforced concrete structure, which, in the first case, became independent from the roof, almost like a transposition of the arches under construction in San Michele di Pagana, while, in the second case, they branched into the ribs of the cross-vaulting.

The sources of natural lighting in both versions constituted a central concern in the design of the elements and the arrangement of the curtain walls. As in the case of the grid geometry in the ribbed shells, Nervi also sought an ornamental design for the *tavellone* perforations to be used in the lighting strips. And thus another type of prefabricated *tavellone* was created and another chapter opened in Nervi's ornamental work, born out of the need to perforate the ribbed shell.

[29] *Deposito sale Tortona. Solai a fungo Fabbricato A*, 23 March 1949, AA.

242-245. P. L. Nervi, *Salone C*, *Torino Esposizioni* complex, Via F. Petrarca 39, Turin, 1949-50.

242. Study perspective drawing, 1949 (BSTA, 024-2801).

243. Study perspective drawing, 1949 (BSTA, 024-2788).

244. Study perspective drawing, 1949 (BSTA, 024-2803).

245. "Armatura arco lato lungo (Reinforcement for long side arch)," 1949 (BSTA, 024-7430).

The main reinforced concrete structure of *Salone C* built by Nervi & Bartoli was made up of four inclined arches with the same taut profile as that of the *Tebaide*, along with mighty ribs that sprung from the "posts" in the corners between the arches before running along the edges of the pavilion vault to join along the ridge. As in the buildings in Tortona, the thickness of the ribs meant that the stiffeners that had not played any role in the web pattern of the prefabricated geodesic structure of the hangars were now exposed.

The ribbed shells of the four vault webbing, positioned above the arches and stretched between the ribs, were built out of special precast ferrocement *tavelloni* that gave rise to the usual ribs to be cast in place, woven in accordance with a geodesic geometry. For the lighting system, Nervi created strips continuous with the abutment of the four vault webbing, using what he referred to as "open reticular areas."[30] The gallery running behind the arches was covered by prefabricated waved beams, similar to those used in the *Tebaide*.

Building work started at the end of 1949. The arches were built using cast-in-place reinforced concrete in timber formwork. The surfaces exposed immediately after stripping revealed the rough criteria used to construct the formwork and carry out the pouring, which had been adopted with a view to finishing with a plaster coating as had probably been planned since the development stage; Nervi had already used the same technique with the spurs in the adjacent *Salone B*.

Prefabrication of the waved beams and the special *tavelloni* took place under cover in spaces below *Salone B*. A procedure designed to improve on the technique tried out in San Michele di Pagana was adopted for the waved beams. This was based on the ferrocement moulds invented in Torre in Pietra. Nervi ordered a series of ferrocement moulds to be shaped on a matrix with a polished surface, perhaps made out of masonry, and these were then secured to special supports so that the workers could prefabricate the waved beams, again in ferrocement. The composition of the cement mortar had to be changed from that of the original mortar used in the first tests in Via della Magliana or that used in the Turin building project, given that the workers now used ladles rather than trowels to lay it. Once removed from the moulds, the beams had now achieved extremely slender dimensions, approximately 2 centimetres, and had come increasingly to resemble bent plates. These were also fitted with stiffening "diaphragms" in the cavities, similar to those of the *Tebaide* floors.

Nervi invented a special type of precast ferrocement *tavelloni* to accommodate his plans for the natural lighting strips. Although the wave ashlars could be easily opened at the side, the *tavelloni* required radical modification of their manufacture so that light could potentially pass through them to al-

[30] Pier Luigi Nervi, patent IT 465636, *Procedimento di costruzione per la realizzazione di superfici resistenti piane o curve costituite da reticolati di nervature in cemento armato, completate o meno da solette di collegamento tra le nervature*, 19 May 1950, p. 3, ACS.

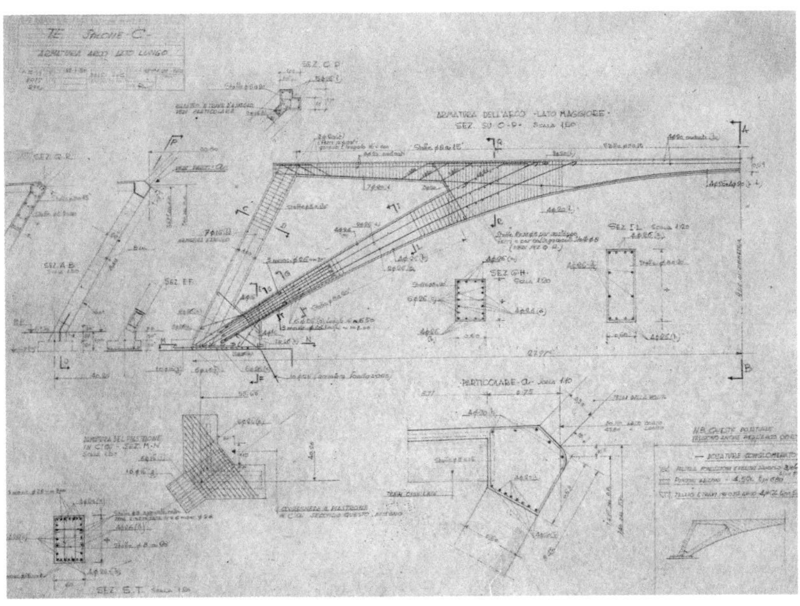

246-250. P. L. Nervi, *Salone C, Torino Esposizioni* complex, Via F. Petrarca 39, Turin, 1949-50.

246. Prefabrication of the beams, 1949-50 (photograph by A. Moisio; Maxxi - PLN, F08729).

247. Prefabrication of the *tavelloni*, 1949-50 (photograph by R. Moncalvo; Maxxi - PLN, F08692).

248. Prefabrication of the beams, 1949-50 (photograph by R. Moncalvo; Maxxi - PLN, F08728).

249. Prefabrication of the *tavelloni*, 1949-50 (photograph by A. Moisio; Maxxi - PLN, F08731).

250. Prefabrication of the *tavelloni*, 1949-50 (photograph by R. Moncalvo; Maxxi - PLN, F08694).

low their use in monumental vaults such as that of *Salone C* – and not merely in industrial roofs. To do this, Nervi broke down his usual *tavellone* into two parts, the four sides and the lid, each always manufactured using the same ferro-cementate procedure. Installation of the sides created a grid of cavities in which the reinforced concrete ribs were, as usual, cast in place. The spaces enclosed by windows created a strip of skylights in the vault. The *tavelloni* and waved beams in precast ferrocement were left unpainted, even though they still showed signs of staining and patchy colouration, like those in *Salone B* or San Michele di Pagana.

Laying the *tavelloni* and perforations in the vaults of *Salone C* posed challenges similar to those raised by the grids of the Torre in Pietra vault and the half-dome of *Salone B*, due to the margin of freedom in the choice of geometric pattern. A lack of decisive laws governing the make-up of the waved shell again highlighted the unresolved formal traits of Nervi's creativity.

When the *Salone C* building site was still in full spate, Nervi filed a patent in May 1950 for the new kind of *tavellone* installed for that roof. In his lexicon, the sides defining the skylights were defined as "*elementi-forma*" (shaping elements) that generate a "grid of ribs in reinforced concrete" and may or may not be "accompanied by connecting slabs." The patent text is another valuable document that allows us to reconstruct Nervi's predilection for a cement surface free of any trace left by the manufacturing process. "The *elementi-forma* (prepared in moulds made out of cement or another suitable material), offer exposed surfaces of great regularity and perfection unachievable using normal systems," confirmed the builder, who was increasingly set on achieving the polished finish of sheet metal for his cement products. The *tavelloni*, with or without a "small slab," reduced to "triangular or diamond-

251-252. P. L. Nervi, *Salone C*, *Torino Esposizioni* complex, Via F. Petrarca 39, Turin, 1949-50.

251. Assembling the vault, 1950 (photograph by R. Moncalvo; BSTA, 024-7569).

252. View (Maxxi - PLN, F01076).

shaped frames with straight or curved sides," act as proper "shuttering," Nervi explained, or "cement *elementi-forma*" that make it possible to "achieve flat or curved surfaces consisting of grids of reinforced concrete ribs." "[...] unlike shuttering in timber or iron, which is removed when the work is complete, these cement *elementi-forma* are left attached to the work and contribute to the static result."

In the patent, Nervi did not limit himself to considering technical aspects of the vault of *Salone C*, but also described the procedure, designed in the Torre in Pietra building site and replicated in that of *Salone B*, for the manufacture of curved *tavelloni* to be used for the construction of "curved structures." He wrote of "barrel vaults, pavilion vaults, vaults with a spherical dome, and any other shape." Perhaps influenced by projects such as the Terni church, Nervi envisaged how he would be able to obtain matrices based on a curved surface shaped on the building site to resemble the shape of the roof, although no longer inverted, as was the case in Torre in Pietra and Turin. Now he envisaged constructing "moulds in cement or plaster," also referred to as "counter-forms," directly on a "template made out of masonry or another material with its surface finished with cement, reproducing a true representation of some or all of the surface being constructed." These moulds therefore give rise to "*elementi-forma*," namely the *tavelloni* to be laid. Nervi did not specify how he would shape the mould to give it the curved shape that he had obtained through direct contact with the inverted surface when manufacturing the *tavelloni* in Torre in Pietra. It was no mere chance that he had to change the mould configuration when making his *tavelloni* in accordance with the criterion set out in the patent.

Although the waved shell had shown that it was capable of producing the greatest number of variants with a specific aesthetic quality, the patent text contains a clue to the extent to which the rigid geometry of the ribbed shell – as installed by Nervi in Torre in Pietra, *Salone B*, Tortona, and *Salone C* itself – had now been radically reconsidered in a way that allowed it to develop along unforeseen aesthetic lines, also waved in accordance with a rigorous static approach. One adjective encapsulates the spirit of the new research in progress, which was able to guide that dubious creative freedom involved in weaving the grids: "isostatic."

chapter eight
strength through form and structural prefabrication

Arcangeli's floor and Nervi's formwork

Over the coming decades, Nervi's obsessive repetition of an invention that had arisen within the engineering department of his own firm Nervi & Bartoli at the end of the 1940s is inexplicable unless we examine the significance that he attributed to the grid of lines on a ribbed shell in prefabricated *tavelloni* made out of ferrocement. He believed it was a new manifestation of a form of ornamentation that was able to simultaneously follow and determine the forces at play by means of the structure's monolithic nature, which is consistently put back together when the parts are ultimately assembled and welded with reinforced concrete. One work in particular marks the point at which the geometric taste for a spiral and radial pattern, as practised by Nervi during the 1940s, was superseded by the appearance of a rigorous rational principle that set out to sweep away every trace of that taste, only to be overtaken in turn by Nervi's interplay of form.

Once he had investigated and developed the various devices possible through the prefabrication of *tavelloni* in ferrocement, Nervi was able to reconsider the exceptional structural figures that he had envisaged for the E42 pavilions and dust off their geometrical patterns in light of his recent trials on construction systems. Those shells, which were designed as continuous and then described as grooved almost like the spokes of an umbrella, could now become ribbed by the interwoven lines generated by the assembly of his prefabricated *tavelloni* in any shape and size. They also open themselves up to the phenomenon of light due to the "*elementi-forma*" that emphasised the grid of ribs against the transparency of the skylights. The result was a design for an imaginative spiral pattern expressed by a mushroom-shaped shell that he intended to generate a bright, sheltered space for a market in Foggia, the apotheosis of the "*en réseau*" vaults of the British Gothic discussed by Eugène Emmanuel Viollet-le-Duc. His aim was to transfigure these into metal bars woven in accordance with revolving hyperboloid surfaces that were extraordinarily similar to the cement structures of Anatole de Baudot, again Gothic-

253. P. L. Nervi, design for Foggia marketplace, about 1953 (Maxxi - PLN, F07436).

254. A. Arcangeli, design for structure (AA, folder No. 101, "Progetti varii studiati in India").

255. P. L. Nervi, collection of images of vaulted structures of monuments and cathedrals (Maxxi - PLN, D150-216).

256. P. L. Nervi, collection of photoelastic tests (Maxxi - PLN, J001-039).

inspired and now implemented in almost intangible filiform, cartilaginous cement material by Nervi, who held those same Gothic inspired vaults up as examples in his university lectures together with other Gothic structures. In the design for Foggia, despite the cohesive strength of the result, Nervi's spiral-shaped geometric wave was again guided by a historical model.

During 1949, when Nervi & Bartoli was engaged in the construction of large roofs, the company's engineering department designed a new type of floor that transformed every ornamental grid into a rational configurations of ribs, developed the fundamental attribute of strength through form, exploited the potential of ferrocement for manufacturing *tavelloni* in every shape, and opened up a decisive chapter in the modelling of cast-in-place reinforced concrete without timber formwork. Arcangeli made a crucial contribution to the new course undertaken by Nervi & Bartoli in the construction of floors. He graduated in engineering from Rome University with a thesis on an Olympic Stadium in Rome (the structure of the *Stadio Berta* tiers was his main inspiration). He had previously worked for the company in 1940 and had only recently returned to work in Nervi's engineering department at the end of a period of captivity in India. Arcangeli whiled away his time in captivity by carrying out structural calculations, sometimes of his own invention (beamed floors or vaulted roofs) or by carrying out proper static checks on built works, such as, significantly, the spiral staircase of the *Stadio Berta*.[1] The invention of a type of floor that resulted in two patents and in many constructions, designs and formal variations in Italy and abroad – in France, in South Africa, in Australia and in the United States – was stimulated by his participation in a call for tenders and arose in the context of Nervi's reflections on questions of a structural grid for ornamental purposes, the quality of the concrete surface, and the potentials of prefabrication in ferrocement.

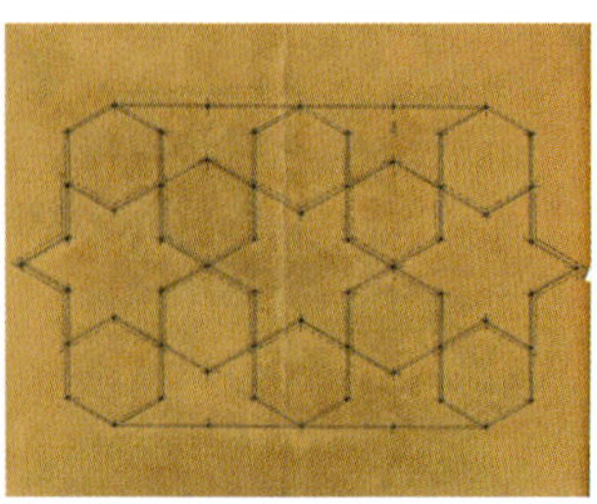

The technical and formal criteria of a roofing system with waved beams were transferred to the manufacture of cast-in-place reinforced concrete. They covered aspects ranging from thicknesses to surface quality, including an industrial approach to processes, doing away with timber, and a static scaffold. It is likely that Nervi's studies for vault roofs with *tavelloni* variously configured to define geometric grids of ribs were the starting point for the transfiguration of that grid approach, from the simple outcome of assem-

[1] AA, folder No. 101 *Progetti vari studiati in India*.

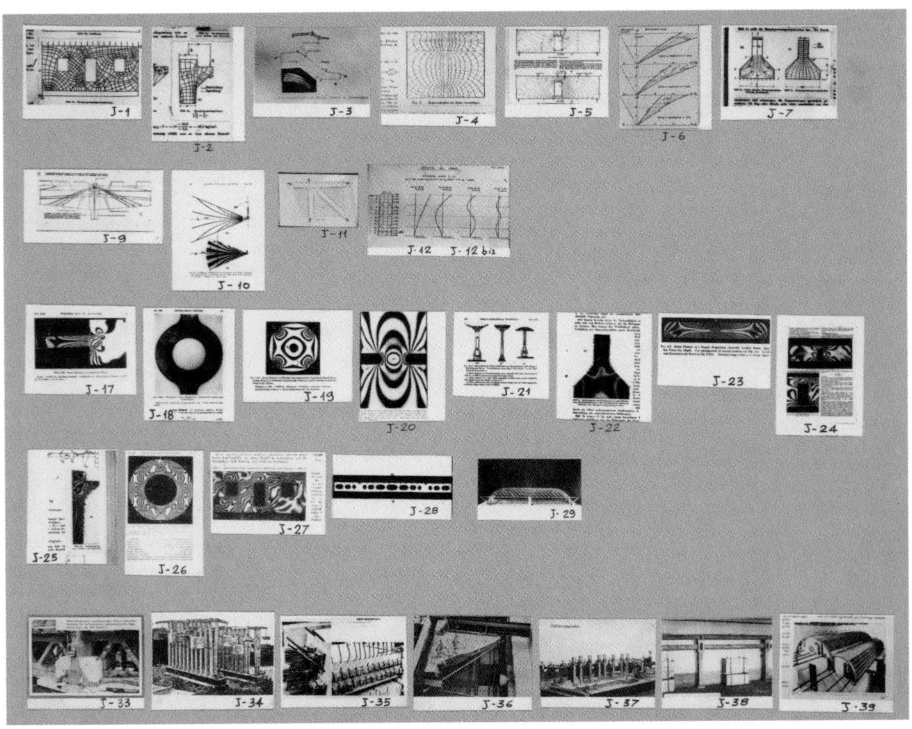

257-258. Nervi & Bartoli (inventor: A. Arcangeli), patent IT 455678, *Perfezionamento nella costruzione di solai, volte, cupole, travi-pareti e strutture portanti in genere a due o tre dimensioni, con disposizione delle nervature resistenti lungo le linee isostatiche dei momenti o degli sforzi normali*, 23 July 1949 (ACS).

259. P. L. Nervi, patent IT 455750, *Perfezionamento nella costruzione delle solette, lastre e altre strutture cementizie armate*, 23 July 1949 (C. Greco, *Pier Luigi Nervi. Dai primi brevetti al Palazzo delle Esposizioni di Torino. 1917-1948*, Lucerna, Quart Edizioni, 2008, p. 297).

bling parts to rib patterns determined by isostatic lines. Only through this crucial transformation was it possible for Nervi to resolve the structural ambiguity that he came up against when he designed or built barrel or pavilion vaults using ferrocement *tavelloni* divided up by ribs in reinforced concrete. After designing a geodesic structure Nervi's interest in discovering a new configuration for those networks is confirmed by a polarised photograph of the photoelastic analysis of a Vierendeel girder, published on the cover of the first issue of "Strutture", and a collection of similar photoelastic analyses collected in one of his albums packed with images summarising his vision of structural systems. During that particular period in the history of Nervi & Bartoli, when they were beginning to experiment with other structural configurations, Nervi's other important collaborator, Musmeci, must have developed his own vision of a structure that was increasingly liberated from Nervi's kind of geometries.

Having the insight to translate isostatic lines into ribs bent in an unusual way requires extreme freedom of invention and a repertoire of motifs that is evidently partly dependent on knowledge of a construction culture based on complicated geometrical patterns, exemplified by the Indian ones that Arcangeli saw during his captivity and transcribed into diagrams for a star-shaped web structure.[2] These structural sketches are valuable in documenting the methods he used to discover new patterns for a reinforced concrete structure. Arcangeli's isostatic lines essentially harnessed the forces in accordance with their own geometry, which already contained the premises for an ornamental pattern.

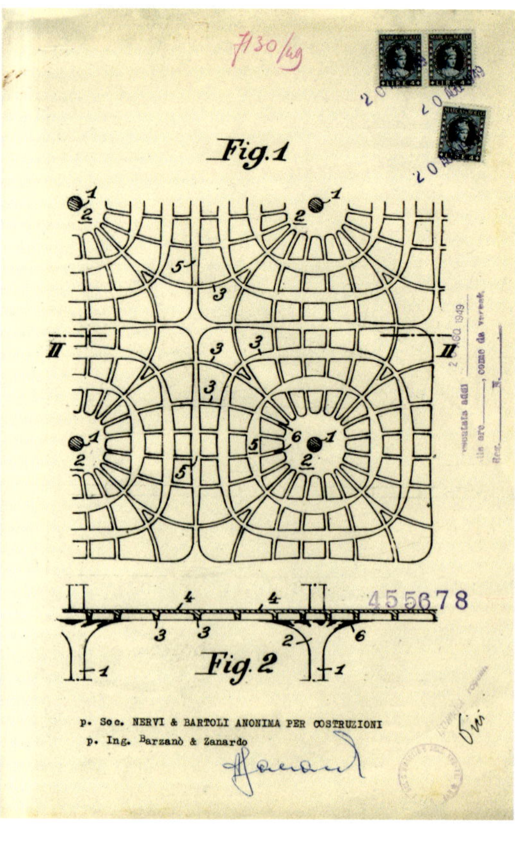

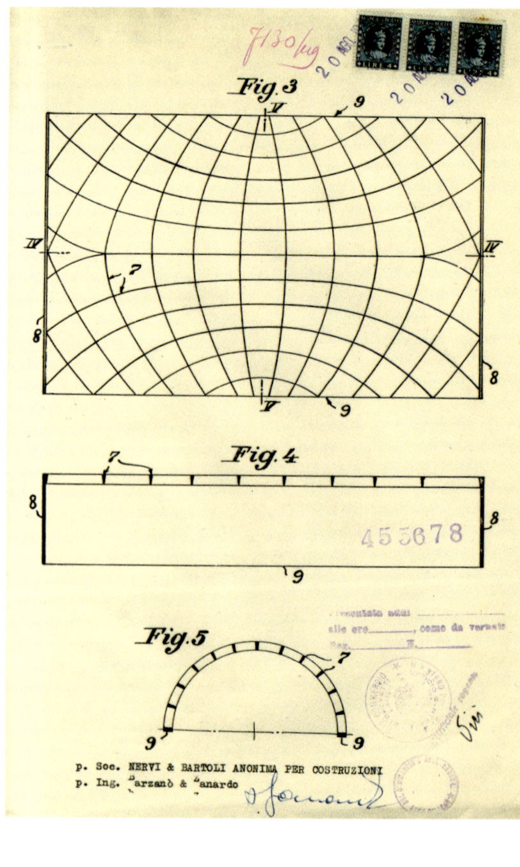

The most evident manifestation of the spirit of experimentation that drove the research and studies carried out in the Nervi & Bartoli engineering department at the end of the 1940s was the invention of a floor with ribs bent in accordance with isostatic lines. The company filed a patent for this floor on 23 July 1949 indicating the name of its inventor, Arcangeli. During this period, Arcangeli carried out inspections at the building site on the floor made out of waved beams for the *Tebaide* and did calculations for the mushroom-shaped floor for the buildings in Tortona. The probable origin of that type of floor in Nervi & Bartoli's ribbed roof systems seems to be confirmed by the title of the patent, which did not restrict itself to indicating the case of the floor: "Improvement in the construction of floors, vaults, domes, wall beams, and load-bearing structures in general in two or three dimensions with arrangement of the resilient ribs along the isostatic lines of moments or right-angled stresses."[3]

Identification of the isostatic lines corresponded to a twofold desire – characteristic of Nervi's thought expressed in *Scienza o arte del costruire?* – to create a form of logical and economic structure for his construction that was also of powerful aesthetic impact. Terragni had in any case sensed the opportunity for creating a new architecture by highlighting invisible isostatic lines, which were the lines that were being discovered at that time through the first photoelastic analyses. Nervi was able to appreciate the consequences in the creative process of those isostatic lines. He saw them as a means of automatically generating an almost natural or organic structure to achieve the blend

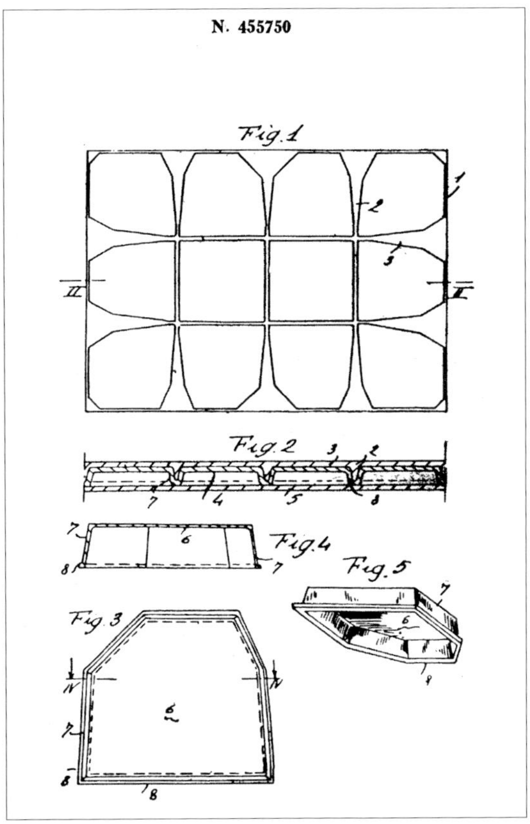

[2] Untitled, undated drawing, *ibid.*

[3] Società Ingg. Nervi & Bartoli, Anonima per Costruzioni (inventor: Aldo Arcangeli), patent IT 455678, *Perfezionamento nella costruzione di solai, volte, cupole, travi-pareti e strutture portanti in genere a due o tre dimensioni, con disposizione delle nervature resistenti lungo le linee isostatiche dei momenti o degli sforzi normali,* 23 July 1949, ACS.

that he had always struggled to find in the imponderable alchemy between building site and calculation.

The structure with isostatic ribs patented in 1949 made it possible for the very fine thicknesses typical of waved beams and ferrocement *tavelloni* to become the benchmark for cast-in-place reinforced concrete construction. Further confirmation of the inspirational part played by ferro-cementate in the design of a structure with isostatic ribs is the fact that this material remained essential to the construction of the new structure. It endowed it with lightness but also the finish level of its perfectly polished surfaces; cast-in-place concrete could be moulded in accordance with flowing isostatic lines due to the formwork shaped in the ferrocement. Nervi's singular insight and understanding of the nature of ferrocement and how to fully exploit its ability to be moulded at will made it possible to conceive of an inexpensive way to build a structure with isostatic lines.

The text of the patent, which was also written by Arcangeli, was drawn up to explain the static nature of the invention of a "discontinuous system of ribs arranged along isostatic lines," in the same way as a "filamentous structure," providing an example of the "trabecular bone" studied by Karl Culmann. The stated purpose was to "achieve a considerable savings in materials without changing the play of internal forces." The appearance of Culmann's example in the explanation of the isostatic lines contributed to an understanding of the emergence – in one accomplished theoretical package – of Nervi's concept of strength through form.

The patent explains the applications of the discontinuous system described to various types of structures, from a "membrane" ("thin vault, dome, box-section structure, wall beam, etc."), to a "spatial structure" ("towers, antennae supporting cables and wiring, etc."), floors with or without main beams, "flat foundation slabs" or "foundations with reversed arches," "bridge decks," "vaulted roofs of any shape," "tanks and silos with flat or curved walls," and lastly "box-section structures for roofs, bridges, canal bridges, walkways, etc." The list made it clear that any membrane structure could be impacted by logical filaments of form-resistant isostatic lines.

In order to document the characteristics of the invention, Arcangeli used the example of the structure of a floor and included schematic drawings of a "self-supporting cylindrical vault." The device used to be able to give free rein to the continuous flow of isostatic lines along the entire thickness of the floor was mushroom-shaped pillar endings. In the patent, the mushroom-shaped pillar takes the form of a paraboloid, with the aim of obtaining a disc with isostatic lines radiating from its top. The pillar consequently assumes a circular cross-section. The diagrams attached to the patent document an essential and consistent vision in every part of the structure with isostatic ribs, as envisaged by Arcangeli.

Arcangeli and Nervi addressed the problem of building a structure with isostatic ribs out of reinforced concrete cast in place and prefabricated. The patent explained the type of formwork and the way it was assembled: "the practical implementation of the structures is particularly easy with reinforced concrete and is achieved by arranging the formwork in the shape of a box, which may or may not remain incorporated in the structure, with or without a base, made out of any material, such as wood, sheet metal, cement, or other materials, and pouring the concrete mixture between the forms themselves in order to create the ribs and, if necessary, above or below these in order to form the continuous slab. The construction may also be carried out by prefabricating rib elements, to be connected in situ by pouring the cement mixture into the nodes and then welding the metal reinforcement if necessary."

Even though the idea for a structure with isostatic lines was unmistakably inspired by the activities of Nervi & Bartoli, Nervi planned a strict separation of merits and skills between the invention, of which he acknowledged that Arcangeli was the author, and his construction, over which he wished to maintain control. Otherwise it is difficult to explain the absence in the patent text of any reference to the possibility of building the formwork using ferrocement and the filing of another patent on 23 July, this time only in Nervi's name, designed to protect the invention of ferrocement formwork, now referred to also as "*ferrocemento*," for the construction of cast-in-place reinforced concrete structures and proposed as a supplement to the patent of April 1943 on ferro-cementitious felt. The invention consisted of the "construction of shuttering for the pouring of reinforced concrete structures, prepared using slabs reinforced with multiple layers of netting instead of with the usual boards [...]."[4]

After using ferrocement moulds at the Torre in Pietra or Turin building sites, Nervi could now claim, against the backdrop of Arcangeli's invention, that he was able to use that material to achieve "constructive economy," by means of the system of matrices and moulds, "executive perfection" due to the polished surfaces of the prefabricated parts and, last but not least, "design freedom," which meant being able to envisage structural figures that were no longer conditional upon the ability of carpenters to assemble wooden formwork. The text of Nervi's patent was a veritable manifesto of a future structure in cast-in-place reinforced concrete, which went beyond anything that Arcangeli himself had envisaged (and even Nervi would never dare to put to the test). The structure's forms were released from any geometry of mathematical origin, with surfaces now perfectly polished – "perfect regularity of the exposed faces of concrete mixtures," as specified in the patent. As an entrepreneur who was aware of the constraints imposed on creation by every step in the construction process, he yet again returned to emphasise what he felt as a personal and revolutionary conquest in his concrete technique: "Complete freedom of form in the structure being built in reinforced concrete, as a consequence of being able to adapt the ferrocement shuttering to any curved, slanted or otherwise shaped surface, a freedom that is in sharp contrast to the strict limitation to prismatic shapes and developable surfaces characteristic of shuttering made out of boards."

In this patent, Nervi succeeded in explaining his own view of concrete cast in place and left exposed, virtually admitting his personal failure to monitor the finish of surfaces in his monumental works, from the *Stadio Berta* to the arches under construction in *Salone C*. The perfectly polished ferrocement formwork was able to guarantee that "perfect regularity of the exposed faces of the concrete mixtures" that he had been unable to achieve at his sites, except by working on the material after stripping off the formwork. Now the surface of the cast-in-place concrete could be of the same quality as that of the ferrocement *tavelloni* and waved beams without further need for the cladding he had been forced to resort to in the two Turin *Saloni*. "[...] there is no further need for plaster to even up the surface, and the problem, unresolved until now, of ensuring a good finish for the surfaces of reinforced concrete

[4] Pier Luigi Nervi, patent IT 455750, *Perfezionamento nella costruzione di solette, lastre ed altre strutture cementizie armate*, 23 July 1949, quoted in, Greco, *Pier Luigi Nervi* cit., pp. 296-297.

structure has been resolved in the most effective way," wrote Nervi. It is worth remembering again the opposing approaches of Nervi and Le Corbusier in their view of the ideal qualities of concrete, polished to perfection for the former, *brut* to the point of *malfaçon* for the latter. It should nevertheless be emphasised that Nervi's formwork, at least as described in the patent, was still designed for the manufacture of individual elements or structures characterised by the repetition of elements such as "coffered" floors, confirming that his system was derived from the criteria of prefabrication that had now been extended to works poured on the building site. The phrase that Nervi used to introduce the structure involving Arcangeli's type of floor should be interpreted in this sense. "Ferro-cementate shuttering is applicable to the widest array of reinforced concrete structures featuring a sufficient number of identical repeated elements, and they are particularly economical and convenient for the construction of large floors for industrial or warehouse use."

Just as the Arcangeli patent did not refer to ferrocement, in Nervi's patent was never mentioned the term "isostatic" (which was to make its first appearance in the May 1950 patent), and when Nervi referred to that type of floor, he spoke of "ribs that may be straight or curved in one or two directions or crossing one another in several directions, according to the static requirements [...]." It is in any case significant that in order to represent his formwork graphically, Nervi prepared a diagram of a floor with polygonal and even irregular caissons but never using the curved lines apparent in Arcangeli's diagram.

Nervi did not limit himself to presenting ferrocement moulds. Against the backdrop of Arcangeli's invention, he envisaged a mobile scaffolding system or "*ponte*" (bridge), to be used to construct floors. This made it possible to construct the various floor sectors between the spans using a limited number of moulds, with the aid of a device for raising and lowering the level holding the moulds in order to make the various casts. Nervi specified measures that he practised on his building site, such as treating the ferrocement moulds with greases, oils, or products such as Bormeol before pouring the concrete, and emphasised the quality of the moulds because they could easily be reused and were waterproof to prevent the concrete mixture from emerging during the pouring.

The changeover from Nervi's spiral and radial grid to Arcangeli's isostatic lines was decisive in generating a form of ornamentation that almost appeared to be an expression of static laws, no longer filtered by sophisticated historical geometries but created automatically. Those lines seemed to do away with the ample creative margins that Nervi always stated and practised when he designed the grids of his ribbed shells at Torre in Pietra, Turin, or Foggia. A new form of ornamentation could now be moulded in the reinforced concrete to be left exposed. Although it appeared to be determined by logic, it still boiled down to appearance.

Sliding formwork in ferrocement for ribbed floors

The filing of two complementary patents on isostatic ribbed structure and ferrocement formwork marked the start of a series of studies and works carried out by Nervi & Bartoli for experimental purposes between 1949 and 1951 in order to check and demonstrate the qualities of the new system. Alongside the vaulted roofs and floors in *tavelloni* and waved beams prefabricated in ferrocement, Nervi began to establish himself as the creator of an exceptional ribbed structure in cast-in-place reinforced concrete. Building sites in Leghorn

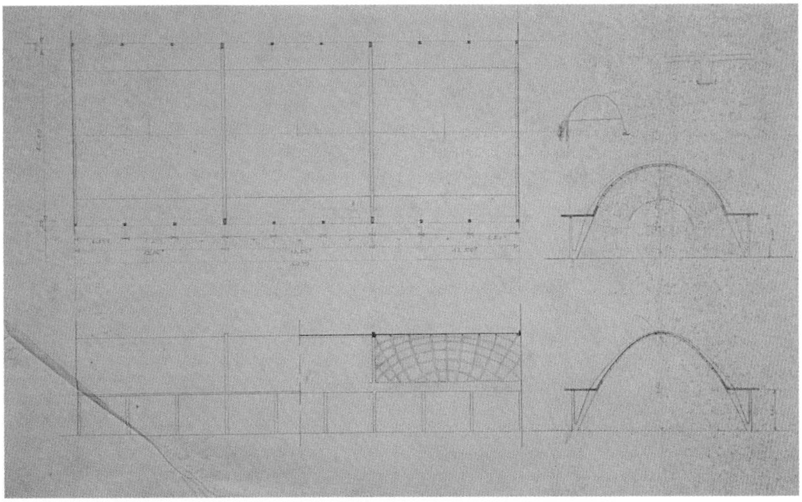

260. P. L. Nervi, swimming pool of the *Accademia Navale*, Viale Italia 72, Leghorn, 1949.
Option with vault with isostatic ribs (CSAC, n. inv. pra 15, n. id. 12712, coll. 148/4).

for the swimming pool of the Naval Academy, in Bologna for the *Fabbricato Ballette*, for the *Manifattura Tabacchi*, in Rome for the *Opificio Gatti* and the *Fabbricato Lavorazione*, again for the *Manifattura Tabacchi*, and in Rosignano for the Solvay sanitation system building marked the first opportunities for Nervi & Bartoli to test Arcangeli's structures for floors and vaults as well as a variant prepared by Nervi.

In studies for the Naval Academy swimming pool roof, which was subsequently built out of the usual waved beams, two options were proposed with an isostatic ribbed structure. The first was a series of small elliptical vaults with their intrados woven out of curved lines and a span of 8.50 metres, and the second was a single large roof with semicircular or parabolic arches and vault webbing, again woven out of the same curved lines. In these shapes, the ribbed shell would have looked as though it were decorated with lines surprisingly interwoven in accordance with rules incomprehensible to most people – almost a decorative apotheosis of Islamic geometric taste.

Ideas of applying Arcangeli's invention were abandoned, and Nervi & Bartoli produced a "waved vault," formed from semicircular arches consisting of three curved ashlars prefabricated out of ferrocement, each with a chord line of 5.22 metres and three diaphragms in the cavity (one central and two at the ends, all open to allow the passage of the reinforcements once installed), shaped in moulds, again in ferrocement, with their surfaces polished so that they could be left unclad. "[...] the surface of the vault intrados is without finishing and does not require further plastering," was now the customary annotation made by Nervi & Bartoli.[5] A sliding crane lifted the ashlars and then slid them along a special wooden scaffolding.

After planning the application of several variants of Arcangeli's structure with isostatic lines, Nervi decided to implement the floor proposed in his patent in the *Fabbricato Ballette* (the work was built between 1949 and 1950). With a view to tendering for the work involved in this major project, Nervi & Bartoli began studies focusing on the Arcangeli floor and Nervi's sliding

[5] Undated and unnumbered drawing, CSAC, n. inv. pra 15, n. id. 12712, coll. 148/4.

261-263. P. L. Nervi, swimming pool of the *Accademia Navale*, Viale Italia 72, Leghorn, 1949.

261. Perspective drawing (BSTA, 024-2804).

262. Matrices for the prefabrication of waved beams (Maxxi - PLN, F8625).

263. Installation of waved beams (Maxxi - PLN, F8626).

264. P. L. Nervi, tobacco bale store (referred to as "*Fabbricato Ballette*"), *Manifattura Tabacchi* complex, Via della Manifattura 3, Bologna, 1949-50. Option with floor with isostatic ribs, 1949 (Acm-EPFL, 0172.02.0003.2094).

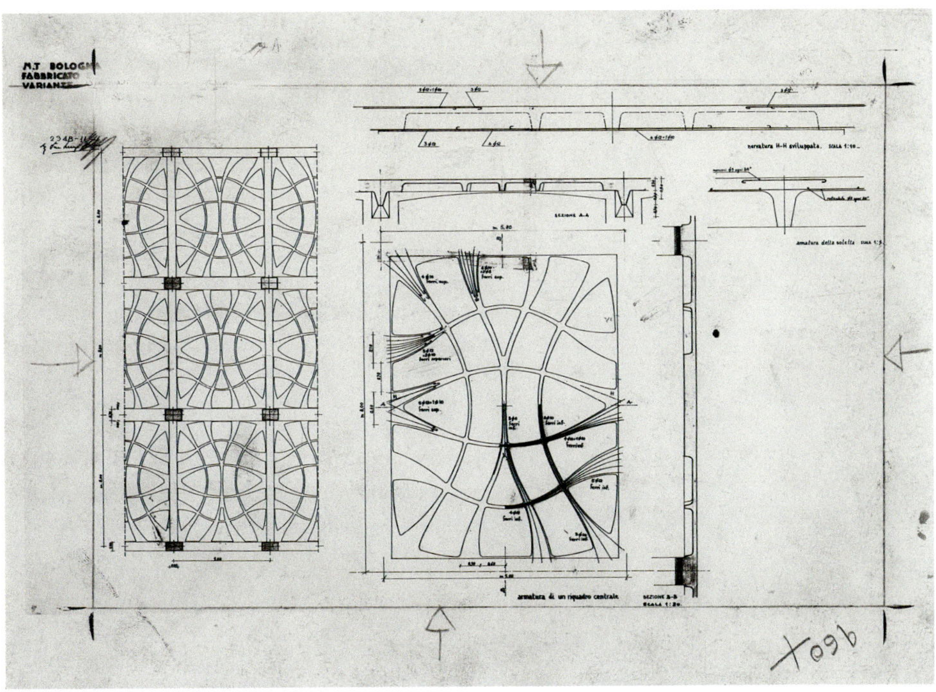
M.T BOLOGNA
FABBRICATO
VARIANTE
2348-11
armatura di un riquadro centrale
SEZIONE B-B
SCALA 1:20
960

265-267. P. L. Nervi, tobacco bale store (referred to as "*Fabbricato Ballette*"), *Manifattura Tabacchi* complex, Via della Manifattura 3, Bologna, 1949-50.

265. "Tavellone per casseforme dei solai (*Tavellone* for floor formwork)," drawing dated 12 June 1950 (CSAC, n. inv. pra 1025, n. id. 14820, coll. 150/1).

266. P. L. Nervi, sliding formwork prototype, laboratory, Via della Magliana 238, Rome (ASCBT, folder 6, doc. 9870/14).

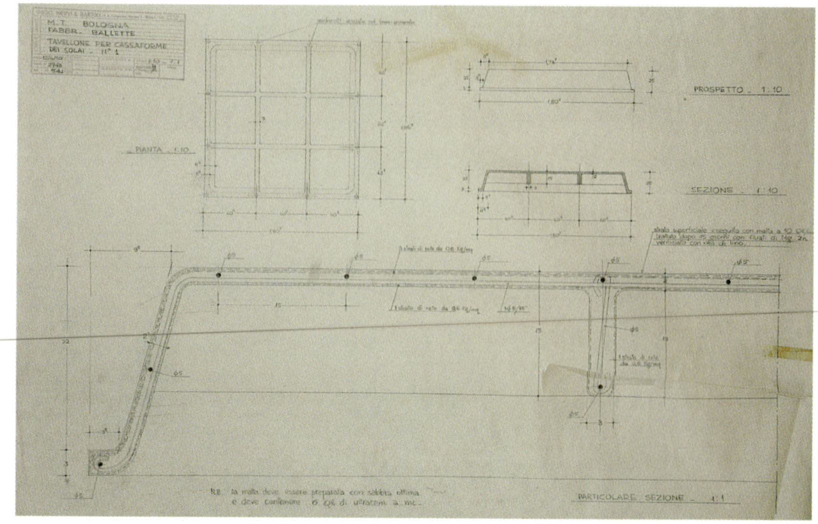

scaffolding with the aim of introducing principles of prefabrication for cast-in-place reinforced concrete floors and reducing their construction costs. The colossal size of the building, which had five stories and was approximately 130 metres long, was ideal for the application of formwork systems with a sliding scaffold and ferrocement moulds for a construction involving sets of four-floored sectors (the roofing had a conventional floor).

All parts of the skeleton, starting from the foundations, were examined with a view to pouring the concrete, with details demonstrating the outline of a real breakthrough in Nervi's ideas on cast-in-place reinforced concrete. The drive to perfection achievable by constructing the floors using his formwork system prompted him to undertake to improve the other parts as well. For the foundations, Nervi invented an amplified plinth, always making use of his ferrocement formwork. For the pillars, he developed a formwork system with wooden boards reinforced by an Innocenti pipe structure that could be dismantled and reused on the different storeys with the aid of a crane. The fact that Nervi now included "planed boards" in his drawings for moulding the pillars confirmed the degree of concrete finish he desired for all parts. "The formwork," it is specified in the drawing, "must be prepared on the ground with the aid of clamps. Then it is hoisted, tied, and sealed. The tubular frame to which weights may be secured if necessary is used to maintain a fixed shape during the pouring, which takes place from above with the aid of cranes. [...] The shuttering and the frame are both moved and positioned by the crane."[6]

The precise descriptions in Nervi's patents of a formwork system designed for "large floors for industrial or warehouse use" are reflected in the devices designed for the Bologna building site. For the scaffolding, Nervi combined a vertical structure of Innocenti poles and a horizontal timber frame with compartments corresponding to the formwork, stiffened to prevent bending during pouring. The scaffolding was installed on four carriages sliding along rails; it could be raised and lowered through 110 centimetres

[6] Drawing No. 2948-57e, CSAC, n. inv. pra 1025, n. id. 14820, coll. 150/1.

by means of hydraulic jacks so that it could be moved once the concrete had set. The formwork of the ribbed floor, each one stiffened by diaphragms, was made out of ferrocement shaped on plaster moulds and bolted to the wooden frame. While Nervi used the terms "*cassero*" (shuttering) or "*cassaforma*" (formwork) in the patent, in the working drawings he used the term "*tavellone*," which was usually used to describe the elements he had used to construct the vaults and half-domes. The ferrocement formwork also moulded the two opposing edge beams obtained, which could be made slender and easily take up an arched profile in the joints in order to follow the direction of the moments. Nervi prepared special formwork to construct the other beams. These were always made out of ferrocement, installed using special scaffolding designed in accordance with that principle (structure in Innocenti poles and timber). A special formwork in ferrocement was also designed for the creation of the three expansion joints for the building structure, whose positions corresponded to those of the beams.

The ferrocement sliding formwork system for Nervi's floor was tested beforehand by constructing the prototype of one span in the Via della Magliana laboratory. In view of the chronological coincidences between the preparation of the patents and the Bologna project, the prototype may have been constructed to test the invention even before the patents were filed.

Nervi & Bartoli planned all the pouring operations to be carried out using its formwork system with sliding scaffolding. The stages began with occupation of the six spans so that it was possible to gradually move the outer

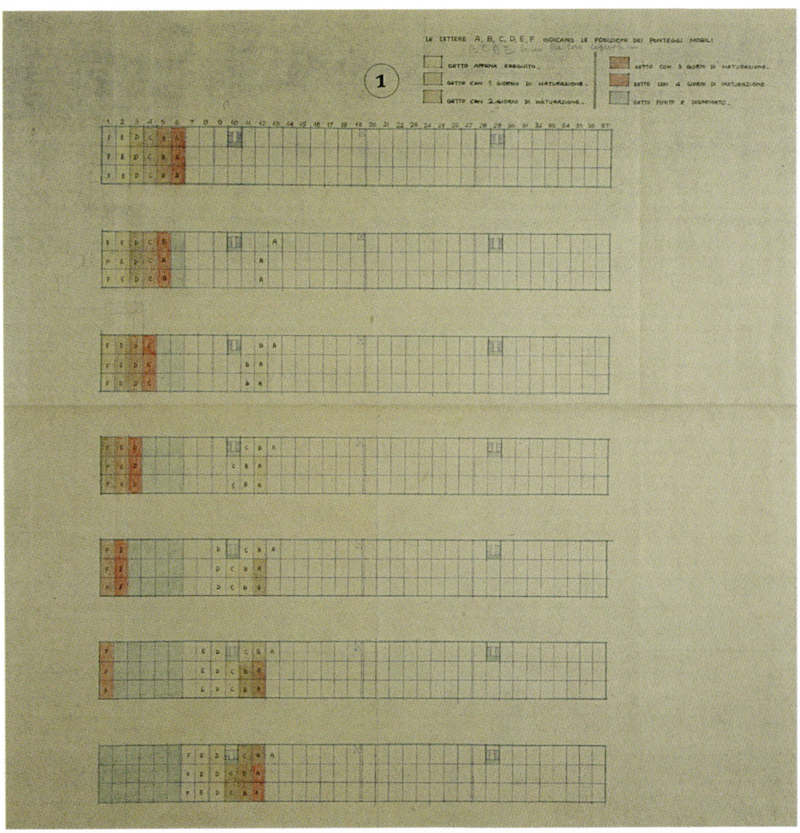

267. Drawing showing the various floor pouring stages (CSAC, n. inv. pra 1025, n. id. 14820, coll. 150/1).

268-270. P. L. Nervi, tobacco bale store (referred to as "*Fabbricato Ballette*"), *Manifattura Tabacchi* complex, Via della Manifattura 3, Bologna, 1949-50.

268. "Cassaforme per i pilastri interni (Formwork for internal pillars)," drawing dated 15 June 1950 (CSAC, n. inv. pra 1025, n. id. 14820, coll. 150/1).

269. Moving the sliding formwork, 1950 (BSTA, 024-4437).

270. Installing the slab and beam reinforcement, 1950 (BSTA, 024-5128).

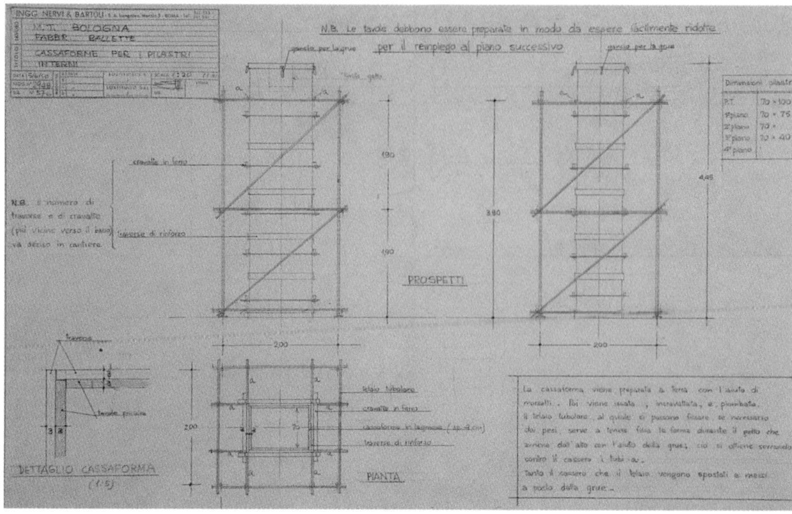

row of scaffolding four days after pouring. A Wolff crane on rails was used to position the pillar formwork and carry the concrete for pouring. The workers pushed the scaffolding by hand.

The anonymous industrial architecture of the *Fabbricato Ballette* conceals the existence of an exposed reinforced concrete skeleton manufactured in accordance with a sophisticated system able to transfer the full potential of the perfectly executed ferrocement structures and their exposed surfaces when the concrete was cast in place. Nervi and his firm extended the structural prefabrication of specific shapes even to the casting in situ, all as a result of Nervi's obsessive drive since the 1930s to eliminate the conventional timber formwork. Identification of a single span and its identical repetition throughout the structure were the premises for the financial success of his ferrocement formwork system.

Following the Bologna experience, Nervi & Bartoli were awarded contracts to build other buildings, again belonging to the *Manifattura Tabacchi* – in Rome in 1951 (the floors of the *Fabbricato Lavorazione* were described as "isostatic ribs" and designed by Arcangeli) and in Chiaravalle in 1954 (the floors were said to be "cross-beamed"). They were also characterised by the same type of floor constructed using the ferrocement sliding form system. Nervi and Musmeci designed a weave of beams crossed in accordance with a pattern of equilateral triangles with each side measuring 3.70 metres for the Roman building basement floor.

It was not until 1951 that Nervi & Bartoli started testing the structure of isostatic ribs patented by Arcangeli for the first time. The occasion arose with the construction of a design by the engineer Carlo Cestelli Guidi for the headquarters of Giuseppe Gatti's *Opificio* in Via Prenestina, Rome. This was a two-storey building with a storage basement punctuated by the Arcangeli floor structure, the production area without pillars, and covered by a system of saw-tooth roofs and portal frames in cast-in-place and prestressed concrete.

The ribbed floor, termed an "isostatic mushroom-shaped floor," formed a structure made up of continuous lines. It was devoid of main beams and stood on pillars with a square cross-section and a mushroom-shaped ending distributed into square spans measuring 5 x 5 metres to form a rectangle of 6 x 15 spans. As with the cross-section of the pillars built for the *Manifatture Tabacchi*, the choice of a square pillar cross-section involved a substantial change in the movement of the isostatic ribs, compared to that envisaged by Arcangeli in the patent. The whole system was simplified, reducing the number of ribs and hence the number of construction moulds, with a consequent increase in the number of gaps.

271-273. P. L. Nervi, tobacco bale store (referred to as "*Fabbricato Ballette*"), *Manifattura Tabacchi* complex, Via della Manifattura 3, Bologna, 1949-50.

271. View of the floor and pillars without finishing, 1950 (ASCBT, folder 6, doc. 9870/11).

272. View of floor and pillars with finishing (photograph by V. Villani; Acm-EPFL, 0172.02.0003.209).

273. View (BSTA, 024-7177).

The manufacture of individual parts for the floor formwork, referred to as "*tavelloni*" or "*forme*" (forms) in the drawings, usually occurred on the building site, using ferrocement moulded on polished matrices, carried out using the same material in all cases. In the manufacture of *tavelloni* with curved sides that varied in accordance with the isostatic line geometries, ferrocement proved itself to be effective for the construction of those structures – moulded by invisible static forces, envisaged by Nervi in all cases. The *tavelloni* used for the *Opificio Gatti* differed from those used for the *Manifattura Tabacchi* because they sprouted bars around their edges that were to be used for the various systems of anchorage on the scaffolding.

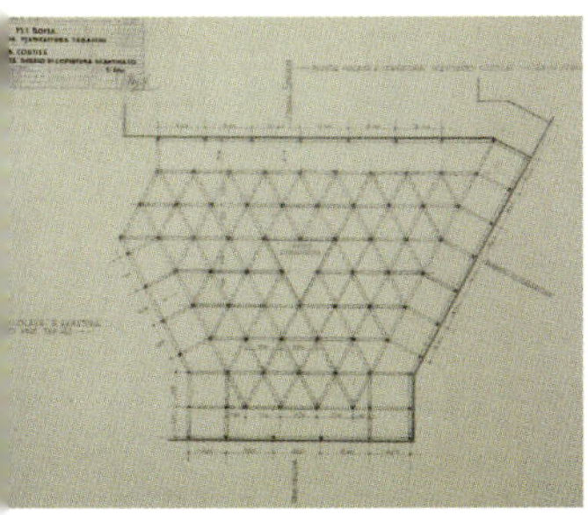

274. P. L. Nervi (with the cooperation of S. Musmeci), *Manifattura Tabacchi* complex, Circonvallazione Ostiense 191, Rome. "Zona cortile. Pianta solaio di copertura scantinato (Courtyard area. Plan of roofing floor for cellar)," drawing dated 13 September 1951 (CSAC, n. inv. pra 594, n. id. 14156, coll. 287/1).

During the construction stage, the continuity of isostatic ribs was broken down into two types of formwork, "form A" and "form B," the former square with curved sides like an astroid, to be arranged in the middle of the spans and the second rectangular, again with curved sides, but conversely to be positioned between the pillars. In order to install the formwork, Nervi & Bartoli prepared six sliding scaffolding for form A, of the same kind as those tested in Bologna. These slid along Decauville rails, but with a "continuous panel" instead of a frame. The design complexity of the individual *tavelloni* involved changing the upper structure of the scaffolding and thus inventing a system of connection to the *tavelloni* that stiffened the formwork while pouring. The *tavelloni* were anchored to the panel by means of special protruding bars, bolted hooks, and a layer of mortar to be "finished with a template" in order to mould the lower faces of the ribs.

In order to set up form B, Nervi developed a separate device, still consisting of a wooden panel with ferrocement *tavelloni* above. This was secured to the edges of form A, with plaster connections constructed in situ to obtain polished ribs that were supported by Innocenti poles, but with special *tavelloni*. Each of these had a hole for the application of a screw with a bolt that could be used, once the concrete had set, to hold the form in position during stripping. The workers standing on the floor lowered the form to the ground using steel cables applied to the screws in place of the bolts. The executive drawings show the configurations of the mushroom-shaped capitals specially prefabricated in reinforced concrete with cavities to hold reinforcements and for the concrete to be poured once the parts had been positioned.

To facilitate stripping, the *tavelloni* were sprinkled with Sformeol, an emulsion of oil and water, which still stained the surfaces as had occurred with the ferrocement manufactured on the previous Nervi & Bartoli building sites. Nervi's concern for finishing all the concrete in situ, even the concrete due to emerge from the timber formwork of the edging beams, was so great that he was prompted, as in the building site of the *Manifattura Tabacchi* in Bologna, to advise technical measures to polish the surfaces of the boards that

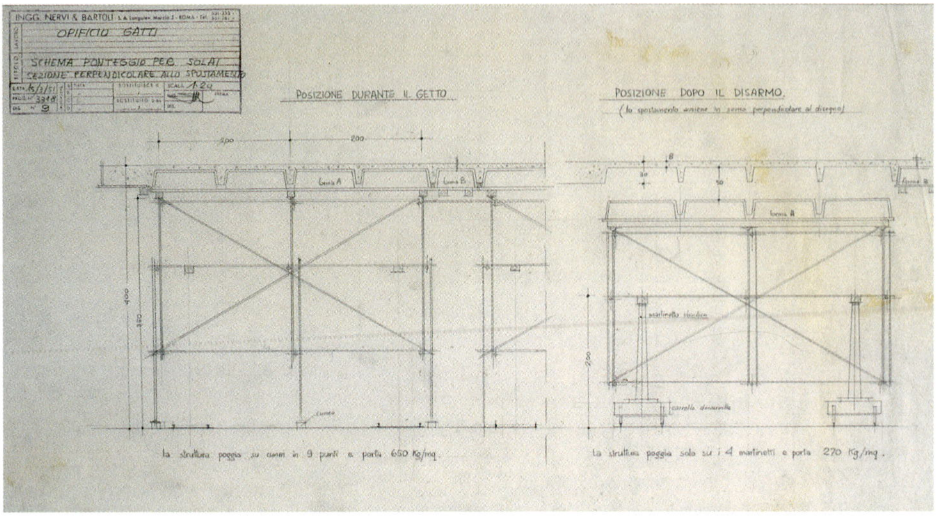

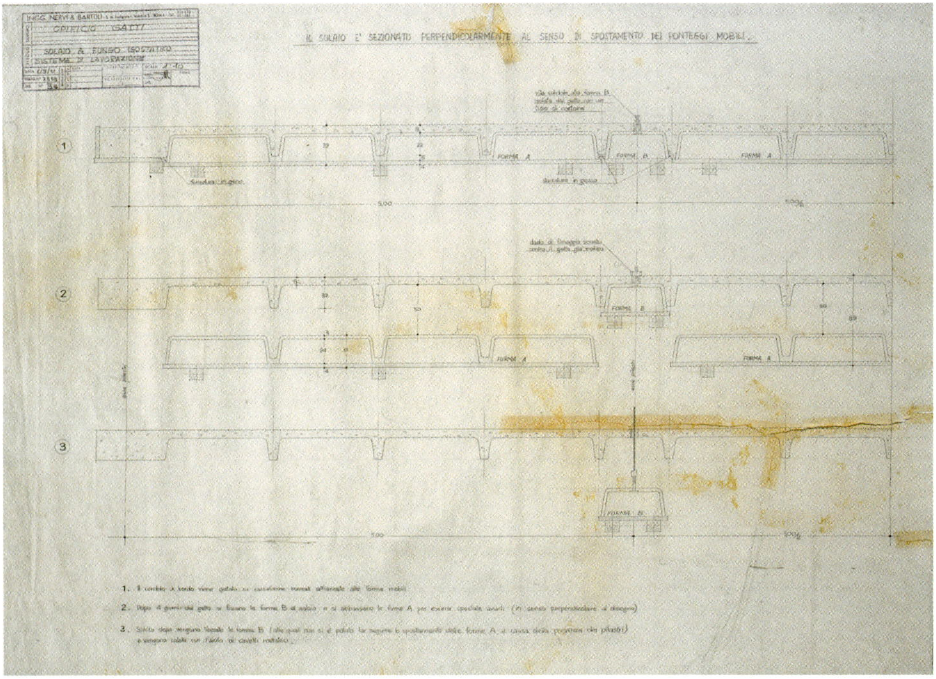

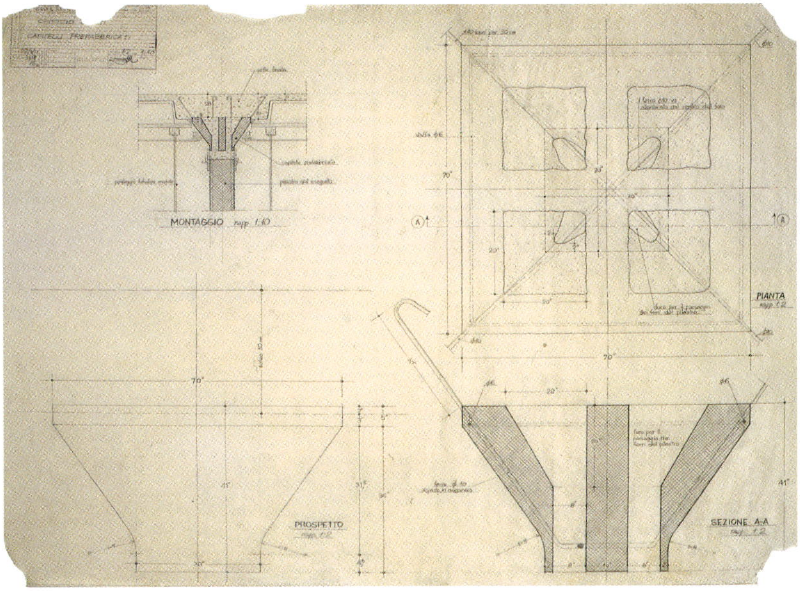

275-277. P. L. Nervi, structure for *Opificio G. Gatti*, designed by C. Cestelli Guidi, Via Prenestina 940, Rome, 1951-53.

275. "Schema ponteggio per solai. Sezione perpendicolare allo spostamento (Diagram of scaffolding for floors. Cross-section perpendicular to movement)," drawing dated 15 March 1951 (CSAC, n. inv. pra 891, n. id. 14580, coll. 155/5).

276. "Solaio a fungo isostatico. Sistema di lavorazione (Isostatic mushroom-shaped floor. Working system)," drawing dated 6 March 1951 (CSAC, n. inv. pra 891, n. id. 14580, coll. 155/5).

277. "Capitelli prefabbricati (Prefabricated capitals)," drawing dated 27 March 1951 (CSAC, n. inv. pra 891, n. id. 14580, coll. 155/5).

went beyond planning. He specified "plaster filling." For the concrete pouring, Nervi also gave advice to prevent the formation of honeycomb patterns on the surface after stripping. For this purpose, he specified "curing the pouring of the lower part of the narrow ribs, especially at the crosses, to prevent the entry of coarse gravel elements."[7] The pouring of each span was left to set for four days. The surface operations were limited to filling and cladding.

Despite the flagrant conflict between the prismatic shape of the pillars and the flowing weave of the continuous isostatic rib structure – a conflict that was not present in the floor patented by Arcangeli – the floor of the *Opificio Gatti* became one of the international masterpieces exemplifying a structural system that could elevate matters of construction statics and science to form the basis of a new aesthetic for reinforced concrete. It succeeded in transforming the ordinary flat patterns into a bony membrane similar to that generated by the waved beams of *Salone B* in Turin and was evidence of the creative efforts expended by Nervi and his collaborators in the Nervi & Bartoli engineering department. The floor became internationally renowned due to the spectacular photography of Oscar Savio, commissioned by Nervi and as usual expertly retouched to remove residual construction defects and convey an effect of perfection and brightness that sometimes appears unreal.

By this point the floor with waved beams of the *Tebaide*, the floor with and without isostatic ribs of the *Fabbricato Ballette* and of the *Opificio Gatti*, and designs such as that of the triangular-meshed floor for the *Manifattura Tabacchi* in Rome revealed that between 1949 and 1951 Nervi was the main proponent of the quest for new plastic configurations for floor structures. He became established as the precursor of what was soon destined to become the innovative model for the structure of an exposed reinforced concrete floor, the Tetrahedron Floor System of Kahn's Yale Art Gallery.[8]

[7] Drawing No. 3318-30e, CSAC, n. inv. pra 891, n. id. 14580, coll. 155/5.

[8] See Roberto Gargiani, *Louis I. Kahn. Exposed concrete and hollow stones*, Lausanne, EPFL Press-Routledge, 2014, pp. 29-57.

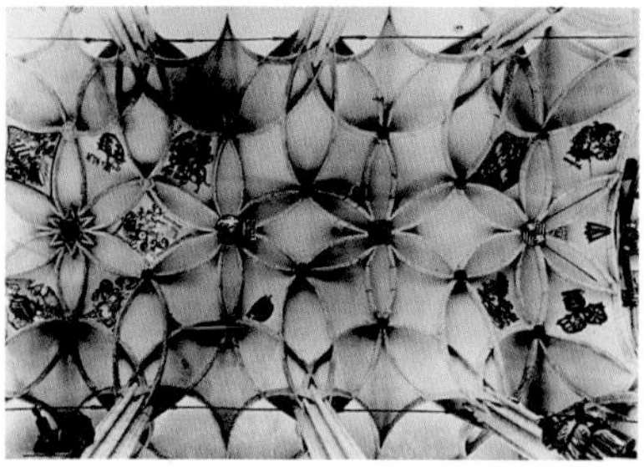

278. Photographic collection of P. L. Nervi, photograph of vault of the church of Santa Barbara, Kutná Hora, Czech Republic, about 16th century (Maxxi - PLN, F04619).

279-282. P. L. Nervi, structure for *Opificio G. Gatti*, designed by C. Cestelli Guidi, Via Prenestina 940, Rome, 1951-53.

279. View of floor with sliding formwork in situ (Maxxi - PLN, F08753).

280. View of ferrocement contoured *tavelloni* of the sliding formwork (Maxxi - PLN, F08750).

281. View of basement floor (photograph by O. Savio; Acm-EPFL 0172.02.0003.2102).

282. View of weaving room (photograph by O. Savio; Acm-EPFL 0172.02.0003.2101).

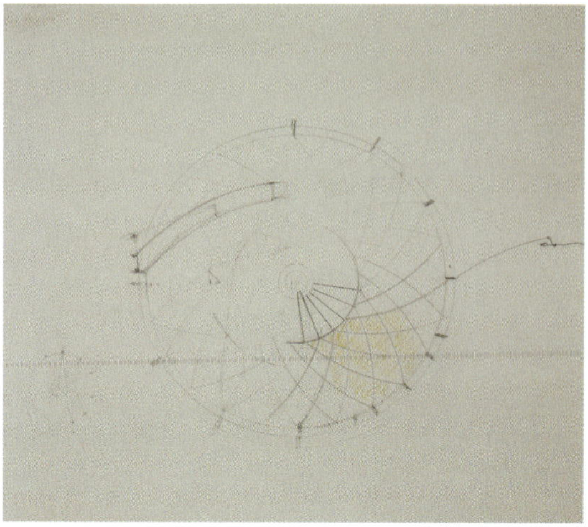

283. P. L. Nervi, design for roof of the *Augusteo* and reset of the *Ara Pacis*, Rome, 1951 (CSAC, n. inv. pra 783, n. id. 14452, coll. 290/6).

Partly as a result of the elaborate process of implementing forms A and B and the prefabricated capitals, Nervi further modified the Arcangeli floor, limiting himself to reproducing it in the version tried out with the *Opificio Gatti* in only a few other industrial buildings, such as the washroom building of Società Solvay in Rosignano and the raw tobacco depot of the *Manifattura Tabacchi* in Lecce, as he tried to find ways to make the most of the aesthetic power of isostatic ribs for the design he was about to propose for the Unesco headquarters in Paris. The construction of the floor for the *Opificio Gatti* might have been the very action that prompted Arcangeli to explore a radical variant of this system, where the form B structures became actual flat beams with their geometry independent from that of the isostatic ribs.

Advent of the variable cross-section pillar

As long ago as his first structures, designed during the 1920s, Nervi recognised the discovery and exposure of the flow of forces in the moulding of a plastic material such as reinforced concrete as decisive creative moments in the advent of a new product aesthetic. The waved shells, ribbed shells, and isostatic lines that later carpeted monumental and industrial spaces, harnessing the phenomenon of light with their ferrocement and reinforced concrete weaves, seem to be the logical consequence of a move toward the discovery of the invisible dynamic action of forces in matter. In terms of his own research, which Nervi was now able to evaluate objectively with several decades of hindsight, he realised that the interpretive excesses of the dynamic forces of reinforced concrete that had fascinated him during the 1930s had been influenced by the political climate of fascism, although he was still capable of including them in his impassioned examination of the unexplored natural potentials of that product.[9] With his new aim of exploring the invisible forces at play – even structures such as floors that, from the Hennebique system onward, had become established due to their imposing trilithic presence – now featured

[9] For this critical view of Nervi's about his own work see Pier Luigi Nervi, *Preface*, in, *The Works of Pier Luigi Nervi* cit., p. VI.

284-285. P. L. Nervi, structure of the Kursaal beach establishment, designed by A. Lapadula, Lungomare L. Catulo 36/40, Ostia, 1950.

284. Construction of the restaurant roof (Maxxi - PLN, F00886).

285. View of the restaurant (Acm-EPFL, 0172.02.0003.2105).

essential materials and flowing lines due to Arcangeli's discovery. They formed part of the filamentous ribbed approach that enlivened the shells of vaults and domes. Even the laying planes, out of which the arches with their taut lines were carved, were angled along guidelines imposed by the vaults for which they acted as abutments – and every other support of those roofs had already been transformed into a post or spur. The only thing that had so far escaped Nervi's creative drive were the pillars, still standing vertically in their reinforced concrete skeletons, even when the isostatic lines of the floors flowed down to the ground in a way that was not necessarily channelled through those vertical static conduits. There is no doubt that, partly due to the universe revealed by Arcangeli with his isostatic lines, the time was ripe in 1950 or thereabouts for re-examining the static and aesthetic canons of the reinforced concrete pillar. As was always the case in Nervi's work, however, the reasons that led him to invent configurations for his structures could not be easily traced back to a single static approach, and the formal excesses that he had admitted were a defect in the creative stage of the 1930s had actually not disappeared but found a different outlet in the new cultural climate. They were now being fed by his meeting, direct and unprecedented for Nervi, with the work of exponents of international standing. His intuitive drive to configure reinforced concrete pillars in an unusual manner, which was decisive for achieving the formal fruition of his idea of structure, all permeated by "aesthetic feeling" and "aesthetic expressiveness," was crucially impacted by his realisation of the plastic and sculptural strength of Le Corbusier's exposed concrete.

That the isostatic lines raised new questions over pillar configurations amongst the employees and managers of the Nervi & Bartoli engineering department is evident in studies carried out by Arcangeli himself following his patent. He sought to establish the consequences of his invention and sketched some solutions for walls and roofs grooved by isostatic ribs. Sometimes he imagined a row of pillars generated by the characteristics of his structure and obtained a triangular-meshed geometry vaguely similar to the star-shaped grid of lines that he had drawn while imprisoned in India. In other cases, the isostatic ribs on the walls of a suspended box generated a pillar that was no longer mushroom-shaped, like the one shown in the patent, but branched into three or even four sections so that it could link up more effectively to the descent of multiple forces. This was a problem of structural geometry that also fascinated Nervi during the years around 1950. It was to become a significant corollary of his search for a pillar shape that was capable of offering maximum strength through a combination of its lines and surfaces, and that could capture the forces and disperse them into the foundations without buckling, each time adapting its configuration to the different ways in which the forces acted.

The matter of variable cross-sections had already cropped up with all its plastic and creative impact in the curved profiles of the *Stadio Berta* roof forks and then reappeared in full force in cases were the profiles could easily be moulded through the use of formwork that could be folded at will, such as that made out of ferrocement, and manufactured on a large scale by the prefabrication of parts, as in the case of the *Salone B* gallery beams. The abutment sections of the taut arches of the *Tebaide*, newly cast in place like the cantilever beams of the *Stadio Berta*, also had a variable cross-section. The origins of the invention of a variable cross-section pillar can also be traced back to Nervi's quest for an ideal shape for each component of his structures. It is no coincidence that one of the first pillars – destined to become a veritable genealogy of the different configurations of Nervi's columns – copied the mushroom-shape

286. P. L. Nervi, structure of the Kursaal beach establishment, designed by A. Lapadula, Lungomare L. Catulo 36/40, Ostia, 1950.
Steel reinforcement of the diving board (photograph by A. Lapadula; AL).

that Arcangeli claimed as the ideal connection for pillars with a shell traversed by isostatic lines (the description he used in the patent was "flared head" when discussing the pillar of the "mushroom-shaped floor").[10]

When Nervi & Bartoli built the hall of the restaurant in the *Kursaal* beach resort in Castel Fusano near Rome, designed by Attilio La Padula, Nervi significantly modified the usual system of prefabrication using ferrocement *tavelloni* in terms of the movement of the lines of ribbing, which reflected the effect exercised by the sinuous movement discovered by Arcangeli. Due to its upturned umbrella vaulted roof cantilevered on a single central column, the hall became a small-scale prototype of the visionary and grand structures designed by Nervi for the E42 buildings and the market at Foggia. The roof and column formed a continuous profile that set out to express the natural unbroken flow of the many forces into the single foundation. The umbrella-shaped shell was made out of prefabricated *tavelloni* in ferrocement while the column was made out of cast-in-place reinforced concrete and clad in plaster (at its centre it housed a concrete asbestos pipe carrying rainwater downward). It is significant that the lines of ribbing generated by the *tavelloni* were curved and no longer the overall result of straight segments arranged in a spiral, an evident sign of Nervi's transposition of Arcangeli's isostatic moulding into his ribbed shells. The very act of folding the *tavelloni* also represents the sublimation into logical lines of Nervi's ornamental geometric grid, betraying at the same time Arcangeli's isostatic lines and revealing their decorative aspect. The column is reabsorbed in the interplay of the vault's lines and can even forms part of it since the section of vault abutment, with its smooth intrados, is made out of cast-in-place reinforced concrete, as is that of the stem. The column therefore loses its independence as a vertical roof support and assumes a variable profile, generated by a diameter that gradually increases from 35 centimetres at the base to 65 at the ring where the column ends and passes over into the vault.

Nervi created the ribbed disc of the overhang using a series of U-shaped beams with a slightly arched profile, all prefabricated out of ferrocement and stitched in place by a curb at the jetting end. The prefabricated elements were also painted in the hall, as usual, to conceal the manufacturing defects. The circle shape was even rendered monumental in La Padula's springboard, which Nervi built with a reduced 6-centimetre cross-section using his ferrocement technique.

The authoritative and decisive starting point of Nervi's research into variable cross-section pillars with the aim of determining configurations that were ideal for transmitting loads, may nevertheless be identified in a design dating back to the first half of 1951 for a colossal building designed to contain homes, offices, and auditoriums as well as the radio and television headquarters called Taba Guajanazes, which was due to be built in São Paulo. Nervi found himself working with Bardi's wife, Lina Bo Bardi, and had the opportunity to travel to Rio de Janeiro, where he was impressed by a visit to the headquarters of the Ministry of Education and Public Health.[11]

The Taba Guajanazes is a critical version of the *Unité d'Habitation* in Marseille, replicating the latter's platform on monumental pilotis and devices

[10] Nervi & Bartoli, patent IT 455678 cit., pp. 7-8.
[11] His impression of that visit was revealed by an interview with Nervi, see Libero de Libero, *Saper costruire*, in, "Pirelli. Rivista d'informazione edi tecnica," IV, July-August 1951, No. 4, (pp. 8-10, 12), p. 12.

287-289. P. L. Nervi and L. Bo Bardi (with the cooperation of A. Nervi), design for the *Taba Guajanazes* building, São Paulo, 1951.

287. Study for variable cross-section pillars (CSAC, n. inv. pra 958, n. id. 14697, coll. 157/4).

288. Preparatory model for variable cross-section pillars ("Habitat: revista das artes no Brasil," No. 14, 1954, p. 5).

289. Preparatory model for variable cross-section pillars ("Habitat: revista das artes no Brasil," No. 14, 1954, p. 7).

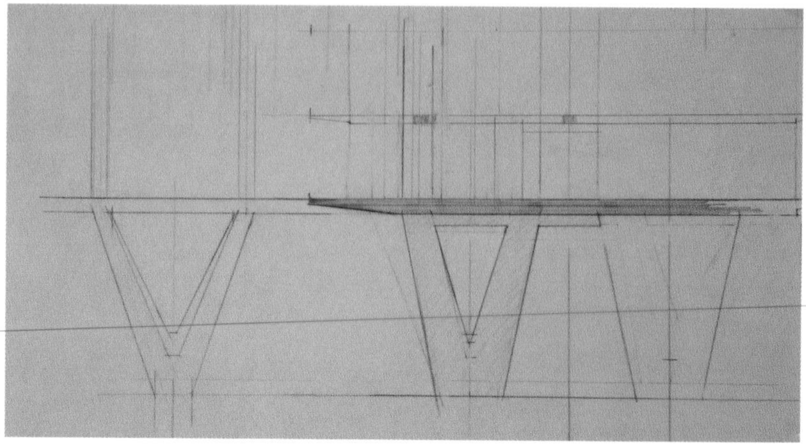

such as the *rues intérieures*. Bo Bardi and Nervi went to great lengths to give their reinforced concrete pilotis their own particular character, based on those of Le Corbusier and the variants proposed by Niemeyer. The study drawings reveal their concern to resolve, through a static and geometrical approach, the plastic impetus expressed by Le Corbusier in his pilotis, which are nevertheless clearly reflected in the profiles explored by Nervi. A V-branched design was selected for the Taba Guajanazes pilotis with the aim of connecting the various framework patterns at the different levels, though not without taking into account the profile of the Marseille pilotis. The pilotis designed by Nervi grew from a base with a round cross-section (he also examined a version with a rectangular cross-section) to connect to the pillars of the lower levels. They then branched out to form two gradually-tapering branches that ended in a rectangular cross-section against the beams. The choice of a V-profile made it possible to concentrate more loads in a smaller number of points, while the volume was already shaped for simple construction, preferring ruled surfaces because this made it easier to lay the boards of the formwork. One sketch by Nervi actually reveals this concern: when he drew the generating lines of the figure, he may even have envisaged the pattern of ridges to be impressed during the concrete-pouring stage through careful laying of the boards.

When they submitted their final design, Nervi and Bo Bardi used a special model made up of two wooden supports to stretch the wires and ensure that the geometrical lines generating the figure were visible. In Nervi's eyes, those wires possessed a scientific value comparable with that of the isostatic lines in identifying strict principles for the design of the pilotis. He may also have intended them to intercept the descending load lines, just as he believed that the grids of the geodesic structure and the ribbed shells were able to dominate the forces at play. All Nervi's subsequent geometric constructions of cast-in-place reinforced concrete pillars obeyed the same creative principles as his São Paulo pilotis. Nervi and Bo Bardi represented the plastic and sculptural configuration produced by the geometric construction lines in another model. In the eyes of Nervi, the Taba Guajanazes pilotis therefore took on a meaning of its own that can be traced back to the contemporary ruled surface figures that were emerging in reinforced concrete construction, becoming a tangible sign of an ideal strength through form.

Terminological issues raised by the new formwork

Construction of the *Fabbricato Ballette* at the Manifattura Tabacchi in Bologna and the *Opificio Gatti* in Rome demonstrated that the technical system, and the process of prefabrication and casting in place, of ferrocement and reinforced concrete implemented by Nervi & Bartoli tended to blur the boundaries of the various means of concrete production due to radical changes in the way the formwork was manufactured. The terminology used by Nervi in documents and drawings is emblematic of the continually-evolving technical processes in his business. The same element of his production system could be described using different terms, because its very meaning was the complex result of its changing role in the installation process. Thus, the *tavellone* that was created to take the place of the brick block to remain in place during the casting-in-place of a floor, may still have been described as a *tavellone* when it was used as a mould to shape the cavity of a floor, again cast in place, but was also described using terms that clarified its new role. Hence, it might equally have been referred to as a "*forma*" (form), "*cassero*" (shuttering), "*cassaforma*" (formwork), or even "*elemento a cassaforma*" (formwork element).[12]

When the formwork became a sliding system, as in Bologna or Rome, the term "*forma*" became ambiguous and was used to indicate both the individual prefabricated element in ferrocement as well as the installation of several elements, firmly anchored together, to create a device – in other words the "*forma*" itself, which made it possible to create a floor span in situ. The Nervi & Bartoli documents clearly reveal Nervi's tendency to use the term *cassaforma* to describe the traditional system of moulding concrete using wooden boards, even though his terminology revealed that he hardly used individual words unequivocally. Nervi often resorted to an expedient that was symptomatic of the complex technical steps performed using the same element and the consequences on the terminology when the company had to use several different formwork systems on the building site. Such was the case with the *Opificio Gatti*. To make sure he was understood by his workers, he added significant details, such as when he defined formwork made out of wooden boards used for floor edging as "*cassaforme normali*" (normal formwork) to distinguish them from the "*forme mobili*" (sliding forms) used for spans of the same floor.[13] Sometimes he used expressions that clearly reveal his continuous struggle with the carpenters in his firm, when he cites their "reluctance to give in" as a reason for the difficulty of using "timber formwork," "except at the cost of unsustainable labour charges and wastage of timber," for the con-

[12] See the terminology used in Nervi, patent IT 455750 cit.

[13] Drawing No. 3318-3e, CSAC, n. inv. pra 891, n. id. 14580, coll. 155/5.

struction of "curved surfaces" and "perforated surfaces" or for the "hollowing of areas that were statically unnecessary."[14]

It is clear that throwing out all the technical and theoretical structure of traditional timber formwork opened up a series of questions that Nervi left unresolved for the time being, fudging the issue by using ambiguous terminology that was still understandable to his workforce, even though he would be forced to be clearer when his system went international. Nervi eventually provided the first exhaustive explanation of the special prefabrication process that he had introduced through his firm Nervi & Bartoli when he described the processes followed for the waved beams of *Salone C* in Turin: a "counter-form" in plaster, "modelled with all the surface accuracy and precision one could want," that could be used to create "cement forms" onto which the waved beams could be poured. Those same "cement forms" were also more appropriately defined as "matrices."[15]

The terminology used to define the new types of floors with isostatic lines also provided significant clues, making it possible to grasp the depth of creative imagination, including in historical terms, that led Nervi to invent the structural configurations best suited to reinforced concrete and construction science. Nervi used the adjective "*cassettonato*" (coffered) to explain his variant of the Arcangeli floor, indicating the value he attributed to his quadrangular compartments with their unequivocally conventional shapes. Nervi's arrival at that kind of coffering in order to make the process of constructing Arcangeli's floor simpler and cheaper does not rule out that his method still betrayed a fondness for historical configurations, if they could rationally be adapted to the new construction processes. From this viewpoint, it is significant that Arcangeli, despite his freedom in imagining the movement of the isostatic ribs, almost refused, or at least knowingly avoided, his master's terminology when he described his own floor, using the vague adjective "boxed,"[16] and certainly never employing an image of coffering. Arcangeli was more precise than Nervi in his use of terminology to indicate the new moulds for the construction of his structures. For example, when he used the term "formwork," he was aware of the need to clarify that it was a "box-shaped" element. Arcangeli again wrote of "forms" or "ribbing elements" without invoking *tavellone* configurations, partly because his structure arose first and foremost out of his vision and calculation of isostatic lines and not, as in the case of Nervi, out of building site construction processes.

When, in 1953, Nervi explained to a wide readership the role of the formwork materials that was so crucial to his concept of structure, he did so with the clarity of a builder who has been relentlessly pursuing building site economies for the manufacture of reinforced concrete for decades. Although the topic cropped up in his writings of the 1940s, he now introduced it with an effective summary of the issues at stake. It should nevertheless also be noted that Nervi made his opening pronouncement just when he was about to mould reinforced concrete using wooden formwork alone, in an extension of its uses that he had never previously tried. His opening words

[14] Pier Luigi Nervi, *Possibilità costruttive ed architettoniche della prefabbricazione strutturale*, in, "Architettura Cantiere," 1952, No. 1, (pp. 33-42), pp. 33-34.

[15] Nervi, *Costruire Correttamente* cit., p. 63.

[16] Nervi & Bartoli, patent IT 455678 cit.

therefore ultimately resonate as a succint coda to the impulse that drove him to a systematic liberation of the timber that he was only now beginning to appreciate. It had a quality that he had never previously considered in all his building sites: the interplay and significance of the impressions it left behind.

"With the spread of reinforced concrete constructions and, above all, through their applications to new and more rational building solutions, it has become increasingly evident that normal wooden formwork constitutes a technical and financial obstacle, since by its very nature (unless one wishes to face unacceptable expense) it requires a design based on a flat or simple curved surface with prismatic elements. It is easy to imagine that placing such restrictions on the shape of a structure, because the building technique is not up to the intrinsic possibilities of reinforced concrete, stands in the way of fully developing a true architecture of reinforced concrete and particularly runs counter to solutions that would be more effective from a static, as well as an architectural, viewpoint; such as double-curved or undulating surfaces, load-bearing elements, beams, or pillars with cross-sections that vary according to the internal stresses."[17]

Even though Nervi's main undertaking in the manufacture of cement structures boiled down to obtaining the best quality of ferrocement when building structures and formwork, the timber he continuously criticised until 1953 nevertheless retained a role in the repertoire of main materials used in the formwork manufactured in Nervi & Bartoli's building sites. Still, its use was limited to the concreting of structures intended to be clad, in monumental works, or left in its raw state in industrial storage sheds. When Nervi succeeded in controlling the finish of the concrete used for his floors through a system of sliding formwork with ferrocement moulds, the timber used to mould the skeleton was confined to the construction of the pillar formwork alone. Until then, Nervi had limited himself to issuing specifications that only concerned surface quality for pillar construction, with the additional requirement of planing the boards used for the reusable formwork in the Bologna building site. Through this solution he intended to reduce every trace of wood on the concrete to obtain a surface that was as polished as that of the floors modelled using ferrocement formwork. This moment was a crucial transition in Nervi's vision of the artistic expressiveness of concrete as a material that was also capable of generating significant impressions, as evidenced by the fact that when he designed the pilotis with Bo Bardi, he considered using wooden boards to be fitted in slats according to the geometry of the ruled surfaces and thus model the surfaces. At this particular juncture, which was decisive for the development of his thinking about the technique and the very concept of cast-in-place reinforced concrete, Nervi's visit to the *Unité d'Habitation* in Marseille, in September 1952,[18] took on a special significance for the genesis of a variable cross-section pilotis and a form of concrete to be left exposed, tattooed with the impressions of the wooden boards.

[17] Pier Luigi Nervi, *L'architecture du béton armé et le problème des coffrages*, in, "L'Architecture d'aujourd'hui," vol. XXIV, July 1953, No. 48, (pp. 68-73), p. 69.

[18] Roberto Gargiani, Anna Rosellini, *Le Corbusier. Béton brut and Ineffable Space, 1940-1965*, Lausanne, EPFL Press-Routledge, 2011, p. 60, note 228.

Principles for a theory of structure

After the numerous experiments that he had now conducted at the building sites of the hangars, at Torre in Pietra, *Salone B* and *Salone C* in Turin, or the *Tebaide*, Nervi had reached a point where he was able to summarise and define the underlying concept of all his works whose structures were the outcome of assembling parts that were manufactured on a large scale on site in reinforced concrete or ferrocement: the "structural prefabrication" he wrote of in 1952.[19] In his view, that kind of prefabrication did much more than offer effective answers to the economic hardships of the reconstruction period. In any case, the start of his research into that kind of building in cement was independent of the matters under debate following the destruction wrought by war from approximately 1943, in which he himself played a part. One of the most significant starting points for grasping his particular idea of structural prefabrication is his study of the technical aspects involved in construction of the *Arco dell'Impero*. Even though the economic issue was crucial in that case, due to autarky, which made it necessary to do without the metal reinforcement for concrete, the real issue for Nervi and his firm was the invention, not so much of a series of effective construction details, but rather of an entire technical process that influenced the manufacture of the material itself and the system of scaffolding used for its execution. He even challenged and criticised the very shape of the arch within whose limits the prefabrication had to be implemented. The second series of hangars translated that visionary drive into a new system of producing cement structures that were always determined by a reticular or geodesic shape. The theoretical and practical experiments that he conducted between the end of the 1930s and the early 1940s were the starting point for a chain of building sites where Nervi developed a particular sense of the concept of "structural prefabrication," as outlined at the beginning of the 1950s after a decade of tests. His form of prefabrication was now no longer shackled by the limitation of transforming a certain structural configuration into an industrial production process, but had become a principle essential to the very discovery of unprecedented formal configurations for roofing structure. That principle now made it possible to devise more and more massive structures characterised by a lightness that was evident in the constituent elements, the building site machinery, the temporary scaffolding for assembly and, consequently, contributed to determining the shape of the structure. Nervi made no bones about specifying the characteristics of his structural prefabrication: "I think I can say that it offers a wealth of opportunities that overcome purely economic or technical considerations, to the point where it will become a crucial element of new structural and architectural solutions." When designing the structure of his system of prefabrication, Nervi emphasised the need to identify a "uniform structural scheme," calculate the measurements, weight, and manoeuvrability of the parts, and then educate the workers on the building site.

In the virtuous cycle linking design and construction, the ferrocement should be considered the crucial invention that made it possible to proceed beyond the usual structural shapes and develop the simple action of on-site manufacture into proper structural prefabrication. It is therefore unsurprising that Nervi's reasoning focused on the materials of moulds for prefabrication and his methods of "welding" parts through the use of special joints in order to obtain a "monolithic structure."[20] It should be noted that Nervi did not include in his concept of structural prefabrication the manufacture of floors with sliding formwork for the large-scale modelling of cast-in-place concrete, as occurred at the *Manifatture Tabacchi* or *Opificio Gatti* building sites.

The virtuous cycle of design and construction implicit in structural prefabrication includes another concept that was fundamental to Nervi's vision for his cement structures, which he was finally able to outline in a cohesive and comprehensive manner due to the very act of prefabrication. Nervi may have already used an expression summarising that concept at the end of the 1930s, but he did not begin to discuss the significance of what he referred to as "strength through form" until the early 1950s when he was defining structural prefabrication.[21] While ferrocement was the crucial material that made it possible to transform the technique of on-site manufacture into the concept of structural prefabrication, Arcangeli's discovery of isostatic lines for reinforced concrete shells should be considered the decisive revelation of a universe of configurations that had never yet been envisaged for structures and that Nervi summarised with the definition of strength through form. Significantly enough, his short and decisive essay on strength through form was published as soon as construction work was finished on the *Opificio Gatti*, in 1951, proving the extent to which the isostatic lines of that floor, which figures among the illustrations to the essay, were crucial to Nervi's concept.

The idea of "form," which resonated in Nervi's definition, concealed his decade-old aspiration to find an alternative to the Hennebique skeleton. It was first identified in reticular or geodesic structures, then discovered in waved or ribbed shells, and in the approach of arranging the cement material and its inherent resilient fibres differently, so that the forces at play no longer flowed in accordance with certain rigid geometrical physiognomies dictated by building site practices determined by a need to construct the moulds used to model the material easily, but instead spread and scattered over the surface. The surface was thereby enlivened, vibrated, and rippled almost as though traversed by their motion – the better to resist destructive impulses, – without ever attempting to oppose those forces by applying the material more thickly in the weakest parts. In Nervi's definition, the shape that had always struggled to find a rational basis for its existence in the 19th and 20th century language of architecture now became the very reason for the structure. In his own theoretical journey, the creative margins inherent in the concept of "nature of form" that he had been expounding since the beginning of the 1930s were now stripped down in order to base the same "form" no longer on an "idea of emotional origin" but on the logical principles and certainties of the static system that gave rise to "strength through form." There is no doubt that the transition from "nature of form" to "strength through form" also reflects the end of an exciting and dramatic political twenty-year period that Nervi was able to interpret through his structures and the advent of the years of economic and industrial development of which he would continue to be one of the leading lights.

In the light of the expression "strength through form," Nervi's terminology revealed another layer of significance in the use of the same term, "form," to describe the mould used to model the parts of his structure. It was possible to bend that form at will by means of the ferrocement that simultaneously acted as the material for the mould and vector product, the quintessence of the very idea of form that embraces everything – structure, mould, and material.

[19] Nervi, *Possibilità costruttive ed architettoniche della prefabbricazione strutturale* cit.
[20] *Ibid.*, p. 35.
[21] Pier Luigi Nervi, *La "resistenza per forma." Caratteristica statico-architettonica del cemento armato*, in, "Pirelli. Rivista d'informazione di tecnica," IV, July-August 1951, No. 4, pp. 11-12.

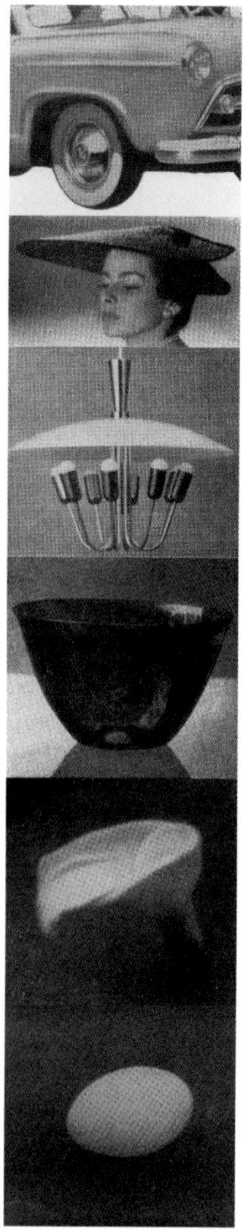

290. Examples of objects illustrating the concept of "strength through form" ("Habitat: revista das artes no Brasil," No. 3, 1951, p. 17).

Nervi's form, which is strong by virtue of its configuration, always retains its prerogative of continuity, even when it appears perforated to achieve lighting effects or is interwoven to an extent that reflects Argan's image of the *tricoter*. Its continuity can be substantiated because of its reduced thickness (Nervi also wrote about "systems with superficial strength"), which had been made possible by ferrocement, the felt-like substance that Nervi had envisaged since the invention of his product and was therefore decisive not only for "structural prefabrication," but also for "strength through form."

Nervi suggested the founding principle of his strength through form: a form that is "not due to the concentration of actions by agents and resilient cross-sections along individual elements." The list of articles that he drew up to illustrate strength through form, "flower calyces, lanceolate leaves, straws, egg and insect shells, shells, fans, lampshades, car bodies, glass vessels and even items of clothing such as hats,"[22] is surprising and sometimes even surreal. Almost as if to provoke a reaction in his astonished reader, he suggests those very objects as examples of possible future structures that could come about as a result of folding his ferrocement at will.

This list, which also includes fuselages and wings of aeroplanes and ship hulls, does not set out to specify individual models but rather to pose the question of what should be the figurative horizon of a creator of structures that follow the principle of strength through form, and how references should be selected to ensure that the "set of unconscious static intuitions,"[23] can give rise to a new static concept. It should not indeed be ruled out that the models of Gothic cathedral vaults that Nervi showed his students and that are reflected in some of his ribbed shells may have seemed to him inadequate to reflect the creative drive stirred within him by the discovery of isostatic lines. The images pasted into his tables of photographic prints, which increasingly come to resemble the Mnemosyne by Aby Warburg in their unconscious genesis of strength through form, appear to have been chosen and mounted with the mounting anxiety of someone who is living through a decisive moment of transition in the creative principles of structure and is preparing the theoretical and cultural ground for the advent of creations such as those of Musmeci. The warning issued in his essay referred first and foremost to Nervi himself, and his greatness lay in being aware of this: "we are not yet accustomed to thinking statically 'by form'."

Nervi tried to list the surfaces that could fulfil the imperative of strength through form, beginning with those with a "simple" shape such as barrel vaults and "conoidal vaults," continuing through those with "more complex shapes" such as "toroidal surfaces" and "translation surfaces" before going on to envisage the surfaces that could be obtained by "combining" different kinds of vaults with "corrugations designed to eliminate the local warping which is the main cause of instability in systems with surface strength."[24] He closed his discussion by explaining the significance of Arcangeli's isostatic lines from which everything stemmed: "Those lines are something absolute that depend solely on the play of forces and resistances in place. The marvellous thing is that by doing so and limiting our role to that of modest interpreters of truth and physical reality, we come to find an unforeseen and ever more expressive harmony of form."

The concept of strength through form, the very different objects that each document their means of manifestation, and that universe of "harmonies of form" that are yet to be discovered nevertheless seem to clash with a vision that is aware of limits imposed by the universal, unchangeable laws that regulate the mechanics of forces and resistances, as though the laws had brought into being certain figures that could only be led along in a relentless quest for perfection. Nervi's idea of combining different kinds of vaults showed that he was channelling the potential of discovering tools of creation, such as that of the isostatic lines, into a fusion not unlike that practised by Felix Candela.

The presence of this twofold drive, visionary and realist, in Nervi's thinking is revealed quite dramatically when we compare his statements on strength through form with other concepts, "*forme-tipo*" (standard forms) or "*soluzione-tipo*" (standard solution), that he expounded at the very same time during his speech at the conference on Divine Proportion organised by the Milan Triennale and held in Milan in October 1951.[25] The subject matter may have prompted him to tone down the complexity unleashed by strength through form to schematic "proportions in technique." What is certain is that Nervi now returned to his role as an engineer builder aware of the rigorous rules of the building site and informed his audience of principles that seemed at times to be a direct reflection of his own way of building, always aiming to optimise each solution and his system of technical reproducibility through prefabrication. These principles guided the entire creative machinery, not merely that of his own works, towards "standard forms" determined on a scientific and experimental basis and perfected in a quest for "maximum economy" to a point where they became "forms that are no longer modifiable," resolved in "standard solutions" that froze invention, and reduced every creative impulse. Words such as "sincerity" and expressions such as "modest adherence and obedience to the laws of the physical world" reveal the level of morality that led Nervi to limit his universe of possible forms and aim for that absolute style of technique that he had envisaged at the beginning of the 1940s. At the very time when Le Corbusier, who was also present at the conference, was rejecting a system based on standard types, Nervi was revealing the part they could play in determining increasingly large structures. Nervi's apodictic conclusion heralded the end of any invention now governed by absolute principles: "Once he has achieved the '*soluzione-tipo*', man becomes a spectator of his own creation." What was not evident, as he expounded on the "proportions" determined by the technique, even as he described Arcangeli's discovery, was the existence of the "completely new" world opened up by "'form-resistant' structures."[26]

"Isostatic lines are [...] preferential lines of the flow of forces within a solid and concentrating the resilient material along those lines gives rise to the maximum static efficiency. Those lines depend only on the interplay of forces acting in the system, and all we can do is find them and exploit them, not change them. Creative freedom has almost completely disappeared and has been reduced to a choice of the most appropriate isostatic lines."[27]

No-one was better than Argan at summarising Nervi's insight on the value of form in the process of creating structures when, to fill the first time with an authoritative critical force penetrating the symbolic folds of his work, Argan revealed the essence of the process of creating Nervi's structures, and his intuition of form as a significant manifestation of the fluid nature of con-

[22] *Ibid.*, p. 11.
[23] *Ibid.*, p. 12.
[24] *Ibid.*
[25] Pier Luigi Nervi, *Le proporzioni nella tecnica*, in, "Domus," December 1951, No. 264-265, pp. 45-48.
[26] Nervi, *La "resistenza per forma"* cit., p. 11.
[27] Nervi, *Le proporzioni nella tecnica* cit., p. 46.

crete that, through its imponderable nature, drives the certainties of figures handed down by history, as happens with contemporary artistic phenomena. Argan solved the puzzle contained in the definition of "strength through form" with these words:

"The awareness that 'technical' architecture must of necessity merge with the overall aesthetic of modern architecture is the starting point of Pier Luigi Nervi's technical research, the primary aim of which is to achieve a formal value: in plastic structures form is not merely the sensitive manifestation or plastic representation of constructive forces but an essential agent of their determination and organisation into systems of balance."[28]

Prefabrication for industrial buildings

In the business activity of Nervi & Bartoli, when Nervi first came up with the concept of structural prefabrication, the idea began to be increasingly adopted for that part of the structure usually constructed using cast-in-place reinforced concrete or made out of brick elements, while the usual ribbed shells in ferrocement *tavelloni* were now increasingly being used to build roofs for industrial buildings in Leghorn or Genoa. Nervi experimented with the prefabrication of colossal pieces in ferrocement, in the form of long ribbed sectors, almost like claw-shaped shells to be fitted opposing one another. Using these he could even include the saw-toothed roofs that were usually made out of brick or cast-in-place concrete in his structural prefabrication. In any case, Nervi & Bartoli had already produced big *tavelloni* in the fans of *Salone B*. On the strength of the structural prefabrication system, the firm was awarded commissions for the quick and inexpensive construction of large buildings after the successful introduction of its systems at the end of the 1940s for the production of large buildings for the *Manifattura Tabacchi*.

The building site of the Azienda Trasporti Municipale factory workshop in Turin, built between 1952 and 1954, is one of the buildings that demonstrates the development of Nervi & Bartoli's structural prefabrication. The matrices of the curved saw-toothed roofs, prefabricated out of ferrocement, were produced using arched brick walls clad in a layer of perfectly smooth plaster. Assembly of the curved segments gave rise to a V-pattern ribbed geometry that was new in the series of those hitherto seen for Nervi's shells. The prefabrication also affected the reinforced concrete frames of the saw-toothed roofs over the large space that could be produced at will on a large scale with an arched profile through the use of moulds modelled out of ferrocement.

Control of structural prefabrication in ferrocement and reinforced concrete was responsible for determining the formal qualities of Nervi's 1952 design for a roof with a parabolic profile and 10-metre span for a building for Piaggio in Pontedera. The claw-shaped *tavelloni* and frames with curved sections tried out in Turin would have created a sophisticated rhythm of in-

291. P. L. Nervi, design of sugar silos for Società Eridania, [Genoa]. Plan, cross-section and detail, drawing dated 11 April 1952 (CSAC, n. inv. pra 388, n. id. 13806, coll. 266/1).

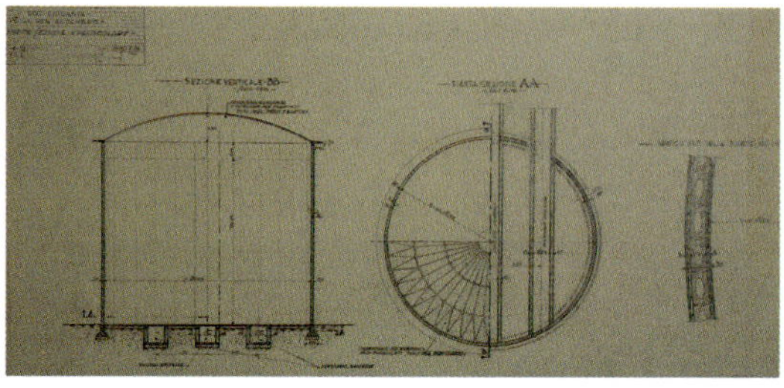

[28] Argan, *Pier Luigi Nervi* cit., cover strip.

292-293. P. L. Nervi, workshop of the Azienda Trasporti Municipale, Corso Tortona 47/a, Turin, 1952-54.

292. Matrices for prefabrication of saw-tooth roofs (Maxxi - PLN, F08852).

293. Saw-tooth roofs (ATMT).

terwoven lines in the roof. The perspective drawing also shows the effect of the cement surfaces through a stippling pattern that reveals Nervi's growing attention for the formal values of the material.

Nervi & Bartoli's technical vocabulary now included configurations and techniques that had previously remained very much at the margins of the company's output, such as vaulted shells with extrados ribs in cast-in-place reinforced concrete, now used in the storage shed of Società Solvay in Via Adamello in Milan (three vaults with an elliptical cross-section) and in the salt storage shed *Capannone Sale Comune* of the *Manifattura Tabacchi* in Bologna (series of vaults with a parabolic cross-section). One particular kind of structure was implemented between 1952 and 1953 for the construction of a cask store, again for the *Manifattura Tabacchi* in Bologna. Nervi & Bartoli transferred their decades of experience in the prefabrication of reinforced cementi-

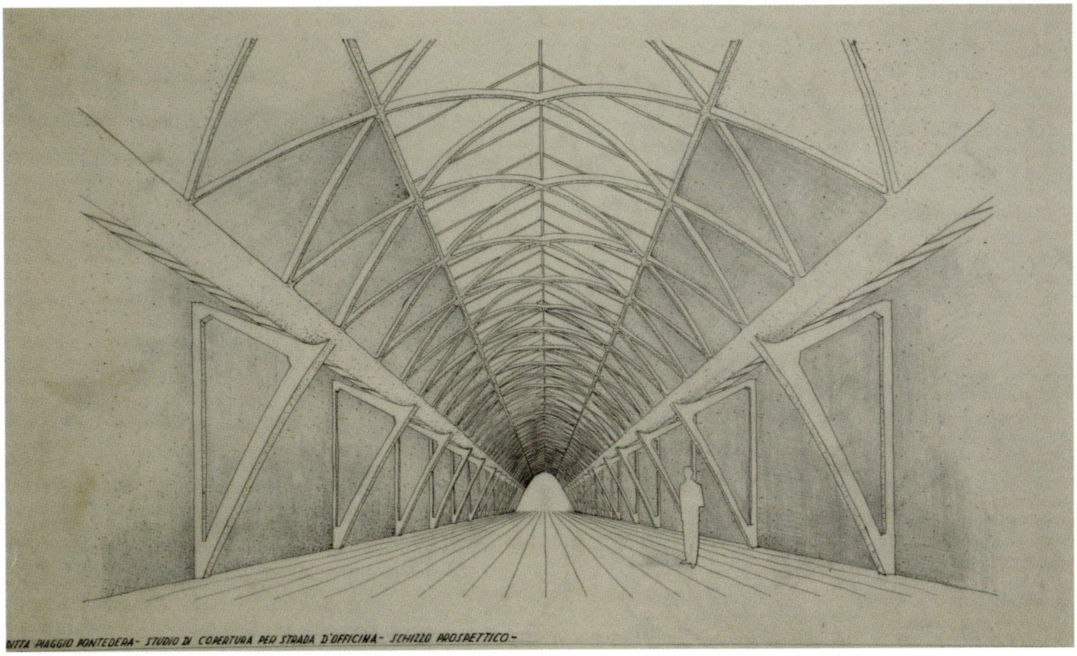

294-95. P. L. Nervi, workshop of the Azienda Trasporti Municipale, Corso Tortona 47/a, Turin, 1952-54.

294. Ferro-cement moulds for prefabrication of saw-tooth roofs (Maxxi - PLN, F08853).

295. View of roof (ATMT).

296. P. L. Nervi, design of "copertura per strada d'officina (roof for workshop road)" for Ditta Piaggio, Pontedera, 1952. Perspective drawing (CSAC, n. inv. pra 949, n. id. 14689, coll. 157/4).

297. P. L. Nervi, design for "studio stazione autobus. Prospettiva (study of bus station. Perspective drawing)" (BSTA, 024-6035).

298. P. L. Nervi, shed for Società Solvay, Via Adamello, Milan, 1951. Demolished. "Schizzo prospettico interno (Interior perspective sketch)" (CSAC, n. inv. pra 29, n. id. 12787, coll. 268/3-4-5).

299. P. L. Nervi, barrel store, *Manifattura Tabacchi* complex, Via della Manifattura 3, Bologna, 1952-53. Construction of the roof (BSTA, 024-7185).

tious structures, including their system of sliding scaffoldings, to the construction of reinforced brick vaults, though still weaved the ribs in accordance with the now traditional geodesic geometries (the extrados vault arches were made out of cast-in-place reinforced concrete, as were the vault abutment walls). The fact that the drawings do not give any guidance on the extrados finish suggests that the weaves generated by the structures and bricks were intended to be left exposed. In order to provide an alternative to the formal quality of that system, Nervi also proposed building the vaults out of cast-in-place reinforced concrete with a "special type" of formwork. He did not specify what kind this was (probably ferrocement), although he described its consequences on the surface finish: "the vault intrados does not require plaster."[29] In order to buttress the vaults, Nervi planned to use chains and also designed a system, which was never realised, for pre-stressing the vertical supports.

In the mid-1950s, again for large industrial buildings, Nervi & Bartoli returned to and modified the procedures for building floors that they had developed and tested for the *Manifattura Tabacchi*. The resulting system was no longer coffered but took the form of long parallel beams spaced close together. One of the most significant results of improving and adapting the pro-

300-302. P. L. Nervi (in cooperation with the Servizio Costruzioni Fiat), *Nuovo Ampliamento Nord* building, Fiat Mirafiori complex, Corso G. Agnelli 200, Turin, 1955.

300. Construction of ferro-cement formwork for prefabricated beams (Maxxi - PLN, F07523).

301. View of building site (Maxxi - PLN, F07588).

302. Prefabricated beams in reinforced concrete (Maxxi - PLN, F07544).

cesses of prefabrication already adopted by the firm was achieved in 1955 for the construction, in just nine months, of the building known as the *Nuovo Ampliamento Nord* to the Fiat building at Mirafiori in Turin, which was three stories high, 20 metres wide, and 640 metres long.[30] While Nervi used his sliding scaffolding with frames fitted with ferrocement forms onto which he could cast in place the structure, for the other storeys, where the beams only rested on the ends, he used T-beams prefabricated through ferrocement moulds and assembled by cranes running on rails. The section of floor between the beams was constructed using a special sliding scaffolding, again fitted with a ferrocement form. For both structural prefabrication methods, Nervi continued to guarantee an effective economy of execution and a polished surface without the need for cladding.

[29] Drawing No. 3657-27, CSAC, n. inv. pra 1025, n. id. 14820, coll. 150/1.
[30] Pier Luigi Nervi, *New Technique for Turin factory. Mass production methods cut erection time*, in, "Prefabrication," May 1957, No. 43, pp. 304-309.

VA 49794
VA 49779

303-305. P. L. Nervi (in cooperation with the Servizio Costruzioni Fiat), *Ampliamento Officine Principali* buildings, Fiat Mirafiori complex, Corso G. Agnelli 200, Turin, 1954.

303. Constructing ground form for pouring prefabricated roof trusses (photograph by T. and G. Vasari; Maxxi - PLN, F08939).

304. Raising a reinforced concrete roof truss (photograph by T. and G. Vasari; Maxxi - PLN, F08949).

305. View of building site (photograph by T. and G. Vasari; Maxxi - PLN, F08936-7-8-B).

Penstocks and canal bridges

In Nervi & Bartoli's multi-faceted business activities, the penstocks for which Nervi invented his construction systems summarised in patents are worth a chapter to themselves. He also used ferrocement for this type of construction, beginning with a 1947 commission to build 80-centimetre diameter pipes for the Naples aqueduct. In order to check the mechanical performance of his material, he carried out tests in the Oberti laboratory on a pipeline made up of several sections measuring 130 centimetres in diameter and 3 centimetres in width. Following these tests, in 1949 Oberti used ferrocemento for a penstock in the rock at Castelbello, in Alto Adige.

In the isostatic line structural version, a ribbed shell in ferrocement made its appearance for the construction, in 1949, by Nervi & Bartoli, of a large pipeline for the canal bridge between the Meschio and Piave rivers near Treviso. The construction of annular ribs fulfilled the principle of structural prefabrication. The words that Nervi wrote about this work of engineering are symptomatic of his approach to Arcangeli's invention, including its unexpected ornamental aspects: "The ribs of a large pipe acting as a canal bridge arranged along the isostatic lines of the main tensions give rise to an interplay of curves that could appear to be the outcome of a sophisticated decorative sensibility."[31]

For the installation of a pipeline carved out of the rock in the Ancipa reservoir in Sicily on behalf of the Ente Siciliano di Elettricità, Nervi invented a machine that was able to produce a cement cladding, irrespective of whether or not it was reinforced, by centrifuging it under pressure against the rock through a rotary device and vibrating it using equipment arranged along the walls of the machine (the device was inspired by the cement gun machinery). Once the cladding had been carried out, special "cement drainage blocks" were installed together with a final waterproof layer, again in ferrocement. The entire process of manufacturing the pipeline, including the machine, became the subject of a patent filed by Nervi in March 1951.[32]

An additional improvement to the pipeline construction introduced by Nervi, again in 1951, involved accelerating the concrete setting by means of a sophisticated circular device made up of different layers of material that, once fitted into the precast ferrocement shell of the pipeline jacket, made it possible to exercise pressure on the concrete pouring that promoted the rapid outflow of water through cracks in the shell. The system allowed him to construct what he referred to as an "actual prestressed pipe,"[33] in this case describing the concrete as prestressed. Nervi also envisaged using heated coils to dry the mixture quickly.[34] Nervi & Bartoli used the procedure perfected using the coils to prefabricate penstocks from as early as 1952–1953.[35]

In order to calculate the sizes of the pipelines, Nervi used tests, which were now carried out by the Istituto Sperimentale Modelli e Strutture (ISMES)

[31] Nervi, *La "resistenza per forma"* cit., p. 12.
[32] Pier Luigi Nervi, patent IT 474615, *Procedimento di costruzione di camicia di tenuta per condotte forzate cementizie in galleria, in stato di precompressione, dispositivo per la realizzazione del procedimento e prodotto ottenuto,* 24 March 1951, ACS, preceded by Pier Luigi Nervi, Patent IT 474148, *Sistema di prova di grandi tubazioni per convogliamento di fluidi sottoposti a sovrapressione interna o meno e relativo dispositivo,* 25 February 1951, ACS.
[33] Pier Luigi Nervi, patent IT 483330, *Procedimento di fabbricazione di condotte forzate in conglomerato precompresso, e relativo dispositivo,* 27 November 1951, p. 11, ACS.
[34] Pier Luigi Nervi, patent IT 495328, *Procedimento di fabbricazione di condotte forzate in conglomerato precompresso, e relativo dispositivo,* 7 November 1952, ACS.
[35] *The works of Pier Luigi Nervi* cit., pp. 82-83, and Pier Luigi Nervi, US patent 2,771,655, *Process of manufacture of forced pipings made of precompressed concrete, and relative equipment,* 25 November 1952.
[36] Neri, *op. cit.*, pp. 69-72.
[37] *Ibid.*, p. 131.

of Bergamo, of which Danusso was president and Oberti was director. This was set up in 1951 as a development of the Milan Polytechnic Testing and Model Laboratory.[36] Nervi continued to use ISMES to define the efficacy of his complex structures using models, and he also became president from 1964.[37]

Nervi & Bartoli's foray into penstocks and canal bridges ended halfway through the 1950s and was enhanced by other inventions, such as a pipeline with a cavity for drainage (this was built in 1954 for the Valico tunnel, in San Floriano, in Trentino Alto Adige), improved compaction of concrete under pressure (this was implemented for the construction, again in 1954, of the pipelines of the Somplago power station near Udine), or a new system for building canal bridges on large pillars, with semi-cylindrical pipelines. The latter were prefabricated using a mould dug in the ground, finished using Nervi's usual methods, and made to slide on a cart along the finished canal section before being installed using a mobile crane and temporary trusses (the system was implemented for the construction, between 1954 and 1955, of the S.A.D.E. canal bridge, between the Meschio and Piave rivers, in Vittorio Veneto).

306. P. L. Nervi, prestressed reinforced concrete pipes (BSTA, 024-6941).

307. P. L. Nervi, design for canal bridge for aqueduct connecting the Meschio and Piave rivers. "Campata tipo del ponte-canale (Typical canal bridge span)," drawing dated 22 October 1949 (CSAC, n. inv. pra 1209, n. id. 15183, coll. 41/6).

308. P. L. Nervi, design for canal bridge for aqueduct connecting the Meschio and Piave rivers, 1954 (P. L. Nervi, *Buildings, Projects, Structures. 1953-1963,* New York, Frederick A. Preager, Publisher, 1963, p. 18).

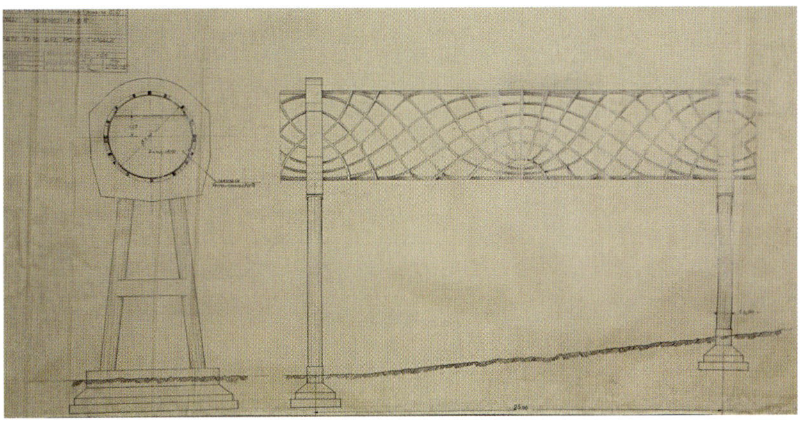

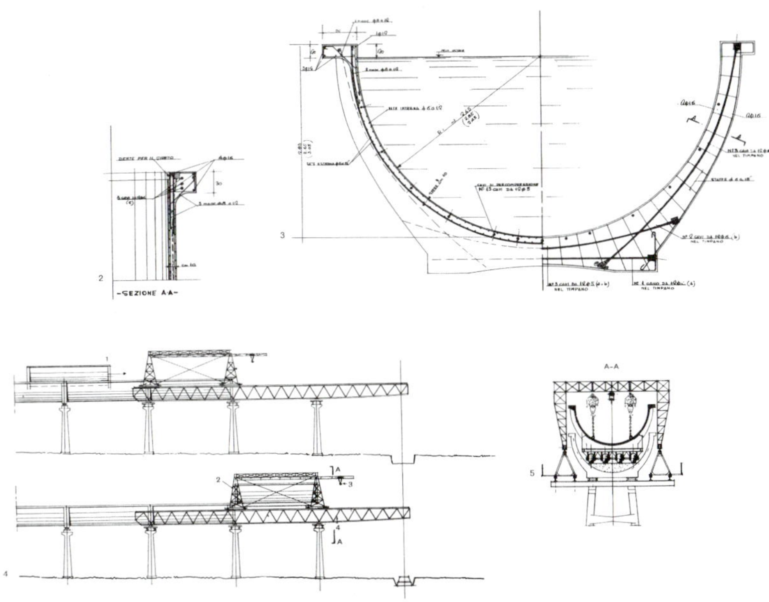

chapter nine
the unesco headquarters in paris: perfect *béton brut*

Structural icons

Not long after the publication of *Oeuvre Complète*, in which Le Corbusier extolled the beauties of *béton brut* and all its unexpected defects, work started on the construction of the prestigious headquarters building of Unesco in Paris. The type of concrete used for this featured sophisticated technical and manufacturing qualities that left no room for the *malfaçon* that had spread following the success of the *Unité d'Habitation* in Marseille. Nervi saw the Unesco buildings and their execution as marking a turning point in perceptions of the plastic qualities of reinforced concrete. Following the experience he had in that building site, he gained confidence with a series of techniques for finishing cast-in-place concrete in wooden formwork and left exposed. This filled him with such enthusiasm that he became established in Italy through his firm Nervi & Bartoli as the leading exponent of a particular kind of *béton brut*, even if he had previously preached the virtues of structural prefabrication, promoted the polished surface quality guaranteed by ferrocement, and been the most critical constructor against the use of wooden formwork.

The design of the three main Unesco headquarters buildings – the Secretariat, the *Bâtiment des Conférences,* and *Bâtiment III*, also known as the Annexe Building – had been entrusted since 1952 to Marcel Breuer, Pier Luigi Nervi, and Bernard Zehrfuss who worked under the control of the *Comité des Cinq*, made up of Le Corbusier, Walter Gropius, Ernesto Nathan Rogers, Sven Markelius, and Lúcio Costa (Eero Saarinen acted as consultant). During the construction work, which began in November 1955 and ended in 1958, a fundamental part in the technical execution of reinforced concrete was played by the construction company, set up for this purpose through a merger between Etablissements Fourré & Rhodes and Société Dumez, combined in the Groupement des Sociétés, E.ts Fourré & Rhodes, Société Dumez. Fourré & Rhodes et Dumez was to make available qualified workers to achieve "high-

309. M. Breuer, P. L. Nervi, and B. H. Zehrfuss, Unesco Headquarters, Place de Fontenoy 7, Paris, 1951-58. Architectural model of complex, 1953 (Unesco).

quality work," particularly for the exposed concrete.[1] A shipyard carpenter was also hired to produce the formwork with their special geometries. As with the Ronchamp chapel roof, this illustrated the decisive role of this kind of work for the success of *béton brut*. In some of the most significant parts of the exposed concrete in the Unesco complex, the same firm developed the measurements and arrangement of the formwork panels, but always working in accordance with guidelines supplied by Breuer, Nervi and Zehrfuss.

Nervi's entrance into the drafting of the design resulted in the intrusion, from the first version, of some structural motifs that clearly had their origin in engineering, including the grand canopy placed in front of the Secretariat, in the form of a hyperbolic paraboloid, and the unequivocal image, now a trademark of Nervi & Bartoli, of the Arcangeli floor that was intended to be used to draw grids of exposed isostatic lines on all the floors of the Secretariat during the initial stages of the design. Even when the design of the headquarters was amended due to the decision to construct the buildings on the land in Place de Fontenoy, various documents attest to the fundamental role played by Nervi in defining the formal nature of the most significant parts of the structure.

In the final project, the design of the reinforced concrete structures still revealed the crucial contribution made by Nervi, ranging from the corrugated shell of the roof and walls of the *Bâtiment des Conférences* to the pilotis of the Secretariat. The two canopies of the Secretariat, known as "auvent Piazza" and "auvent Fontenoy," became further outstanding motifs of Nervi's reinforced concrete.

The entire design of the corrugated shell on the *Bâtiment des Conférences* shell translated into cast-in-place concrete reinforced the quest for strength through form that Nervi staged with his waved shells in ferrocement. The power of the corrugation also reveals an echo of Freyssinet's grand structure for the Orly hangars. Nervi's folds follow the variations of the moments and were reinforced, in the roof resting on the middle row of tall pillars, by a slab that was alternately positioned low down or high up in relation to the waves, in all cases to accommodate changes in the moments. "[...] we happily allowed ourselves to be guided by the unerring compass of constructiveness and statics," he confirmed.[2]

After the initial studies for the branching V-shaped pilotis of the Taba Guajanazes, performed with Bo Bardi, with the design for the pilotis of the Secretariat and the pillars of the *Bâtiment des Conférences*, Nervi was ready to erect his first variable cross-section columns in exposed reinforced concrete. Significantly, Nervi's visit to the *Unité d'Habitation* in Marseilles in 1952 fell at the very time when he was defining the formal characteristics of the structures for the Unesco headquarters. The unusual motif of the Secretariat pilotis, its angle, the elliptical cross-section of its base, and its cantilevered ending should be seen as a critical interpretation of the formal qualities of Le Corbusier's pilotis. Nervi's intervention was to render logical, in terms of static function, what must have struck him as the fascinating outcome of a creative plastic impulse. Every detail of Nervi's pilotis is intended to justify its own existence in the mechanics of the forces at work. The ellipse expanded the base to hold the figure firm against lateral forces (the longest axis measured 1.30 metres, while the shortest axis measured 70 centimetres) – and to create "a cross-section with an efficient moment of inertia."[3] The change in cross-section to the rectangular configuration at the top meant that the structure could be developed into the cantilevered protuberance supporting the platform. Lastly, the angle meant that it could be positioned in order to minimise the lines of supports required to raise the building off the ground, while its angle in relation to the vertical took into account a calculated eccentricity of the loads at the base.

The twist imposed on the cross-section of the *Bâtiment des Conférences* pilotis also served to establish, at the base and summit, a proper relationship with other structures in order to be able to erect statically perfect columns enlivened by the dynamic drive of the forces traversing them, as was the case with the cantilever beams of the *Stadio Berta* roof, the waves of the corrugated shells, and the beams with their expanded ends.

Guidelines for exposed concrete, *brut* and *bouchardé*

The concrete of the Unesco headquarters was defined as "*béton apparent*," subdivided into "*brut*" and "*bouchardé*," in the various documents by Breuer, Nervi, and Zehrfuss that preceded the start of the building site. The joint presence of both finishes revealed their caution over a material that was becoming popular partly due to the artistic properties of its imperfections in the wake of the masterful *Unité d'Habitation* in Marseille. All this suggests that construction of the Unesco headquarters was conceived from the outset as the model for a form of concrete different from the celebrated *béton brut* of Marseille and more in line with the rigorous formal parameters of US Architectural Concrete. The Unesco headquarters building site can be said to mark an initial significant evolution from the Marseille kind of *béton brut* to *béton apparent* with different levels of finish.

In documents dating from 1953, the reinforced concrete structure of the Unesco headquarters was described in general terms without entering into specifics of the way the various surfaces would be treated. Breuer, Nervi, and Zehrfuss were nevertheless already raising questions about the finish of the exposed concrete in terms of the sophisticated work that would go on for certain parts following the stripping. This was summarised briefly as "bush hammered concrete with special aggregates for the exposed part of the structure."[4] The envisaged bush hammering process not only contradicted the principle of *béton brut*, but was also a departure from the inexpensive working processes imposed in post-war building sites. Even Auguste Perret himself did nothing else to the stripped concrete in his most recent works. In the context of the

[1] Groupement des Sociétés, E.ts Fourré & Rhodes, Société Dumez, letter to Bureau du Siège, 8 November 1955, Unesco, Bureau du Siège, HQ/9.

[2] Pier Luigi Nervi, *Le strutture dell'Unesco*, in, "Casabella Continuità," XXIII, April 1959, No. 226, (pp. 17-20), p. 17.

[3] *Ibid.*, p. 18.

[4] Marcel Breuer, Pier Luigi Nervi, Bernard Zehrfuss, "Explanatory notes," 2 April 1953, typewritten, Unesco, Bureau des Architectes, BNZ/7.

construction of the Unesco headquarters, his latest work embodying this sophisticated kind of concrete, the *Musée des Travaux Publics*, now represented the formal goal for exposed concrete.

Until the Unesco building site, Nervi and his firm had never been engaged in producing concrete works with very fine finishes, and Zehrfuss did not have any significant experience in this area. Nervi's only connection was that he had used hammering for the stones of the *Stadio Berta*[5] and was aware of the monumental effects of chiselling concrete as envisaged by Libera for the *Arco dell'Impero*. It is likely that *béton bouchardé* was introduced to the Unesco building site at the suggestion of Breuer, who had been engaged since 1951 in experimenting with the composition of concrete and bush hammering to achieve special sculpted effects in his works in the United States. Between the specification of bush hammering in the 1953 documents and the finishing techniques actually implemented, a long series of tests were carried out by the workers of Fourré & Rhodes et Dumez while Breuer, Nervi, and Zehrfuss reappraised the treatment of the various surfaces of the complex in order to define a new kind of *béton apparent*.

Development of the formwork

During 1954, various types of processes for exposed reinforced concrete surfaces were specified in drawings and reports drawn up with a view to awarding the work contract. Some technical guidelines had to be reconsidered in light of proposals put forward by the firm or changes in the opinion of Breuer, Nervi, and Zehrfuss. The firm's engineers were required to clarify many of the technical details for laying the concrete.

An initial detailed explanation of the structure and the working methods is given in a technical report entitled *Structure en béton armé* (Structure in reinforced concrete) drawn up partly as an outcome of a meeting with Nervi on 13 May 1954, before the contracts were awarded.[6] The opening section of the report described the role of the "structure in reinforced concrete" as not only "static"; it "also performs a very important architectural function," and the winning contractor was called on to demonstrate particular commitment, a "sincere and effective technical and executive cooperation" to achieve the goal of a "good aesthetic outcome for these structures."

Some technical details for perfectly finished concrete are given under the heading "*Béton armé brut*," where it is explained that a special reinforcement must be arranged along the consolidated pouring lines, "to ensure a connection between successive layers and prevent cracking." It is also worth remembering that in order to resolve the same technical detail, so crucial to exposed concrete, Kahn had opted to use grooves in the cylindrical wall of the Yale Art Gallery – a solution that was also to appear in successive phases of the manufacture of the Unesco concrete. The same possible use of prestressed concrete was also advised in order to "reduce the tendency of concrete to crack due to shrinkage and thermal fluctuations." Nervi's technical contribution is discernible in this kind of concern. The same vibration operations are also described in relation to the finish of *béton brut*, in order to "improve the compactness and uniformity of concrete" ("previbrate the concrete by means of light, even

[5] "The company Ingg. Nervi & Bartoli finished work on the fine bush hammering of the stairs in strong stone [*pietra forte*] […]," *Svincolo cauzione Ingg. Nervi & Bartoli*, 21 March 1934, ASCF, pack No. 7473.

[6] No author, "Structure en béton armé," undated [May 1954], Unesco, Bureau du Siège, HQ/2.

vibration"). The specification of a special secondary reinforcement to prevent surface cracking also reflected Nervi's experience and the characteristics of his ferrocement, now transferred to reinforced concrete.[7] On rereading the advice that Nervi wrote after the construction of the *Stadio Berta*, there is no doubt that his intention for the Unesco headquarters was to build those "monolithic and waterproof surfaces" that he had stated were an essential technical condition for doing without reinforced concrete cladding.

Confirming his concerns over the formal quality of concrete worked by bush hammering, Breuer, Nervi, and Zehrfuss warned that the aggregate grading would have to be calculated based on the surface finish (this cautionary note was probably struck by Breuer), while the specification that the mixture should be hand-compacted reveals other technical suggestions by Nervi: "The concrete of the *béton brut* structures [...] shall be compacted with particular care to prevent surface damage due to lack of compactness and to obtain well-defined edges. The size of the aggregate particles shall be determined by preliminary testing in order to obtain the best surface appearance and the most appropriate quality for the bush hammering."

The matter of *béton brut* cannot be addressed without considering the execution of the formwork. The report for the structure of the Unesco headquarters therefore describes how these should be constructed, with guidelines including the type of component, the board joint and woodworking, and their measurements (not greater than 6-8 centimetres). The fact that boards were chosen with a ban on using plywood or metal sheets is a significant indication of the wish to pursue the aesthetics of *béton brut* started by the *Unité d'Habitation* in Marseille. Detailed specifications such as these engineered the new turn in the manufacture of cast-in-place reinforced concrete that was to characterise Nervi & Bartoli's future work.

While the report simply indicated that the firm had to supply drawings of the formwork, following the general guidelines of *béton apparent* for the pilotis of the Secretariat, the specifications went into details for the waved walls of the *Bâtiment des Conférences*. The report clearly indicates the intention to give the concrete its own special character and Nervi's wish to avoid possible cracks between the roof and walls. Breuer, Nervi, and Zehrfuss also advocated a kind of *béton banché* with bolts replacing the usual wire connections with the aim of avoiding rust spots. They intended to achieve a formal characterisation of the waved wall surfaces according to two different methods, raw concrete for the exterior and bush hammered concrete for the interior.

Nervi must have given information about the type of surface to be achieved for the Fontenoy canopy, to be executed as a fragment of the floor with isostatic lines given that, because ferrocement was not being used, the specifications for the concrete finish state that the formwork, "shall be finished with a plaster coating because of the short profile of the ribs and the need to achieve perfectly smooth concrete surfaces."[8]

The drawing of the cross-section of the Secretariat, dated 31 May 1954, is one of the ones that best represents the preliminary studies of the concrete finish. The notes demonstrate the intention to convert the platform into a sophisticated interpretation of Le Corbusier's *béton brut* by using techniques

See also the handwritten note with the explanation "Retour de Mr. Nervi le 20-5-54."

[7] *Ibid.*
[8] *Ibid.*

and expedients able to produce the effect of perfect concrete. All the "*portiques*" (portal frames), formed by pairs of pilotis with cantilevered endings became sculptural elements that could be worked using *béton bouchardé*, in a desire to avoid the *malfaçons* visible in the Marseille pilotis. The finish of the floor intrados differed according to the sectors. Surprisingly enough, in order to achieve the goal of perfection, in the open sector of the platform, the middle strip is designed as an actual representation of *béton brut* to be achieved by means of a cladding. The fact that the wooden boards were spliced male-female to reduce the effect of bleeding, in accordance with the practices of Architectural Concrete, as Kahn also did at the Yale Art Gallery, suggests that the "*coffrage spéciale*" (special formwork) was introduced at the instigation of Breuer. The *brise soleil* along the facades were in precast reinforced concrete, but a "*surface bouchardée*" (bush-hammered surface)[9] was specified for their heads, while the floor edging beam was designed to be "*brute de décoffrage*" (left raw after stripping).

Between the end of 1954 and the beginning of 1955, Fourré & Rhodes et Dumez develop the technical details for executing the reinforced concrete structure. In the case of the Yale Art Gallery, as well as in the case of the Unesco headquarters, the contractor suggested components for the formwork that were more practical and inexpensive than the boards desired by Breuer, Nervi, and Zehrfuss – undoubtedly because they had become a reflection of *béton brut* after they had been used in the *Unité d'Habitation* in Marseille. However, it cannot be ruled out that, at least for the pilotis, the insistence of Breuer, Nervi, and Zehrfuss on boards may have been determined by a wish to emphasise the ruled surfaces generated by those structures through the subtle interplay of impressions. The terms used by Fourré & Rhodes et Dumez to explain the choice of components for the formwork are significant. "We envisaged using metal moulds for the formwork of the ground floor columns (of the Secretariat)," they explained, "but in accordance with your wishes, we could make the following formwork: wooden battens or unplaned planks, leaving the vertical generating lines exposed after stripping."[10]

The study on the construction of the waved walls and the roof of *Bâtiment des Conférences* marks an important chapter in the development of the construction details, partly due to the matter of the cracks. In this case too, Fourré & Rhodes et Dumez adapted to the guidelines given by Breuer, Nervi, and Zehrfuss, amending their proposals. The explanations reveal the nature of the discussions between the firm and the designers, and the taste of Breuer, Nervi, and Zehrfuss for certain *béton brut* type effects in the style of Le Corbusier.

"For the formwork of these vertical walls, because of your wish to obtain perfect sharp edges, we wish to change the formwork system shown in our drawings. The joints between the formwork panels must never be located on an edge, but rather on a flat surface, and the formwork for the edges must be made out of a piece of wood at the required angle without any sheet metal parts. The panels must be made out of unplaned plank with horizontal slats

[9] Drawing No. A26 "Secretariat. Detail des façades, Suffren Segur," 31 May 1954, SCRC, T1086_006.
[10] Groupement des Sociétés, E.ts Fourré & Rhodes, Société Dumez, letter to Bureau du Siège, 22 January 1955, Unesco, Bureau du Siège, HQ/8.
[11] *Ibid.*
[12] See for example, Bernard Zehrfuss, letter to Pier Luigi Nervi, 22 May 1954, Unesco, Bureau des Architectes, BNZ/4.
[13] Pier Luigi Nervi, letter to Eugène H. Callison, 1 April 1955, Unesco,

to produce a slight architectural flashing effect when the joint of each lath is stripped away."[11]

Nervi was responsible for checking the reinforced concrete studies drawn up by the Fourré & Rhodes et Dumez technicians, from the structural calculations to the drawing showing the arrangement of the reinforcements, sometimes also including drawings of the formwork.[12] He was the main person that the firm's technicians turned to during the building stages for any decision over the composition, casting, and treatment of concrete. Nervi controlled the calculations and drawings produced by the firm for the reinforced concrete structure,[13] recommended studying various compositions of aggregate grading for the concrete mixtures, "both from the viewpoint of the appearance of the finished concrete as well as of its mechanical strength,"[14] gave instructions about the nature of the cement (the one that ensures the least shrinkage and is as light as possible),[15] and was called on to examine the results of the various tests. No change in the technical details of the structure proposed by the firm could deviate from Nervi's instructions, as those responsible were brusquely reminded by Eugène H. Callison, chief engineer of the *Bureau du Siège* and responsible for monitoring the progress of work on behalf of Unesco: "We wish to inform you categorically that as far as Unesco is concerned, Mr Nervi's competence cannot be questioned [...]."[16]

Tests on the building site

The start of work was preceded by tests on the formwork and the processing of the concrete after stripping. In May 1955, Fourré & Rhodes et Dumez constructed the prototype of a pilotis from the Secretariat, a waved wall segment and a pillar segment from the *Bâtiment des Conférences* that they used to carry out various bush hammering, sand-blasting, and painting tests. The firm examined a system of *banchage* using *fourreaux* in cemented asbestos to take into account the ban on using wire imposed by Breuer, Nervi, and Zehrfuss. The waved wall concrete was laid in three different ways to check the effects on the surface shape (vibrated, semi-vibrated, and not vibrated).

The results of the sample inspection were summarised in an official document of 26 May, "*Essais de béton armé*," drawn up by Pierre Marcel on behalf of the *Bureau des Architectes* (Breuer, Nervi, and Zehrfuss). The test hinged around the formwork system, the colour and composition of the concrete aggregates, and the finish. The verdict on the Secretariat and the *Bâtiment des Conférences* pillars was good. The view of the regular formwork board impressions must have impressed the examiners because, in order to define the type of finish of the Secretariat pillars – which Breuer, Nervi, and Zehrfuss had determined was to be *bouchardé* – tests were requested that had to be carried out by means of "sand-blasting" and with "paint or liquids fixing the raw concrete surface,"[17] with the aim of preserving traces of the manufacturing process. For the concrete on the *Bâtiment des Conférences* pillar segment, which also underwent bush hammering, a recommendation was given to reduce the gravel size on the advice of Nervi. Execution of the waved

Bureau du Siège, HQ/2.
[14] Eugène H. Callison, letter to Groupement des Sociétés, E.ts Fourré & Rhodes, Société Dumez, 26 April 1955, Unesco, Bureau du Siège, HQ/8.
[15] See, for example, Pierre Marcel, letter to Bureau du Siège, 26 April 1955, Unesco, Bureau du Siège, HQ/2.
[16] Callison, letter to Groupement des Sociétés, E.ts Fourré & Rhodes, Société Dumez, 26 April 1955, cit.
[17] Pierre Marcel, letter to Bureau du Siège, 26 May 1955, Unesco, Bureau du Siège, HQ/2.

310-315. M. Breuer, P. L. Nervi, and B. H. Zehrfuss, Unesco Headquarters, Place de Fontenoy 7, Paris, 1951-58.

310. Pouring tests for waved walls of the *Bâtiment des Conférences*, May 1955 (Unesco).

311. Bush hammering test, May 1955 (Unesco).

312. Sample pouring, May 1955 (Unesco).

313. Sample pouring, May 1955 (Unesco).

314. Prototype of Secretariat pilotis, May 1955 (Unesco).

315. Prototype of Secretariat pilotis, May 1955. Bush hammering test (Unesco).

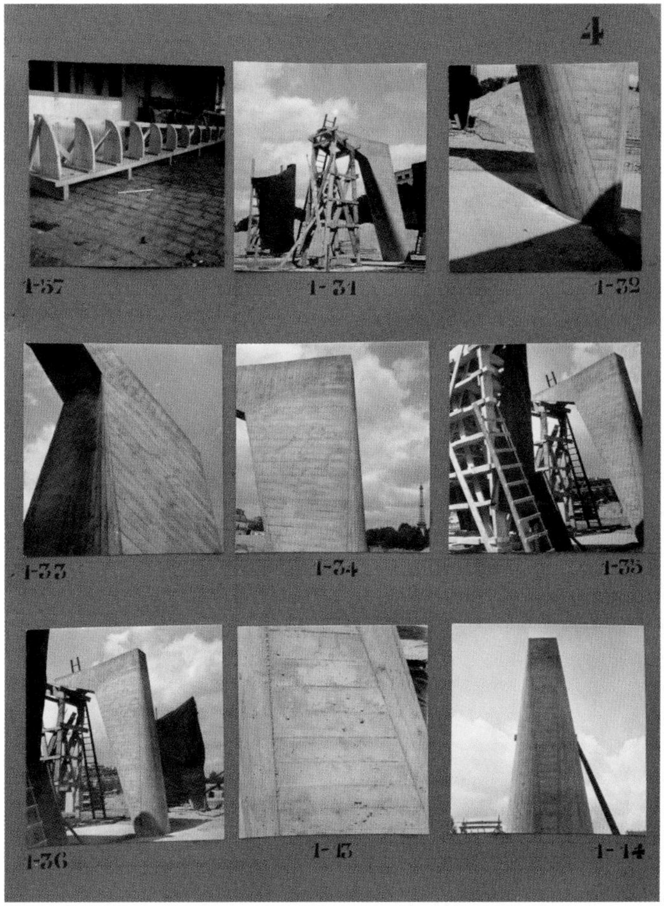

316-319. M. Breuer, P. L. Nervi, and B. H. Zehrfuss, Unesco Headquarters, Place de Fontenoy 7, Paris, 1951-58.

316. Building sites tests, July 1955 (Unesco).

317. Constructing the first portal frame of the Secretariat. Photograph of pouring defects, dated 18 August 1955 (Unesco).

318-319. Constructing the first portal frame of the Secretariat. Photographs of pouring defects, dated 18 August 1955 (Unesco).

wall sample, however, raised some concerns. "We believe that the concrete, which is made out of washed sand, is not fine enough."[18] Some defects were attributed to "compacting" or "*vibration*"; the angled wall sectors were also judged to be unsatisfactory: "I would also ask you to review the horizontal construction joints and the arrangement of the formwork laths on the slanting surfaces (horizontal joints)."[19]

Breuer seemed to have played a crucial role when questions arose concerning details of the formwork component quality due to the surface finish. He undertook to order from the United States a type of cone used for formwork, in view of the unsatisfactory results achieved with the exposed cement asbestos *fourreaux*.[20] In the meantime, Fourré & Rhodes et Dumez encountered the first difficulties in pouring the exposed concrete of the underground structures, some of which had to be demolished, while they worked on the *béton brut* surfaces with concrete castings disputed by the *Bureau des Architectes*.

Alongside the waved wall test, a small, angled wall segment was built to check the composition of the concrete mixture. The other tests to define the quality of the concrete concerned the type of surface processing following the stripping. On the waved wall sample, first the form spacer holes were filled and then the surface was bush hammered with sufficient intensity to make the imperfections, concrete castings, and formwork impressions less evident, yet without achieving significant effects of roughness. Sand-blasting was applied to a sector of the Secretariat pillar and also to the waved wall.

The different levels of processing revealed an intention to leave the formwork board impressions visible under all circumstances. Only rarely were the hammering tests pushed to the point of rendering the stage of pouring the material in the formwork unrecognisable. In general, it the bush hammering was eventually interpreted as a solution for mitigating the pouring defects and smoothing the surface texture, in other words to correct the *malfaçons* of the *béton brut*. It was no longer used to give the concrete the appearance of artificial stone, as was envisaged at the design stage and as happened on Breuer's US building sites. This interpretation of bush hammering also explains the decision, which came about during the development of the building site, to sand-blast some surfaces instead. During the test stage, Nervi spent some days in Paris between the end of June and 4 July,[21] when decisions had to be made about the concrete, and his opinion was decisive.[22]

When Nervi was in Paris, Fourré & Rhodes et Dumez had just carried out the "sand-blasting of an area" and "painting of the stripping raw surface" on the Secretariat pilotis sample. They had built two "panels of about 1 square metre, using metal cones fitted at both ends of a spacer sleeve made out of fibre cement, clamped in all cases by means of a threaded rod."[23] The bush hammering tests on the waved wall surfaces were also ready.

[18] *Ibid.*
[19] Eugène H. Callison, letter to Groupement des Sociétés, E.ts Fourré & Rhodes, Société Dumez, 2 June 1955, Unesco, Bureau du Siège, HQ/8.
[20] Marcel, letter to Bureau du Siège, 26 May 1955, cit.
[21] Bernard Zehrfuss, letter to Marcel Breuer, 6 July 1955, Unesco, Bureau des Architectes, BNZ/4.
[22] Bernard Zehrfuss, letter to Pier Luigi Nervi, 21 June 1955, *ibid.*
[23] Groupement des Sociétés, E.ts Fourré & Rhodes, Société Dumez, letter to Bureau du Siège, 27 June 1955, Unesco, Bureau du Siège, HQ/8.

After examining the samples and carrying out discussions with the various technicians, on 4 July Nervi drafted a letter on behalf of himself, Breuer, and Zehrfuss, addressed to the *Bureau du Siège*, on the subject of the "*Echantillons de Béton armé*" (Reinforced Concrete Samples). His remarks do not contain any reference to bush hammering, but they did reveal a preference for sand-blasting the surfaces. His various tips on the composition of the concrete mixture came about following observations that he had made after examining the samples in May. His advice to use "*pilonnage*" instead of vibration to remove the honeycomb patterns reflects his experiences accrued on the Nervi & Bartoli building sites. A passage from the letter confirms the emergence of the decision to seal the holes in the form spacers without applying the concrete mixture flush with the wall, thereby leaving their presence exposed, given that it was impossible to cancel out the concrete castings. Nervi's letter is the first significant document demonstrating the criteria he followed when appreciating the qualities of the concrete surfaces:

"*Finish quality*. At the moment it is difficult to establish the finish quality of surfces that are required to be left 'brut de décroffage'. We nevertheless believe that these surfaces will generally be sandblasted, but this treatment (one or two applications) will depend greatly on the quality of concrete obtained. On this subject, we ask you again to notify the firm of the need to achieve a better result by improving the quality of the concrete, which must be more even with more regular surfaces. We believe that the concrete can be improved by increasing the percentage of gravel and incorporating plasticising agents. *Concrete appearance*. We have already given you our opinions on the matter of gravel uniformity. We ask you whether it is possible, for exposed concretes, to use the cement that was used beforehand to pour the 'Conference Centre' test pillar. This cement seems to give a better appearance. It must be understood that the cement is to fulfill all the required guarantees. *Cracking*. We have observed that the concrete shows a particular tendency to crack. Samples must therefore be taken of this cement and shrinkage tests carried out. Similarly, the company must not forget to water the concrete for about 10 days after pouring and safeguard against drying out too quickly. *Conference Centre wall* (*voile*). We have carried out a test to seal the formwork fastening tie holes that clearly shows the position. We believe that this solution is acceptable, but it must be finalised with the firm and receive the final go-ahead from the architects. To summarise, surface faults always depend on the quality of the concrete and can only be removed by improving the concrete. With regard to the numerous air bubbles that appear at the time of stripping, we believe that these can only be removed by replacing vibration with properly executed compacting, carried out by hand."[24]

When the formwork and the reinforcement of the first pilotis of the Secretariat were in place, by mid-July, and the firm was about to pour the

[24] Pier Luigi Nervi, letter to Bureau du Siège, 4 July 1955, Unesco, Bureau du Siège, HQ/2.
[25] Groupement des Sociétés, E.ts Fourré & Rhodes, Société Dumez, letter to Bureau du Siège, 15 July 1955, Unesco, Bureau du Siège, HQ/8.
[26] Jean-Louis Lalonde, letter to Marcel Breuer, 22 July 1955, Unesco, Bureau des Architectes, BNZ/4.
[27] *Ibid.*
[28] Eugène H. Callison, letter to Groupement des Sociétés, E.ts Fourré & Rhodes, Société Dumez, 30 July 1955, Unesco, Bureau du Siège, HQ/8.
[29] See typewritten note, dated 4 August [1955], Unesco, Bureau du Siège, HQ/2.
[30] [Groupement des Sociétés, E.ts Fourré & Rhodes, Société Dumez], letter to Bureau du Siège, 6 August 1955, Unesco, Bureau du Siège, HQ/8.

concrete,[25] Breuer suggested some crucial technical solutions for the waved wall surfaces of the *Bâtiment des Conférences*. He suggested introducing the grooves that would divide the walls up into deep lines of shade. On 13 July, he told Nervi about his idea of applying a "wood strip at top line of pouring" to the formwork.[26] This solution became part of the technical vocabulary of future works by Nervi & Bartoli. To test out Breuer's solution, it was decided to build a test wall segment, which was ready by 23 July, with this type of strip and also using a "new type of tie," which was cheaper and had been suggested by the contractor, "using asbestos tubes and steel cones on the outer surface, which give a similar effect to what we would get from standard tie rods."[27]

Before undertaking construction of the waved wall, the firm was asked to carry out other tests using "*coffrage spécial*" (special formwork) and without mechanical vibration but "*en pilonnant le béton*" (compacting the concrete).[28] The need to obtain Nervi's opinion, under all circumstances, on all technical solutions planned by the firm for construction of the waved walls was emphasised.[29] During an inspection visit to the building site at the beginning of August, Nervi himself, by agreement with the firm, advised carrying out a simple "*voile plan*" as a new test, using the same inclination as that of the waved walls of the *Bâtiment des Conférences*.[30] It was then specified that a "*coffrage spécial*" should be used for the external face, the pouring should be laid in horizontal layers, and the concrete should be compacted manually before vibrating it gently.[31] Nervi again made known his opinion of the kind of concrete that should be used for the "*voiles des Conférences*." He was now quite familiar with the fashionable terminology that had arisen around Le Corbusier's *béton brut*.

"[…] we believe that the most important matter is to obtain the best possible quality of *béton brut* for all parts that are bush hammered or left rough after stripping. For this reason, more tests must be carried out by increasing the quantity of gravel still further. […] the most effective precaution to prevent cracking is plenty of well-arranged reinforcements, preferably made out of hard, notched steel (Tor or something similar). The other general precautions are as follows: use of a good cement supply; good concrete quality; pouring and compacting of the concrete carried out by hand with light vibration; protection of stripped concrete for about 15 days with wetted coverings; arrangement of numerous small iron bars in the construction joint surfaces; as far as possible, the walls (*voiles*) should be poured at a wet time of year (October, November)."[32]

After advising the introduction of grooves, Breuer intervened yet again to confer a specific formal character on the waved walls by distributing the cones and casting strips to make up a pattern, even though the *Bureau des Architectes* considered that "these changes are determined *a priori* for reasons that are far more technical than aesthetic."[33]

[31] Jacques Vitrier, letter to Groupement des Sociétés, E.ts Fourré & Rhodes, Société Dumez, 16 August 1955, *ibid.*
[32] Pier Luigi Nervi, letter to Bureau du Siège, 5 August 1955, Unesco, Bureau du Siège, HQ/2.
[33] Pierre Marcel, letter to Marcel Breuer, 19 August 1955, Unesco, Bureau des Architectes, BNZ/4.

The pilotis of the Secretariat

The first two pilotis of the Secretariat, identified by the code A5-B5, were ready by mid-August 1955. The stripping uncovered a series of severe defects documented by photographs that revealed the resulting large honeycomb patterns, particularly near the bases. The other two adjacent pairs of pilotis, A6-B6 and A7-B7, which were stripped on 23 August, were found, on the other hand, to be "perfectly executed."[34] This success had been achieved as a result of Nervi's advice: "the concrete has been poured with much greater care. As you advised, now we will compact it by hand before the vibration."[35] To rectify the defects in the first pair of pilotis, the firm carried out "finishing tests"[36] – one of the pilotis was ultimately fully sandblasted at the instigation of the *Bureau du Siège*.[37] A decision was nevertheless made to demolish the pair (actually, only the sandblasted pilotis was demolished). Nervi asked the *Bureau du Siège* to delay the demolition until 6 September, the date of his arrival in Paris, because he wished to examine personally the pair found to be defective and compare the concrete that had been executed to perfection in the other two pairs so that he could understand the technical reasons for such different outcomes.[38]

When all the pairs of pilotis of the wing facing the Avenue de Ségur had been completed at the beginning of September, Breuer, Nervi, and Zehrfuss told the firm that they wished to "replace the planned bush hammering by light sandblasting."[39] Sandblasting was carried out on the test pilotis in order to mitigate the effect of the bush hammering required in the specifications and drawings and to expose the impressions of the regular and controlled assembly of the formwork boards. At the time of the sandblasting tests, Breuer, Nervi, and Zehrfuss had, however, postponed their final choice to an examination of the executed work. Their decision of early September had a direct impact on the quality of the concrete: although the first pair of pilotis had been scarred by a variety of honeycomb patterns, the impressions and surfaces were now perfect without significant *malfaçons*. The changeover from bush hammering to light sandblasting was merely another step toward the final decision, which was to do nothing at all to the surface of the pilotis and thereby achieve a form of *béton brut* that could be considered perfect when compared to that of the pilotis of the *Unité d'Habitation* in Marseille. Bush hammering was also stopped in other parts of the Secretariat, which were nevertheless of minor importance.

By mid-September, the decision to give up the idea of light sandblasting seems to have found favour in order to leave the concrete in its *brut* condition, in view of "the excellence of the results obtained."[40] The firm pointed out that, in any case, operations such as sandblasting could not be undertaken when the pouring was already completed, unless it was decided not to give up on the concrete pouring devices, which could be differentiated on the basis of the artistic effects sought: "These devices vary according to the purpose established. It could be to used achieve a uniformly coloured fine-grained film or alternatively to emphasise the gravel in the concrete to obtain a cladding

[34] Pierre Marcel, letter to Pier Luigi Nervi, 23 August 1955, *ibid.*
[35] *Ibid.*
[36] *Ibid.*
[37] Groupement des Sociétés, E.ts Fourré & Rhodes, Société Dumez, letter to Bureau du Siège, 26 September 1955, Unesco, Bureau du Siège, HQ/8.
[38] Jacques Vitrier, letter to Pier Luigi Nervi, 29 August 1955, Unesco, Bureau du Siège, HQ/2.
[39] This expression is set out in several documents; see for example, Pierre Marcel, letter to Bureau du Siège, 1 September 1955, *ibid.*

with different colours and particle sizes. In most cases, the mortar or concrete used near the external facing is different from that used for the skeleton itself. When the concrete is cast, we pour both categories of concrete from either side of a sheet of metal that gradually rises as the formwork is filled."[41] This type of device had never been described in any Nervi & Bartoli construction project documents concerning concrete treatment. In this regard, what Nervi read in the various communications from Fourré & Rhodes et Dumez and saw at the Unesco building site represented for him a kind of initiation to the techniques for achieving a perfect surface for cast-in-place concrete.

Sandblasting was at the root of the initiative to demolish the most defective of the pilotis in the first pair of the Secretariat, which was the one coded A5. Dismantling the pilotis began on 14 September, after Nervi had carried out an inspection. Nervi saw the pilotis after the *Bureau du Siège* had given the order to sandblast it. "We believe that these circumstances influenced Mr Nervi's opinion," the firm noted.[42] Following the final decision to leave the raw concrete, an order was issued to protect the Secretariat pilotis "using boards, kraft paper, or other methods effective against knocks, stains, graffiti etc."[43]

The waved walls of the *Bâtiment des Conférences*

In September 1955, work began on assembling the great scaffolding of metal pipes used to set up the timber board formwork of the *Bâtiment des Conférences* undulations, while the first pillars with a ruled surface forming the load-bearing backbone along the foyer were constructed. For the construction of these pillars, Fourré & Rhodes et Dumez developed a formwork system consisting of four removable sectors that were reused to pour the entire row.[44] Nervi had already tried out a similar device to build the pillars of the *Fabbricato Ballette* in Bologna, but he only began to create variable cross-section pillars to be poured in large, removable building site machines such as the one he used in Paris after he used the system in the *Bâtiment des Conférences*.

In mid-September, in view of the work of pouring the walls and roof that constituted the most complex parts of the *Bâtiment des Conférences*, the firm was again asked to produce a sample of "wall (*voile*) consisting of 6 panels with different particle sizes."[45] At this stage of setting out the technical specifications, the details of the formwork were provided. The "special formwork" was considered to be an ideal type of joint that could also be used for the boards of the inner panels of the "voiles," because the "ordinary formwork" gave rise to a "seal [that] does not meet the rules of good workmanship," and the "leakage of milk of lime and fine aggregate are liable to compromise the quality of the whole."[46] Nervi's orders about the importance of "compacting" the concrete with "special care to avoid surface defects" were also reiterated. "In our opinion, this operation must be supplemented, not replaced, by vibration."[47]

The six "samples of wall (*voile*) elements" with the different dosages were examined on approximately 28 September. The different dosages were tried out with the aim of identifying the best formal effect for a surface to be worked using a bush hammer. Callison informed the firm that for the "bush

[40] Groupement des Sociétés, E.ts Fourré & Rhodes, Société Dumez, letter to Bureau du Siège, 26 September 1955, Unesco, Bureau du Siège, HQ/8.
[41] *Ibid.*
[42] *Ibid.*
[43] Eugène H. Callison, letter to Groupement des Sociétés, E.ts Fourré & Rhodes, Société Dumez, 22 September 1955, *ibd.*
[44] Nervi, *Le strutture dell'UNESCO* cit., p. 19.
[45] Eugène H. Callison, letter to Groupement des Sociétés, E.ts Fourré & Rhodes, Société Dumez, 13 September 1955, Unesco, Bureau du Siège, HQ/8.
[46] *Ibid.*
[47] *Ibid.*

320-323. M. Breuer, P. L. Nervi, and B. H. Zehrfuss, Unesco Headquarters, Place de Fontenoy 7, Paris, 1951-58.

320. Assembling formwork for waved walls of the *Bâtiment des Conférences*, September 1955 (Unesco).

321. Assembling formwork for roof of the *Bâtiment des Conférences* (Unesco).

322. View of building site area, 1956 (Unesco).

323. *Bâtiment des Conférences* roof under construction, 1956 (Unesco).

324-326. M. Breuer, P. L. Nervi, and B. H. Zehrfuss, Unesco Headquarters, Place de Fontenoy 7, Paris, 1951-58.

324. View of *Bâtiment des Conférences* (Unesco).

325. Detail of bush hammering tests carried out on waved walls of the *Bâtiment des Conférences* (photograph by R. Gargiani).

326. View of *Bâtiment des Conférences* (Unesco).

hammered exposed reinforced concrete," the element coded "d-4" had been chosen, which became the "standard dosage," consisting of sand and aggregates recovered from the foundation excavations.[48]

Some decisions that had already been taken at the beginning of September were finally ratified, following a meeting held on 18 September, and are summarised in a letter of 20 October 1955 (Breuer was present, but not Nervi). This document established that for the pilotis of the Secretariat and the pillars of the *Bâtiment des Conférences*, there would be a "*fini 'brut de décroffrage'*," while any decisions for the other parts were postponed.[49] During the inspection toward the end of October, Nervi met the Fourré & Rhodes et Dumez technicians in order to establish the methods for executing the inclined walls and the roof of the *Bâtiment des Conférences*, above all with the aim of avoiding cracks. The technicians again consulted Nervi to get his opinion on the results of the concrete cracking tests.

When Nervi visited Paris on approximately 17–18 November, he discussed the method for constructing the Piazza canopy with the contractor and Callison, and he also gave his final agreement for the construction method

[48] Eugène H. Callison, letter to Groupement des Sociétés, E.ts Fourré & Rhodes, Société Dumez, 28 September 1955, *ibid.*
[49] Pierre Marcel, letter to Bureau du Siège, 20 October 1955, Unesco, Bureau des Architectes, BNZ/7.
[50] Pier Luigi Nervi, letter to Bureau du Siège, 17 November 1955, Unesco, Bureau du Siège, HQ/2.
[51] Eugène H. Callison, letter to Groupement des Sociétés, E.ts Fourré & Rhodes, Société Dumez, 25 November 1955, Unesco, Bureau du Siège, HQ/9.
[52] Eugène H. Callison, letter to Groupement des Sociétés, E.ts Fourré & Rhodes, Société Dumez, 12 January 1956, *ibid.*
[53] Eugène H. Callison, letter

used to produce the undulations of the *Bâtiment des Conférences*. He made two clarifications that took into consideration the proposals put forward by Breuer over a method for controlling the concrete mixture pouring stages by means of deep grooves: "1) Add a Guttaterna border in the middle of the construction joint grooves, as was discussed when we spoke to the contractor. 2) Construction joints. In general, for all construction joints, the firm must take care to leave plenty of small iron bars."[50] Again for the construction of the *Bâtiment des Conférences*, it was also decided to stop obtaining aggregates from the building site and instead use the sand and gravel provided by the company Parthenay, at Viry Châtillon, in Seine et Oise in order to achieve "control of concrete quality from a structural and aesthetic viewpoint."[51]

In January 1956, work continued on the wing of the Secretariat facing Avenue Lowendal, with the preparation of "*coffrages spéciaux*" for the pilotis,[52] using measures to protect the concrete against frost (in addition to the usual methods, it was also proposed to use "infrared radiation heaters"[53]) when work started on pouring the walls of the *Bâtiment des Conférences* and the protection, in February, of the pilotis of the Secretariat and of the *Bâtiment des Conférences* using timber that was "not suitable for formwork and not reusable."[54] Despite the various inspections, tests carried out and documents exchanged, Fourré & Rhodes et Dumez, the *Bureau du Siège*, and the architects misunderstood one another about the finish of certain fundamental parts in exposed concrete, which is borne out by the fact that even in February 1956, the firm still believed that the pilotis of the Secretariat and of the *Bâtiment des Conférences* had to be bush hammered.[55]

After halting work due to frost on 31 January, when work resumed on 29 February, the first *voiles* of the *Bâtiment des Conférences* had not yet been stripped.[56] During Nervi's latest inspection in March to attend the stripping of the first *voiles* of the *Bâtiment des Conférences*, various matters were addressed, ranging from prefabrication of the brise soleil to development of the Piazza canopy.[57] Stripping of the first *voiles* revealed "serious defects," includ-

to Groupement des Sociétés, E.ts Fourré & Rhodes, Société Dumez, 24 January 1956, *ibid.*
[54] Jacques Vitrier, letter to Groupement des Sociétés, E.ts Fourré & Rhodes, Société Dumez, 3 February 1956, *ibid.*
[55] Groupement des Sociétés, E.ts Fourré & Rhodes, Société Dumez, letter to Bureau du Siège, 9 February 1956, *ibid.*
[56] Pierre Marcel, letter to Pier Luigi Nervi, 5 March 1956, Unesco, Bureau des Architectes, BNZ/4.
[57] Groupement des Sociétés, E.ts Fourré & Rhodes, Société Dumez, letter to Bureau du Siège, 17 March 1956, Unesco, Bureau du Siège, HQ/9.

327-328. M. Breuer, P. L. Nervi, and B. H. Zehrfuss, Unesco Headquarters, Place de Fontenoy 7, Paris, 1951-58.

327. Pillars of *Bâtiment des Conférences* (BSTA, 024-7212).

328. View of the pilotis building site and Secretariat platform (BSTA, 024-7208).

ing chipped corners, the presence of honeycomb patterns at each concrete casting, and cracks near the roof.

Breuer, Nervi, and Zehrfuss were also engaged in identifying a hydrophobic product to be spread on the surface of the concrete in order to protect it against water ingress. Suggestions included a type of silicone that had been submitted for examination by Nervi,[58] with "opaque and clear paint"[59] and US products suggested by Breuer.[60] Nervi provided his own recipe for a special compound, which he claimed to have tried several times: "I am sure that we can make it absolutely waterproof without a protective layer through good workmanship, but it would be better to meet and talk about this matter as well as the matter of the gutters on the upper edges. I am also sending you a formula for a waterproofing solution for concrete surfaces that I have used many times with good results and may be useful for the walls of the conference centre building. The solution looks like pure water and does not change the colour of the concrete at all. I think it would be a good idea to carry out tests."[61]

Even though Breuer, Nervi, and Zehrfuss had had several discussions about the matter of bush hammering the concrete, at the beginning of 1957, the extent of this surface treatment was still the subject of discussions and even misunderstandings. In January, the workers began to carry out the bush

[58] Pierre Marcel, letter to Pier Luigi Nervi, 13 July 1956, Unesco, Bureau des Architectes, BNZ/4.

[59] Bernard Zehrfuss, letter to Marcel Breuer, 13 September 1956, *ibid*.

[60] Marcel Breuer, letter to Bernard Zehrfuss, 18 September 1956, *ibid*.

[61] Pier Luigi Nervi, to Bernard Zehrfuss, 4 October 1956,

hammering of some surfaces of the *Bâtiment des Conférences* and attempted to cancel out every trace of the construction process, including the horizontal impressions left by the boards and the holes left by the form spacers. Not all of the surfaces showed a uniform standard of working. In some parts, small cavities were made, sometimes arranged in accordance with an approximate pattern, to create powerful *chiaroscuro* motifs, almost a form of rough rusticated ashlar work, in order to camouflage the presence of the form spacer holes. These attempts at working the concrete contravened the guidelines that had emerged from the complicated process of establishing the nature of the Unesco concrete. These underscored the importance of making all the concrete laying devices clearly visible, including the holes that Breuer also wished to have laid out in accordance with aesthetic principles. Callison therefore decided to give an order to stop the bush hammering.[62] Even when the building site entered the stage where clarifications were required about the finish, it was still not clear to everyone which surfaces had to be left *brut* and which had to be treated by bush hammering.

At this crucial stage of deciding the final finish of the surfaces, the opinion of Breuer, probably the main proponent of concrete bush hammering, may be inferred from the words of Zehrfuss: "I think you ease off on the bush hammering because the concrete is so well-executed." Zehrfuss believed that it

Unesco, Bureau des Architectes, BNZ/10.

[62] Pierre Marcel, letter to Bureau du Siège, 25 January 1957, Unesco, Bureau du Siège, HQ/9.

was possible to "keep the bush hammering at some points of the foyer, but we can do without in other places, for example the *voiles* of the two large halls."[63]

The halt to the bush hammering decided by Callison must have been carried out with the blessing of the *Bureau des Architectes*, because other finishing techniques were now considered for the treatment of the waved walls of the *Bâtiment des Conférences*. In February, the firm was asked to carry out "tests by carrying out very careful brushing of part of the *Salle des Commissions voiles*" along with "bush hammering tests that have also been carried out."[64] The request by Breuer, Nervi, and Zehrfuss specified a need to use a "hard brush (it can be metal)" to work on the surfaces, following the horizontal direction of the board impressions, to implement the "removal and pumicing of burrs, in the same direction" and then to carry out "leaching using a detergent, solvent product," and lastly to perform a "finish brushing with a soft brush." It was planned to carry out these new tests on the "side of the walkway on the Suffren side and on the nearest central pilotis."[65]

The document that Breuer, Nervi, and Zehrfuss considered to have the last word on all the decisions taken over the "concrete finishing" was a report dated 11 April 1957, drawn up by Zehrfuss.[66] It was now clear that the work in progress at the Secretariat and the *Bâtiment des Conférences* guaranteed a standard of surface finish for some significant parts that made the various treatment solutions following stripping ultimately redundant, whether this involved bush hammering as instructed in the first studies or sandblasting as tested on special prototypes produced on the building site. Traces of these solutions, which were envisaged for fear of the practical results once the concrete had been stripped, were left only in certain parts that were still supposed to be treated after stripping. They were now destined to stand as relics of the original sculptural concrete vision. The ups and downs of the building site reflected uncertainties over gauging the exact level of surface working after stripping. Breuer, Nervi, and Zehrfuss could not decide between a process that gave the surfaces a sculptural value and the desire to leave the impressions of the construction process intact; this was because the firm was ultimately responsible for monitoring the technique.

The interminable series of decisions had not yet come to an end, and the work bore the signs of the continuous second thoughts. Breuer wished to limit the bush hammering of the lift block in the Secretariat to the ground floor and the seventh floor; on the other floors, the walls were to be "left raw as they were after stripping and painted." In the *Bâtiment des Conférences*, the face of the "*voiles latéraux*" looking over the great hall were no longer to be bush hammered but left "raw as they were after stripping." Lastly, probably following the results of the washing test requested on 11 April, it was decided that the foyer pillars had to be bush hammered, even though only light working was advised: "This bush hammering must be as light as possible, especially on the edges of these pillars."[67] The intensity with which the pillars were to be bush hammered was further described to the workers by Callison

[63] Bernard Zehrfuss, letter to Marcel Breuer, 28 January 1957, AAA-SI, Box 11, Reel 5722, Frame 1198-99.

[64] Pierre Marcel, letter to Bureau du Siège, 28 February 1957, AAA-SI, Box 11, Reel 5722, Frame 1278.

[65] Eugène H. Callison, letter to Groupement des Sociétés, E.ts Fourré & Rhodes, Société Dumez, 4 March 1957, Unesco, Bureau du Siège, HQ/49.

[66] Bernard Zehrfuss, letter to Bureau du Siège, 11 April 1957, *ibid.*

[67] Pierre Marcel, letter to Bureau du Siège, 1 July 1957, Unesco, Réunions du Comité du Siège, 5/HQ.

[68] Eugène H. Callison, letter

as he interpreted Breuer's guidelines: "The bush hammering must get rid of the existing dirt (rust, spray, graffiti, etc.) but must allow the formwork joints to pass through and not round off or chip the edges."[68]

Following the tests, Breuer, Nervi, and Zehrfuss decided to replace light bush hammering with this type of working in many parts of the complex. Sandblasting was adopted for the exposed concrete pillars on the seventh floor of the Secretariat and in various parts of the *Bâtiment des Conférences*, with the exception of the *Salle des Commissions* where the *voile* had already been bush hammered and the *plafond* was already polished.

Nature of form of the canopies

In addition to the portal frame pilotis, the two canopies in the Secretariat are also Nervi's personal contribution to the affirmation of exposed reinforced concrete structure. The first to be constructed was the Fontenoy canopy, for which Nervi had recommended, from the design stage, the use of a special formwork to obtain a perfectly smooth concrete surface. Only the pillars showed the marks of the thin, perfectly made formwork boards. "The Fontenoy canopy has been stripped and is absolutely manificent," Zehrfuss wrote to Nervi on 8 February 1957.[69] Nervi had wanted that polished surface, the only one of its kind in the repertoire of "*béton apparent*" solutions in the Unesco complex, as a reminder of the surfaces he usually obtained with the ferrocement moulds and formwork that shaped the isostatic lines of his floors. By the end of the work, he was convinced that the beauty of concrete could also reside in the exposed impressions left by the wooden boards and not merely in the perfection of smoothness. When seen in comparison with the other concrete surfaces, the surface that Nervi wished to leave smooth almost looks like an unreal substance, a literal transfiguration of *béton brut*. It is symptomatic of the effect of the estrangement from the material caused by the working process that Marcel cautioned Nervi that a photograph of the canopy, which he had sent him attached to a letter, "does not do proper justice to the concrete material, which is so different from that of other works."[70] This remark by Marcel is one of the rare examples of how Nervi's contemporaries may have perceived the quest for smoothness relentlessly pursued by Nervi with his ferrocement and helps to understand the ultimate meaning hidden in that almost metallic product, very far from the weave of impressions so inherent to the *béton brut* of the 1950s.

Clarification of the structural details of the other canopy, known as the "Piazza," continued between the end of 1956 and the beginning of 1957, again through cooperation between Nervi and Fourré & Rhodes et Dumez.[71] In mid-November 1957, the work was stripped – the structure of the foundations was poured in April, the formwork of the hyperbolic paraboloids was installed in August, and the ironwork was assembled at the end of September.[72] "The concrete and formwork have turned out well, and everyone admires the finished product," was Marcel's comment to Nervi.[73] In this case, Nervi

to Groupement des Sociétés, E.ts Fourré & Rhodes, Société Dumez, 25 July 1957, Unesco, Bureau du Siège, HQ/49.

[69] Bernard Zehrfuss, letter to Pier Luigi Nervi, 8 February 1957, Unesco, Bureau des Architectes, BNZ/4.

[70] Pierre Marcel, letter to Pier Luigi Nervi, 22 February 1957, *ibid*.

[71] Pierre Marcel, letter to Bureau du Siège, 17 December 1956, Unesco, Bureau du Siège, HQ/9.

[72] Pierre Marcel, letter to Marcel Breuer, 26 April 1957, Unesco, Bureau des Architectes, BNZ/4.

[73] Pierre Marcel, letter to Pier Luigi Nervi, 19 November 1957, *ibid*.

329-330. M. Breuer, P. L. Nervi, and B. H. Zehrfuss, Unesco Headquarters, Place de Fontenoy 7, Paris, 1951-58.

329. "Fontenoy" canopy (Unesco).

330. "Piazza" canopy (Unesco).

designed all the lines of the boards that rendered the geometrical and static construction of the surfaces visible through their impressions. The different mechanical flows of the forces seem almost to be reflected in the different working methods used for the concrete surfaces of the two canopies.

Fourré & Rhodes et Dumez designed a different method of protection for the Piazza canopy to the one suggested by Nervi because they believed it would be subject to cracking in the Parisian climate. Nervi proposed applying a bituminous layer onto which a cement coating would be laid to restore a "pale grey appearance" to the work.[74] When Marcel sent Nervi the photographs taken by Lucien Hervé, the first confirmed to Nervi that the Piazza canopy "is always a great success."[75] But he also warned him about the effects of the photographs: "Hervé's shots are a little aggressive, but the reality is less harsh than the photos suggest."[76]

The canopy lends itself to "aggressive" shots due to the asymmetric clarity of its vault webbing, which Nervi explained in terms that are valuable to an understanding of some subtleties of his idea of reinforced concrete structure. That asymmetry gives the canopy the rush it needs to accommodate visitors and officials, and almost makes it look like a hat with a brim. Nervi described the "structural layout" tersely as "decidedly antistatic" and unfeasible if the canopy had been larger. It is clear that Nervi had taken the liberty of playing

[74] Pierre Marcel, letter to Pier Luigi Nervi, 20 November 1957, *ibid.*
[75] Pierre Marcel, letter to Pier Luigi Nervi, 6 December 1957, *ibid.*
[76] For Hervé's photographs of the Piazza canopy, see AAA-SI, Box 12, Reel 5724, Frame 530.
[77] Nervi, *Le strutture dell'UNESCO* cit., pp. 17-18.

on the precarious visible balance of the two asymmetrical vault webbings, partly as a result of anchorage devices hidden in the foundations, as he had done for the large pillars in *Stadio Berta*, and he wished to demonstrate an artificial and symbolic use of the material that he usually installed in accordance with the rules of strength through form.

The reasoning that led Nervi to explore asymmetry and anti-staticity, using methods similar to those adopted for the *Stadio Berta* should again be sought in his desire to accentuate the expressivity of his structure. The adjective he used to define the effect of the structural forcing is the one he had read in Marcel's letter: "We believed that the opposition of the two curvatures was an expressive element that we should retain even at the cost of structural insincerity; I personally think that this very insincerity helps to give the canopy a slightly aggressive and random nature that contrasts well with the structural rigidity of all the rest."[77]

The pilotis of *Bâtiment III*

There can be no doubt that among all the surfaces produced in the Secretariat and in the *Bâtiment des Conférences*, the surfaces of the pilotis and the pillars clearly demonstrate Nervi's aspiration to create a rational profile for each element of his structure that was able to express its static function through its variable cross-section. For the first time, that profile was now defined in terms of formal expression through the careful shaping of the exposed concrete surfaces. Nervi appreciated the fact that the ruled surfaces of his variable cross-section pillars were furrowed by perfect and logical impressions, expressing the geometrical and static genesis of a kind of column designed to carry the flow of forces in a reinforced concrete structure. When the work was complete, he was able to admire the quality of workmanship achieved by

331. M. Breuer, P. L. Nervi, and B. H. Zehrfuss, Unesco Headquarters, Place de Fontenoy 7, Paris, 1951-58. Pilotis of *Bâtiment III* (photograph by R. Gargiani).

Fourré & Rhodes et Dumez, which from now on became the goal of Nervi & Bartoli. The words used by Nervi to describe the pillars and their execution are a kind of declaration of intent for the construction of his future variable cross-section pillars:

"Obedience to the static requirements (the section interlocking with the architrave has a large transverse moment of inertia, and the base cross-section has a large longitudinal moment of inertia) led in the most spontaneous manner to a ruled, architecturally expressive shape, which characterised the ground floor pillars. It may also be observed that the variable cross-section structures obtained by joining homologous points of the two different geometrical figures with straight lines are those best suited to the static and constructional needs of reinforced concrete. They also constitute, and will to even greater extent in the future, the most characteristic and specific architectural element in that, while the ruled forms can be achieved with relative ease using the straight elements of the wooden formwork, they would pose insuperable difficulties to metals and also to stone or masonry. In this case, the arrangement of the formwork boards was designed beforehand, and the contracting company very brilliantly solved the executive problem of preparing sturdy removable formwork that was used several times to pour all the pillars on the ground floor."

Due to his design for another two buildings, *Bâtiment III*, otherwise known as the Annexe Building, which was built alongside the Secretariat, and *Bâtiment IV*, which was never built, the Unesco complex ended up becoming a veritable catalogue of Nervi's variable cross-section pillars. Between April and May 1957, a design was drawn up for the variable cross-section pilotis arranged at the level of the *Bâtiment III* garden, which was the outcome of an

exchange of ideas between Nervi and Breuer, even though Nervi developed the final solution. The pillar shape was derived from an interplay of ruled surfaces that developed from a cruciform base toward a circular top to be joined to the floor beams. The special cruciform cross-section, with three arms only connected together by curved lines, since it performs a stiffening function like the elliptical shape, could also be seen as a metaphor of the specific floor plan outline of the Secretariat.

Nervi was again responsible for monitoring the studies of the formwork system for the pilotis proposed by the firm. When, following a building site inspection, he informed Breuer, on 27 January 1958, of the work in progress, Nervi was visibly pleased with the exposed concrete finish and also the sand blasting.[78] The Unesco headquarters was now destined to show the world a perfect form of concrete that was produced in an unprecedented way. The first two variable cross-section pillars in the *Bâtiment III* garden were stripped on 14 February 1958 to display a concrete that was "nicely done,"[79] and other sculptural works appeared to confirm the exceptional quality of workmanship.

In 1959 Nervi presented the Unesco buildings as "an important episode in the current laborious gestation of a new, sincere architecture for our times and, if you will forgive me my personal passions, one of the first, if not the first, examples of reinforced concrete mixture clearly and decisively entering the field of architecture."[80] The apostle of *béton brut*, Le Corbusier, admired the finish of Nervi's work, which he held up as a goal of "the perfection most possibly perfect" for concrete to be poured into formwork with narrow boards. "This concrete formwork is extremely elegant and very clean," he was to write of the Unesco headquarters in Paris.[81]

[78] Pier Luigi Nervi, letter to Marcel Breuer, 27 January 1958, Unesco, Bureau des Architectes, BNZ/4.
[79] Pierre Marcel, letter to Marcel Breuer, 14 February 1958, *ibid.*
[80] Pier Luigi Nervi, *Circostanze di una collaborazione*, in, "Casabella Continuità," XXIII, April 1959, No. 226, (pp. 8-9), p. 9. The essay is dated 27 February 1959.
[81] Le Corbusier, letter to José Luis Sert, 29 May 1961, (quoted in, Gargiani, Rosellini, *op. cit.*, p. 533).

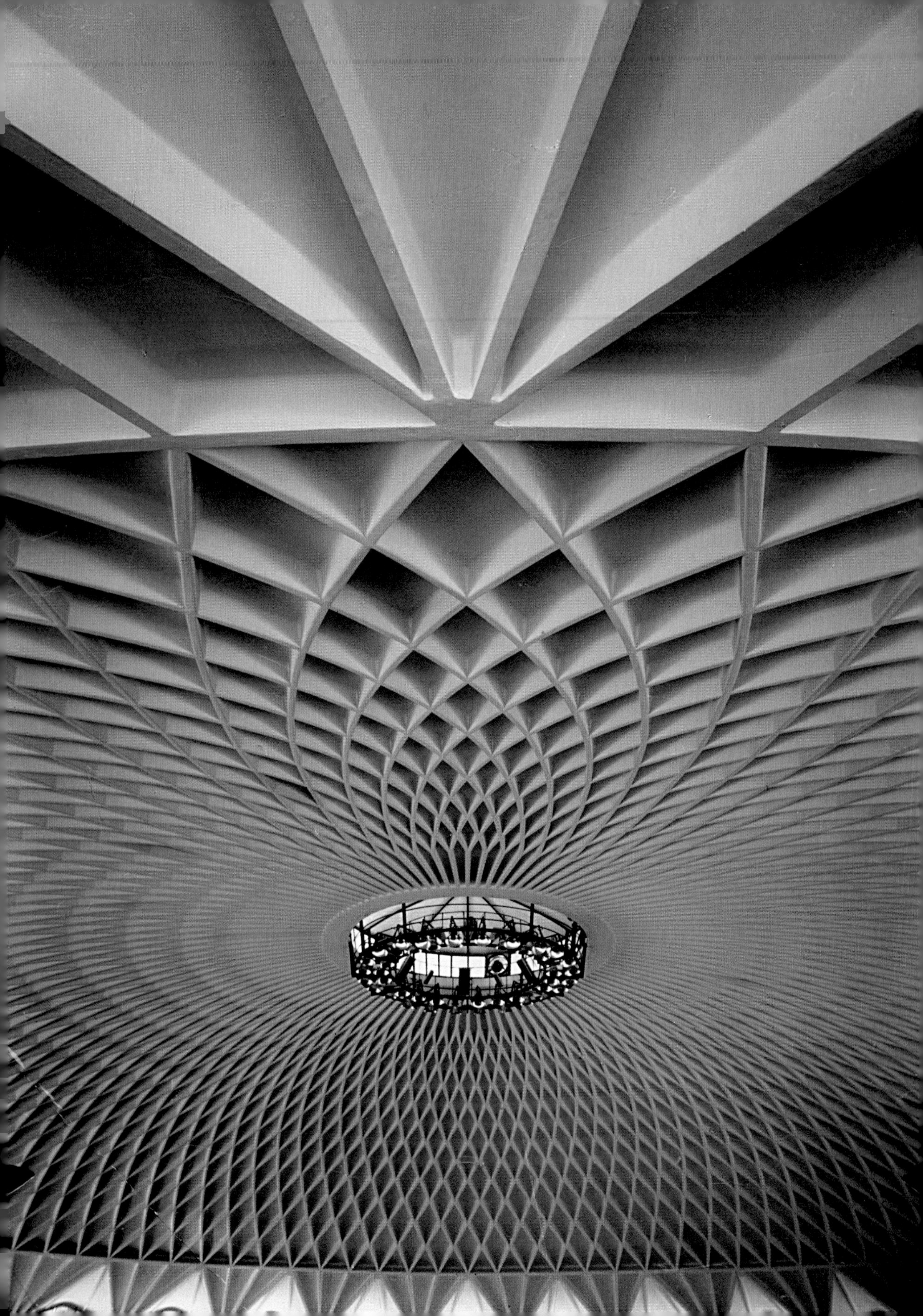

chapter ten
the nervi system

Bone-like skeletons

The commitment to follow the project and building site of the Unesco headquarters led to certain changes in Nervi & Bartoli's professional organisation and opened the way to a series of major international works by Nervi, heralded by the Brazilian experience. In 1953 Pier Luigi Nervi and his son Antonio, graduated in architecture in 1950, founded the Studio Nervi, Architettura e Tecnica Edilizia after the fleeting existence, between 1952 and 1953, of the Studio di Architettura e Tecnica Edilizia, set up by Antonio, and Sergio Musmeci, to which Pier Luigi Nervi acted as consultant and financial backer. In the works of Studio Nervi, structural prefabrication, form-resistant figures, and technical inventions gelled into a repertoire that could be adapted to the design of large contemporary spaces, in the perfect and changeless form of the "Nervi system,"[1] or the absolute style of reinforced concrete perennially pursued by Nervi, when it was necessary for the structure to reflect the configuration of the vaulted roofs. It was as though the prophecy that Nervi made in 1942, when he wrote that "spontaneously," "of necessity," "every statically important construction tends toward easy and simple solutions, and what we could call a 'natural' balance of forces that best interprets the laws of statics and material strength" came true in the 1950s.[2] At the same time, in the process of designing various configurations for different buildings, once Nervi had abandoned his concerns about including his reinforced concrete skeleton within his structural prefabrication system, it seemed destined to undergo continuous changes that revealed Nervi's ability to build genuine organic entities capable of their own intrinsic static strength and eloquence. In the variable cross-section pillar that he had recently discovered was another element that could express

[1] *The works of Pier Luigi Nervi* cit., p. 82, and Pier Luigi Nervi, *Nuove strutture*, Milano, Edizioni di Comunità, 1963, pp. 34, 48.

[2] Nervi, *La tecnica ed i suoi orientamenti estetici* cit., p. 10.

strength through form as well as the isostatic lines; Nervi actually succeeded in defining a sequence of significant figures that became one of the formal characteristics of his reinforced concrete structures.

While strength through form led Nervi to invent roofs with variously ribbed shells, the skeleton itself now underwent a radical overhaul with the aim of imbuing every segment with life, not merely the variable cross-section pillars. In the mid-1950s, Nervi was able to offer his own comprehensive vision of the Hennebique skeleton, transfigured into an organic skeleton where it was possible to identify the function of each component, each different traction or compression action, or the concentration of critical forces at the nodes and the flow of other forces into cantilever parts and the necessary balancing devices. At this point, the segments of the skeleton took on such an identity of their own that Nervis's skeletons almost seem to suggest the animalesque skeletons designed by Viollet-le-Duc for metal structures. In other cases, his skeletons continued to reveal striking similarities with Kahn's structural geometries, due to the continued presence in Nervi's repertoire of forms of the geodesic structure that had been deployed in various guises from the time of his hangars and that he increasingly employed for flat roofs.[3] In Nervi's studies for this type of roof, a triangular-edged geometry often competed with a flowing geometry based on Arcangeli's grid of isostatic lines. Nervi also derived from these different geometries pillars that were configured along lines best suited to meet the forces and aesthetic of the structure, as was also the case in the work of Kahn, Robert Le Ricolais, or Mies van der Rohe.

Various creative impulses drove Nervi's work in the mid-1950s and these led to his skeleton being expressed in a variety of ways: hidden or left exposed. Nervi developed a different device for each building, according to the construction programme and character. After the Unesco headquarters building, Nervi designed another grand structure (although one that was destined to remain virtually invisible) for the Pirelli skyscraper, which was conceived by Ponti in 1952–1953 (the design of the structure, by Nervi with the participation of Danusso, began in 1954; work was completed in 1959; the team included the engineers Giuseppe Valtolina, Egidio Dell'Orto, and Antonio Fornaroli, as well as the architect Alberto Rosselli).[4] Together with the BBPR *Torre Velasca*, for which Danusso designed the skeleton, the Pirelli skyscraper became established as the most significant skyscraper erected in reinforced concrete and not steel. Its structural concept was quite exceptional for the insight that brought it into being and unmistakably revealed the hand of Nervi: he chose to withstand wind loads no longer using a rigid Hennebique-derived type of frame but through the invention of colossal trunks, rooted in the ground with thick bases, and then branching to become more and more slender until they were almost left free to bend at the top. Everything arose from the spindle-shaped floor plan outlined by Ponti and his two triangular points, which inspired Nervi to invent what he defined as wind resistance "through gravity"[5] by means of a structure that was unusual for a skyscraper

[3] See Anne Griswold Tyng, ed., *Louis Kahn to Anne Tyng: the Rome letters 1953-1954*, New York, Rizzoli, 1997, (pp. 69-70), p. 69, and Gargiani, *op. cit.*, pp. 77-83.
[4] See the monographic edition on the Pirelli skyscraper of the review "Edilizia Moderna," December 1960, No. 71.
[5] Pier Luigi Nervi, *L'ossatura*, *ibid.*, (pp. 35-39), p. 35.
[6] Nervi, letter to Miozzi, 20 December 1931 cit.
[7] Pier Luigi Nervi, letter to Vittoriano Viganò, 20 November 1959, Maxxi, private correspondence.
[8] Nervi, *L'ossatura* cit., p. 35.
[9] Drawing, Plate No. 44, 15 April 1955, CSAC, n. inv. pra. 1093, n. id. 14894, coll. 123/2.

– it looked more like a viaduct on large pillars with the openings blocked by the floors. Although they look different, the structure that now resurfaced was essentially the same device that he had invented for the *Monumento alla Bandiera*, that ability to withstand the wind through the "weight force" that he advised Miozzi to apply "to the specific case of high towers."[6]

One of the crucial issues for the static function of the exceptional branching structure of the Pirelli skyscraper is that of the nature of the floors and their level of connection with what Nervi called a "few pillars of appropriate shape,"[7] or "wall pillars,"[8] which varied in terms of mass and elastic behaviour as they grew upwards. It is unsurprising that even in the case of these floors with their indefinable static role, stretching up to 24 metres over empty space, Nervi proposed to use his trademark prefabricated beams in ferrocement with their waved or corrugated profile and envisaged the use of pre-stressing at the top, where the wall pillars became thinner and the joints became less effective. In explaining the design of his structures in ferrocement, he took the opportunity to emphasise the level of formal freedom with which he had always worked with creative awareness: "the profile of the waved beam can be altered and shaped between relatively extensive limits," as well as recalling the hallmark of his product ("prefabrication makes it possible to achieve perfect surfaces without plaster"[9]). Instead of the waved beams, however, Nervi and Danusso examined beams in reinforced concrete with expanding heads in the joints.

Following Nervi's intervention, the entire structure – and not merely the two points, which were made more streamlined and spread out – tended to assert itself and become forcefully wedged in Ponti's curtain wall, which transformed it into an infill, exposing its wall pillars and the way that they thinned markedly toward the top and projected above the roof with a graceful and decisive gesture, like taking off a garment.

Nervi set out his insight about the resistance "through gravity" against wind loads in a diagrammatic drawing of the schematic cross-section that is drawn through a pair of wall pillars in order to expose, in the mass of black

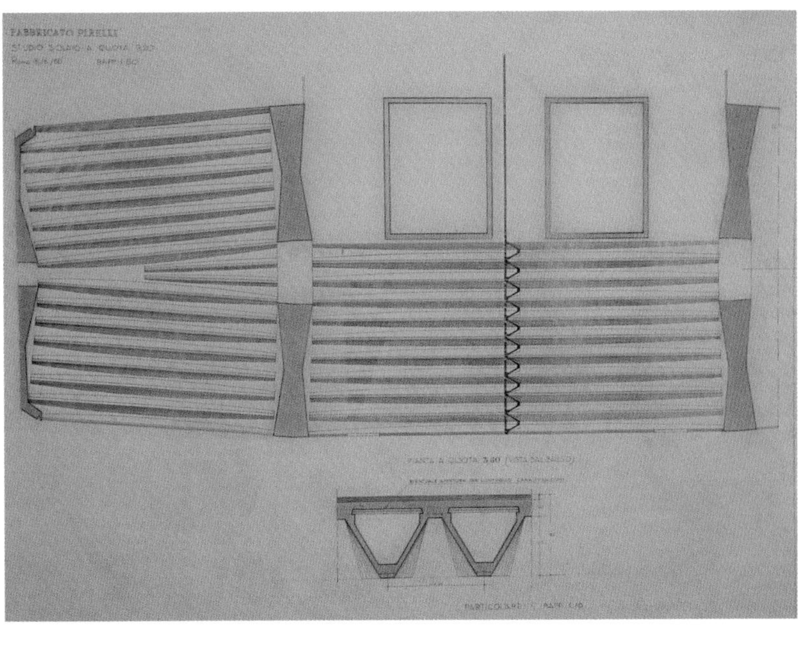

332. P. L. Nervi and A. Danusso, structure for Centro Pirelli, designed by G. Ponti with A. Fornaroli, A. Rosselli, G. Valtolina and E. Dell'Orto, Piazza Duca d'Aosta, Milan, 1955-58. "Studio solaio a quota 9,20 (study for floor at 9,20 metres)," drawing dated 8 June 1955 (CSAC, n. inv. pra 1093, n. id. 14894, coll. 123/2).

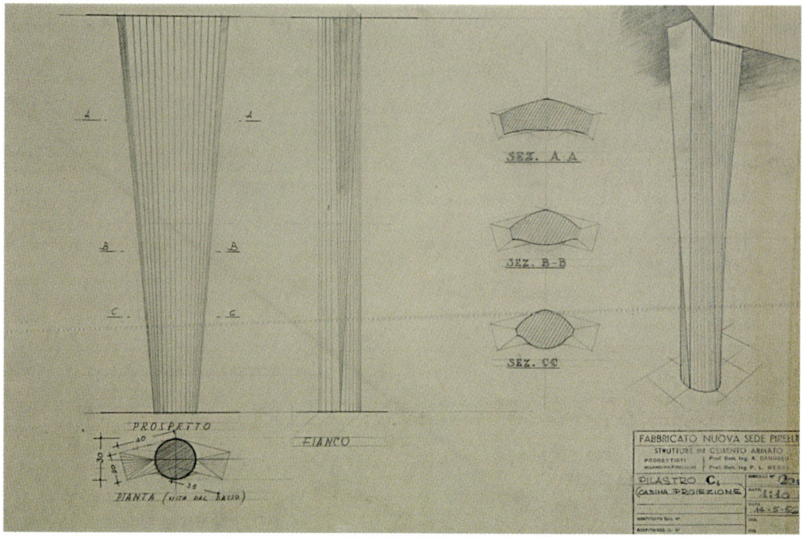

333-335. P. L. Nervi and A. Danusso, structure for Centro Pirelli, designed by G. Ponti with A. Fornaroli, A. Rosselli, G. Valtolina and E. Dell'Orto, Piazza Duca d'Aosta, Milan, 1955-58.

333. Study for variable cross-section pillar for auditorium projection booth, drawing dated 14 May 1956 (CSAC, n. inv. pra 1093, n. id. 14894, coll. 123/2).

334. Cross-section (*The Works of Pier Luigi Nervi,* London, The Architectural Press, 1957, p. 132).

335. Scale model 1:15 produced at the ISMES, Bergamo, 1955 (BSTA, 024-G860).

at the root tapering off toward the foliage, the quantity of reinforced concrete in the colossal foundation structure, which then became thread-like at the top, while the wall pillars separated into two branches from a certain height. Frank Lloyd Wright identified in that diagram a significant confirmation to the mighty foundation root and towering stalk that formed the reinforced concrete structure of his visionary "Illinois Mile-High Cantilever Sky-City to Honor the State of Illinois and City of Chicago" designed in 1956 – the skyscraper was also dedicated to three inventors of reinforced concrete, including Nervi, whom Wright met, also in 1956, in New York.

Once the vital logic of the structural body had been defined, Nervi played around with the device of strength through form, with the aim of designing the individual elements with static flair. For some pillars of the base and the floor beams, he designed profiles with variable cross-sections, partly to obey static rules but also to put his stamp on the work with motifs that were becoming iconic trademarks of his construction method. It is no coincidence that some pillars with a round base and rectangular ending that he designed for the skyscraper were directly derived from those of the Unesco *Bâtiment des Conférences*. In his drawings Nervi depicted the distinctive lines of the ruled surfaces that he probably imagined he could expose using formwork, based on his experience in the Paris building site. In the auditorium on the ground floor backing onto the skyscraper, the structure took on the configuration of portal frames in relief, with interwoven beams, in order to obtain grand diamond-shaped meshes reminiscent of a geometrical pattern dear to Nervi.

In accordance with Nervi's experimental procedure, the skyscraper structure was verified between 1955 and 1956 by building a model to be used for testing by the ISMES. As in the case of the waved ferrocement beams for the *Fiera Campionaria* in Milan, the material used for the Pirelli skyscraper model was the same as would be used in the constructed work, reinforced concrete, but with the strength of the components proportionally reduced in scale so that the mechanical and static function reflected the true situation. The dimensions of the model consequently became significant – it was 9 metres high and made using plaster formwork.[10] This choice, which confirmed a significant change of

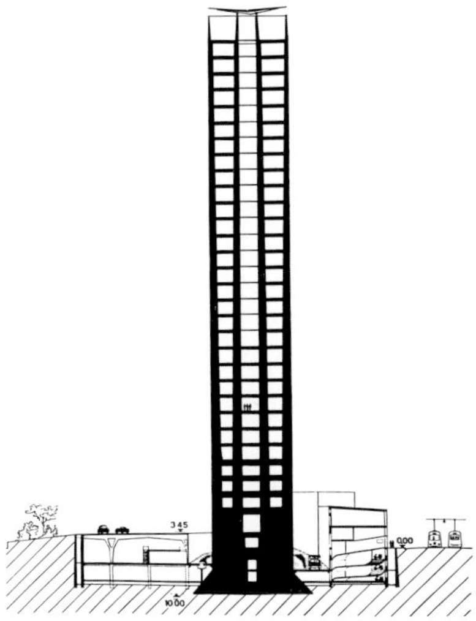

approach in the investigations conducted by the Danusso and Oberti laboratory, reflected Nervi's concerns about the behaviour of reinforced concrete outside the theoretical laws of elasticity, as he had the opportunity to learn from the words of Colonnetti in 1943. The ISMES also used models to test the beams of a typical floor, in view of its importance in determining structural rigidity. The results convinced Nervi to do without pre-stressing.

In order to control the heat expansion effects on the floors, which could have generated critical thrusts against the rigid, sharp end structures, tests conducted at the ISMES allowed Nervi and Danusso to identify the height of the ninth floor as the ideal level for a change in connections between the parts. Up to the ninth floor, where expansion effects could be significant, the floors were left free to deform through the use of sliding supports. The test results revealed "levels of cooperation between the different structural elements that were more considerable than we had expected," in the words of Nervi.[11]

All the reinforced concrete of the skyscraper was shaped using rudimentary formwork and was either buried by cladding or banded with chiseled grey cement mortar, except for the summit structure section, in the plant room, where it was left exposed, showing all the rough impressions of the wooden boards of the formwork on the portal frames, on the pillars, on the beams, and on the mighty cantilevered connections. Nervi gave great prominence in his publications to photographs of the auditorium and the plant room in their rough state, in which he recognised the "plastic richness of concrete cast in place,"[12] confirming his new passion for *béton brut*.

The reinforced concrete skeleton became a rhetorical and imposing motif with every part exposed in the competition design for the Napoli Centrale railway station, drawn up by Nervi with Giuseppe Vaccaro and Mario Campanella in 1954. The sparse diamond-shaped grid of the high, slender beams determined the specific angling of the arms of high cruciform pillars with their stems tapering toward the bottom. The design studies, including those follow-

[10] *Esperienze su modelli del Grattacielo Pirelli,* in, "Edilizia Moderna," December 1960 cit., pp. 40-41, and Neri, *op. cit.*, pp. 82-93.

[11] Nervi, *L'ossatura* cit., p. 38.

[12] Nervi, *Aesthetic and Technology in Building* cit., pp. 22-36, 60, 62.

336-337. P. L. Nervi and A. Danusso, structure for Centro Pirelli, designed by G. Ponti with A. Fornaroli, A. Rosselli, G. Valtolina and E. Dell'Orto, Piazza Duca d'Aosta, Milan, 1955-58.

336. Auditorium under construction, photograph dated 25 June 1957 (BSTA, 024-6857).

337. View of plant room on the top floor (BSTA, 024-6858).

338. P. L. Nervi, G. Vaccaro, and M. Campanella, competition design for the *Napoli Centrale* railway station, 1954-55. Perspective drawing of the concourse (*The Works of Pier Luigi Nervi*, London, The Architectural Press, 1957, p. 97).

ing the first stage of the competition, documented Nervi's desire to define the geometry of the pillars and roof in accordance with symbolic and static lines that sometimes take on traits not dissimilar to those of Kahn's structural geometries. Nervi's pillars were not only cruciform but also V-shaped with a circular base and variable cross-section branches, some again derived from the Unesco buildings pillars with ruled surfaces so that they could, if necessary, be built using formwork made out of wooden boards. Nervi's proposals also included a kind of structure that was destined to pre-empt some of his subsequent work, derived from the floor with isostatic ribs. In this case, the pillar, again tapering downward, is shaped by grooves and juts to create a bundle of ribs continuous with those of the isostatic ribs of the roof. The design of this version confirmed Nervi's interest in Gothic structures and their bundle pillars branching into the ribs of the cross vaults.

In the station built between 1958 and 1960, the roof beams were woven into geometrical meshes made up of equilateral triangles, which gave rise to a special pillar shape branching into three bars. The pillar was inspired by an idea proposed by Enrico Castiglioni, with Edoardo Sianesi and Giorgio Bongioanni, during the first stage of the competition design for the station and was to act as a prototype for Nervi's later branched pillar versions.

As was customary when he designed large bridges and viaducts, when designing the bridge over the Tenza in Salerno, dating from 1955, Nervi designed the structure with the aim of minimising the need for any temporary scaffolding for the construction of the roadway. He therefore proposed a series of large pillars reduced to essential trusses, each branched into four trunks, with variable cruciform cross-section bars and rigid, angled lines, each divided in two – one low base bar and one high bar – to support the roadway, almost as if to reflect the profiles of metal electricity pylon trusses. The large pillars were designed so that they could be built using the same formwork system, while the roadway beams were prefabricated and assembled using a sliding device of the same kind as that developed for the Meschio canal bridge. The height of each base sector of the pillar could be varied according to the topography of the ground, but the presence of a dwarf pillar to support a section of roadway near a bridge abutment laid bare Nervi's business pragmatism.

The skeleton that Nervi envisaged in 1955-56 for the underground Basilica at Lourdes, designed by Pierre Vago, is practically bone-like as a result of the sophisticated moulded folds in its variable cross-section components, giving an impression in the half-light of a presence revealed by a sensitive interplay of light and shade. The most distinguishing feature of that structural device was nevertheless the invention of a narrative of dynamic flows beyond any flight of

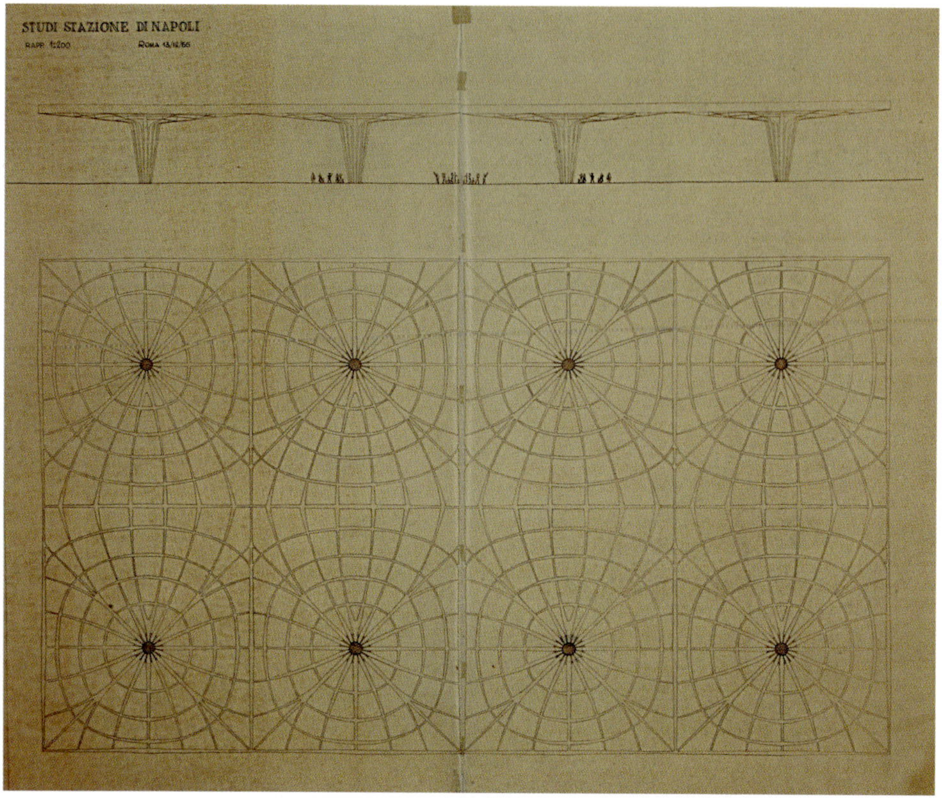
STUDI STAZIONE DI NAPOLI
RAPP. 1:200
ROMA 13/12/55

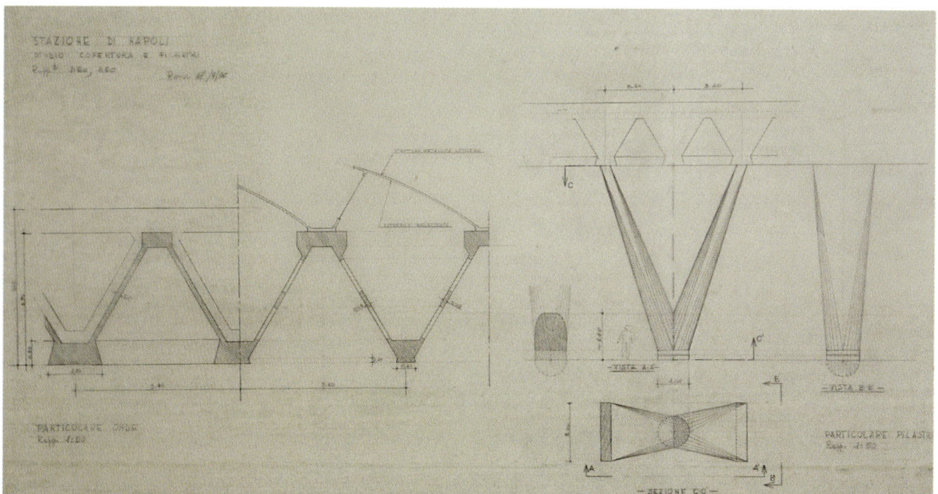
STAZIONE DI NAPOLI
STUDIO COPERTURA E PILASTRI
VISTA A-A
VISTA B-B
SEZIONE C-C
PARTICOLARE ONDE
PARTICOLARE PILASTRI

339-340. P. L. Nervi, G. Vaccaro, and M. Campanella, competition design for the *Napoli Centrale* railway station, 1954-55.

339. "Studi Stazione di Napoli (Studies for Naples Station)," drawing dated 13 December 1955 (CSAC, n. inv. pra 1053, n. id. 14851, coll. 150/4).

340. "Stazione di Napoli. Studio copertura e pilastri (Naples Station. Study for roof and pillars)," drawing dated 27 September 1955 (CSAC, n. inv. pra 1053, n. id. 14851, coll. 150/4).

341. P. L. Nervi, design for bridge over the River Tenza, Salerno, 1955 (*The Works of Pier Luigi Nervi*, London, The Architectural Press, 1957, p. 134).

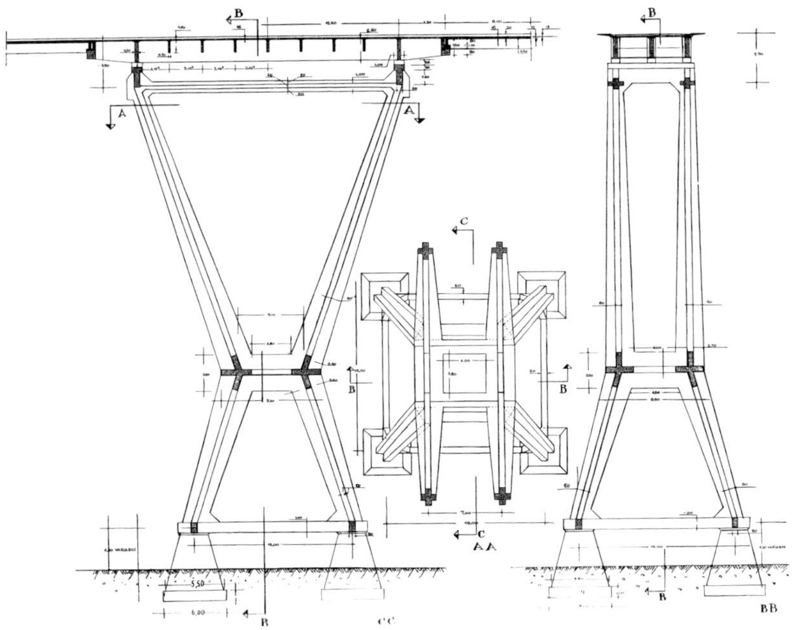

Gothic imagination. Nervi's structural insight was guided by neither arches nor frames, nor a ridge beam pierced by the overhead light, but rather a sophisticated system that set out to stage a series of pressures and tensions that cancel one another out in the phenomenon of light and shade, as if by a miracle, almost as though the symmetry of the radial sweep along the deformed ellipse were not the representation of a contemplative static certainty but the incessant revelation of an interplay of dynamic contrapositions revealing a truth, a balance glimpsed in an instantaneous flash. This idea was also a way of embodying Bernadette's uncertain vision of the Virgin Mary. The pierced beam that cuts the ellipse, the revelation of light and essential balance, conjures up more than a passing resemblance, due to its role in the structural body, to the cut that passes through the long crystal-shaped body of the Pirelli skyscraper. Nervi's contribution ended when the Freyssinet firm joined the project.

The structure was again designed to embody a metaphor, in this case the wings of an aeroplane, in the case of a design for terminals of Fiumicino airport, drawn up between October and November 1957 by Studio Nervi together with Vaccaro, Cesare Ligini, Amedeo Luccichenti, and Dagoberto Ortensi. "Flying stone" was Nervi's enthusiastic description of the wings of an aeroplane made out of prestressed reinforced concrete conceived by Freyssinet:[13] an expression that encapsulates not only his passion for aviation but also the secret attraction for empty space that lay behind his soaring structures. On the floors of the control tower in the shape of an overturned pyramid, on the other hand, the floor pillars radiate from the base of the central core in the form of angled lines and cross the spaces in an unusual and pre-figurative manner in the various versions of the canopy. When designing the supports for his "flying stones," Nervi tested out various structures, includ-

[13] Nervi, *Costruire Correttamente* cit., p. 117.

ing skewed portal frames, to make the system dynamic and a truss made up of bars of different thicknesses to reveal whether they were struts or ties and even give the impression of a Da Vinci-esque petrified aircraft. The purpose of the final solution chosen for the canopy was to identify the ideal combination of struts, ties, cantilever beams, and nodes that would make it possible to cancel out the "bending stresses in the supporting trestles."[14]

Now all components of the structure became members with sections of variable thickness, cleverly outlined with sharply raked lines to make the system even more expressive. Designs such as those for Lourdes or Fiumicino unmistakably reveal the conscious achievement of Nervi's aspiration to create a skeleton that is essential in every part, often soaring into space, which had previously been typical of the rudimentary design of his works in the early 1920s but was now informed by such a mastery of the forces involved that he was able to sublimate every structure into the form of a symbolic and spatial expression.

342-343. P. L. Nervi, structure for underground Basilica in Lourdes, designed by P. Vago, 1955-56.

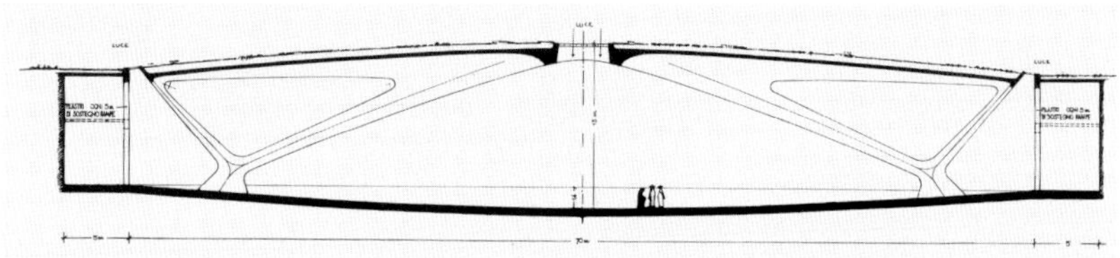

342. Cross-section (P. L. Nervi, *Buildings, Projects, Structures. 1953-1963,* New York, Frederick A. Preager, Publisher, 1963, p. 23).

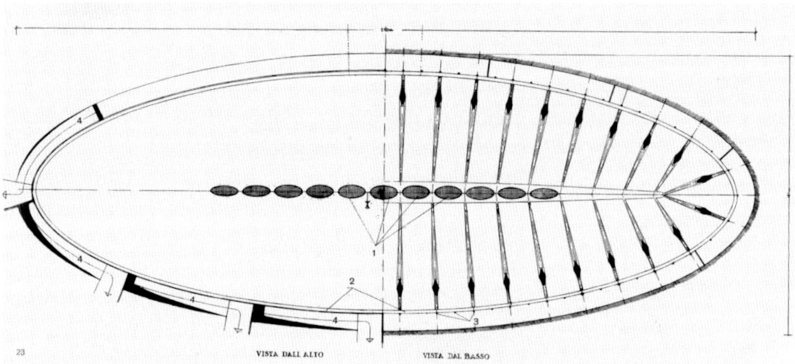

343. Plan (P. L. Nervi, *Buildings, Projects, Structures. 1953-1963,* New York, Frederick A. Preager, Publisher, 1963, p. 23).

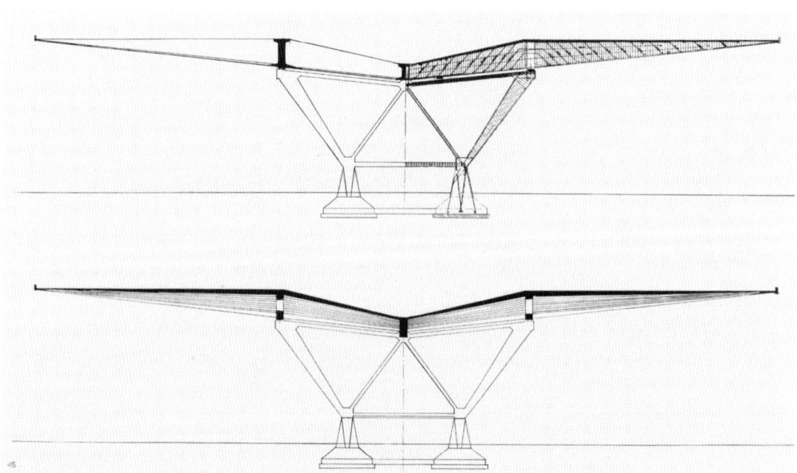

344. P. L. Nervi, structures for Fiumicino International Airport, competition design drawn up with C. Ligini, U. Luccichenti, A. Nervi, D. Ortensi, and G. Vaccaro, 1957. Design for the roof of a terminal (P. L. Nervi, *Buildings, Projects, Structures. 1953-1963,* New York, Frederick A. Preager, Publisher, 1963, p. 45).

Roof shapes, from domes to cross vaults

The success of structural prefabrication in ferrocement decreed by the *Saloni* in Turin won Nervi & Bartoli a series of commissions for the construction of roofs of various shapes and sizes. During the 1950s, Nervi continued to make use of his trademark elements, from waved ashlars to *tavelloni*. Nervi & Bartoli built roofs of various kinds, always made out of prefabricated elements in ferrocement, between 1950 and 1952 in the Parco delle Fonti di Chianciano thermal establishment, designed by Mario Loreti and Mario Marchi. The "waters room" featured a "waved vault" made out of two-centimetre thick variable cross-section arched beams with a lighting frieze made out of a "ribbed grid"; the elliptical roof of the restaurant hall was a version of Nervi's domes made out of ferrocement *tavelloni*. Between 1952 and 1953, Nervi also designed and built yet another canopy, again curved like that of the Kursaal, for the *Emiciclo* hall at the *Fiera di Milano*. This was made out of the usual waved ferrocement beams (with a covering made out of brick blocks) set upon V-shaped pillars in cast-in-place reinforced concrete and plastered: a pared-down echo of the ones he designed with Bo Bardi. In his studies for the profile of the waved beams in ferrocement, Nervi examined the flow of lines and surfaces with the aim of demonstrating the formal power of his system of prefabricated parts using a matrix on the ground that could be shaped as desired. Nervi was now beginning to explore the Mannerist refinements allowed by his system.

A competition design in 1953 for a sports complex in Vienna, drawn up by Pier Luigi and Antonio Nervi, saw the return of Nervi & Bartoli's usual motifs and techniques, expressed together with some recent inventions for the Unesco headquarters (Italo Balletti also contributed to the design). The main building, the sports arena, was the first adaptation of prefabricated waved ashlars in ferrocement pioneered for *Salone B* in Turin to a dome configuration with the aim of obtaining a "corrugated structure."[15] This was to be achieved by positioning the parts along Decauville rails – they were to be raised using a crane positioned in the centre of the dome. Even though the waved ashlars were the same kind as those used in Turin, organised into 10 discs, Nervi actually now intended to change the joint in order to reduce it to a thin band in relief, partly for the purposes of enhancing the convergence effect of the waved beams in the oculus with the stage. The first three series of waved ashlars perforated by skylights created a broad annular band for the lighting system. The same segments were also designed to provide conduits for the air conditioning. On the unperforated waved ashlars, sound-absorbent panels were also added to seal the part facing the outside. The cavities thus enabled the waved beams to become complex structural organic bodies.

For the monumental stand stairway, Nervi designed a majestic symmetrical structure with crossed beams with the intention of replicating the sculptural virtuosity of the *Stadio Berta* stand stairway. Grand portal frames with variable profiles reminiscent of those of the Unesco *Bâtiment des Conférences* were planned for the roof of the car park facing the sports arena. These were to be constructed in reinforced concrete (the option of pre-stressing was also considered), connected by beams, and closed off in each span by a ceiling in the form of a shell with a waved profile determined by the moments. As

[14] Pier Luigi Nervi, *Buildings, Projects, Structures. 1953-1963*, New York, Frederick A. Praeger, Publisher, 1963, p. 45.

[15] Pier Luigi Nervi, Antonio Nervi, Italo Balletti, *Concorso per lo Sportpalast di Vienna. Relazione generale*, in, "Architettura Cantiere," 1953, No. 4, (pp. 37-47), p. 45.

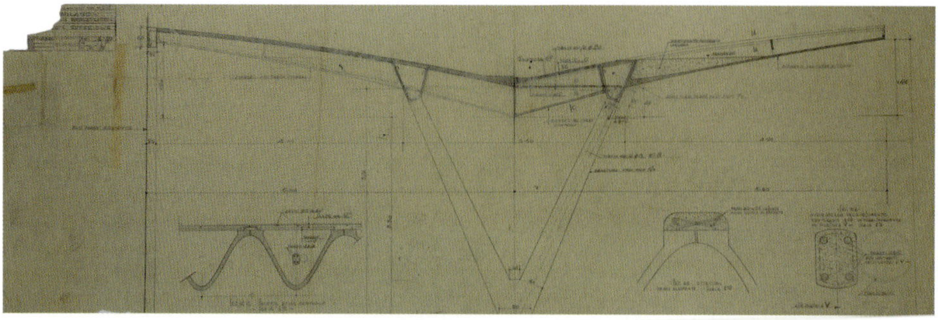

345-346. P. L. Nervi (with the cooperation of A. Nervi), roof of restaurant hall for *Padiglione 25* at the *Fiera di Milano*, Quartiere Fieristico, Milan, 1953. Demolished.

345. Canopy sections, drawing dated 5 September 1952 (CSAC, n. inv. pra 1008, n. id. 14776, coll. 266/3).

346. Matrices for the prefabrication of waved beams (Maxxi - PLN, F08839).

347. P. L. Nervi, structure for headquarters of Società Galbani, designed by G. Pestalozza, E. Soncini, and E. Soncini, Via F. Filzi 25, Milan, 1955-56. Raising the prefabricated floor beams (photograph by A. Villani; Maxxi - PLN, F07663).

348-350. P. L. Nervi, structures for thermal establishment at Parco delle Fonti, designed by M. Loreti and M. Marchi, Piazza Martiri Perugini, Chianciano Terme, 1952.

348. Pre-fabricating the *tavelloni* of the restaurant hall dome (Maxxi - PLN, F08662).

349. Assembling the *tavelloni* of the restaurant hall dome (Maxxi - PLN, F08666).

350. View of the restaurant hall (photograph by V. Villani; Acm-EPFL, 0172.02.0059.0359).

351-352. P. L. Nervi and A. Nervi (with the cooperation of I. Balletti), competition design for sports complex, Vienna, 1953.

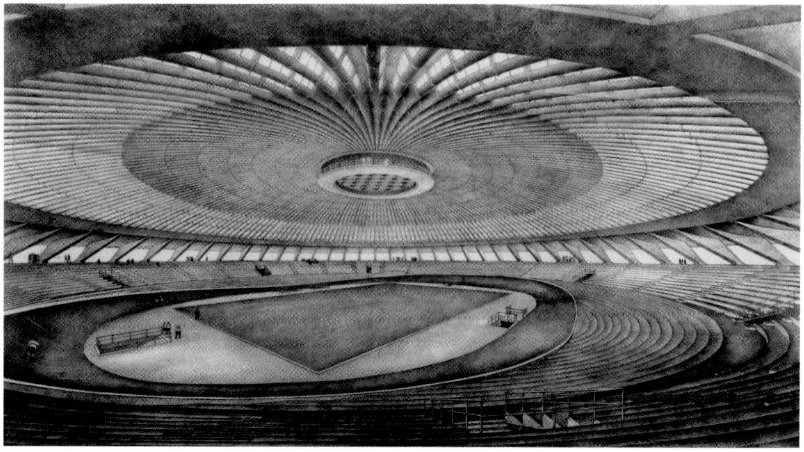

351. Perspective drawing of the sports arena (Acm-EPFL, 0172.02.0003.2061).

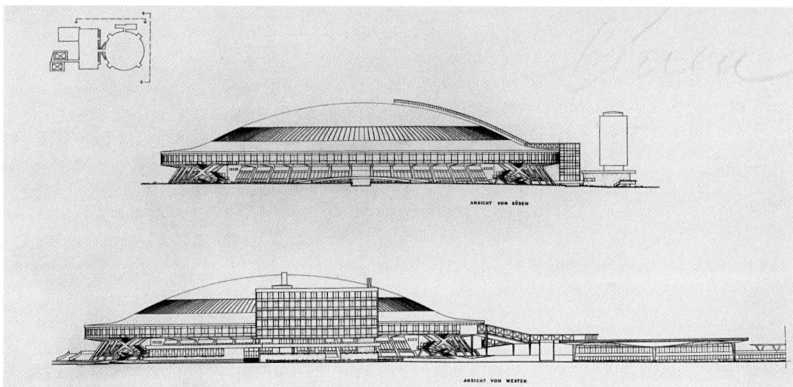

352. Cross-section (Acm-EPFL, 0172.02.0003.2063).

usual, the attributes of the "perfectly smooth" surface of the exposed surface prefabricated elements were emphasised.[16]

After the vault webbing of the hyperbolic paraboloid planned for the Unesco Secretariat Piazza canopy, Nervi again found himself dealing with a similar type of roof when, at the end of 1953, he was called in by Zehrfuss to develop a roof structure for the *Centre des Nouvelles Industries et Technologies* (CNIT) in Paris, designed with Robert Camelot and Jean de Mailly in the form of a vault resting only on three vertices. Even in this case, though, his structure did not take the form of continuous shells. Nervi made his decision after the CNIT had rejected Bernard Lafaille's design for a metal roof and suggested, for financial reasons, that the cement structures of the *Saloni* of the *Palazzo di Torino Esposizioni* were an appropriate example to use as a guide.[17] Nervi broke down the roof into a structure in the form of a triangular geodesic mesh made up of several elements to be prefabricated and cast in place. These were to be made out of reinforced concrete, ferrocement and involved the use of pre-stressing. He displayed a clear desire to create an organic body in which it was possible to recognise the flow of forces along a diversity of lines arising out of the interweaving of parts. As usual, the aim was to allow construction to take place without major scaffolding, in accordance with Nervi & Bartoli's typical practices for large roofs. Partly in the light of the design for the CNIT, the Piazza canopy increasingly appears to be an exception in Nervi's output, which was justifiable due to its small size.

353. P. L. Nervi (with the cooperation of A. Arcangeli), structure of roof for Centre des Nouvelles Industries et Technologies, designed by B. Zehrfuss, R. Camelot, and J. de Mailly (consultant, J. Prouvé), Paris, 1953. Model (*The Works of Pier Luigi Nervi*, London, The Architectural Press, 1957, p. 139).

The interwoven structure of the CNIT vault was created by a hierarchy of elements inspired by the criteria tested in the second series of hangars. This included three ribs converging at the centre of the roof, to be produced using cast-in-place reinforced concrete, triangular meshes made out of parts prefabricated in ferrocement, prestressed bars, and nodes – all in cast-in-place reinforced concrete. Due to their small size, the three edging arches serve to close the triangular grid and appear to be held in place by the central ribs that make their presence felt in the form of a majestic ribbed vault, miraculously skewed. In this regard, it is significant that Arcangeli used the definition "triangular cross vault" when mathematically modelling Nervi's structure.[18] The purpose of this exercise by Arcangeli, who compared the structure to membranes (he therefore envisaged a lack of "bending stresses"),[19] was to test the behaviour of that vault "formed by the intersection of three cylindrical surfaces with horizontal generating lines and a circular leading curve" under the action of its own weight and accidental overload and the means by which the forces were transmitted to the ribs.[20] In 1956 the SOCODEF construction company changed the structural system that he had designed, and Nervi refused to allow his name to appear as a consultant.[21]

The variable cross-section pillar that resulted from studies conducted with Bo Bardi, and that was tested in the Unesco headquarters, also became part of the vaulted structures in two designs dating from 1955-56. Studio Nervi returned to the layout designed for the CNIT cross vault in its design for a cathedral in New Norcia, in Perth, Australia, drawn up with the co-operation of Francesco Vacchini and Carlo Vannoni. In the cathedral cross vault, Nervi again transfigured the motifs that were becoming iconic features of continuous shells for religious architecture into ribbed hyperbolic paraboloids. As usual, these were designed to be built using prefabricated *tavelloni* in ferrocement with fine ribs subordinate to the three edge arches. The three large ribs in reinforced concrete joined in the centre of the vault, as in the CNIT roof. Vault webbing emerged beyond the edge arches to protect the stained glass. Exactly as in the Unesco pillar designs, Nervi proposed that the feet of the three cathedral vault abutments should also feature

[16] *Ibid.*
[17] Centre des Nouvelles Industries et Technologies, letter to Bernard Lafaille, 12 November 1953, Maxxi, P25.
[18] See handwritten calculations, p. 1, AA, folder No. 237.
[19] *Ibid.*, p. 2.
[20] *Ibid.*, p. 1.
[21] Pier Luigi Nervi, letter to Robert Camelot, Jean de Mailly, Bernard Zehrfuss, 5 June 1956, Maxxi, P25.

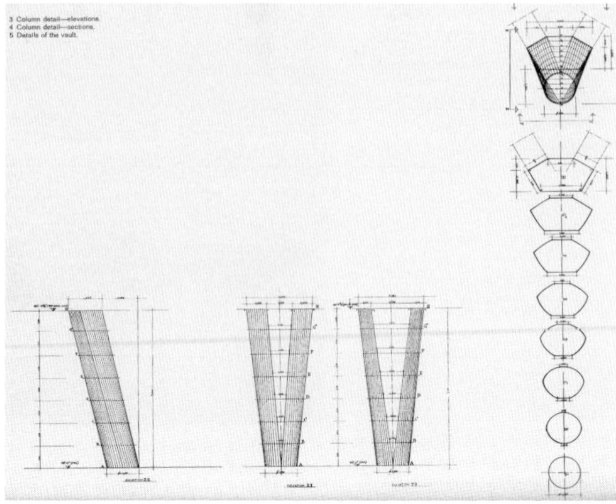

354-355. P. L. Nervi, A. Nervi, F. Vacchini, C. Vannoni, design for New Norcia cathedral, Perth, Australia, 1955.

354. Model (P. L. Nervi, *Buildings, Projects, Structures. 1953-1963,* New York, Frederick A. Preager, Publisher, 1963, p. 49).

355. Variable cross-section pillar (P. L. Nervi, *Buildings, Projects, Structures. 1953-1963,* New York, Frederick A. Preager, Publisher, 1963, p. 50).

a variable cross-section pillar with a round base that he gradually changed first to ellipses, then to irregular hexagons with curved sides, and lastly to a hexagon that was still irregular but with straight lines that flowed uninterruptedly into the vault arches and ribs. In this case, the variable cross-section pillar was directly generated by the vault ribbing system, revealing the extent to which the design of Nervi's structure faithfully reflected the lesson learned from Gothic cathedrals.

Nervi's studies into variable cross-section pillars also affected the structures that, until then, had acted as support spurs for his arched roofs made out of prefabricated waved beams, introduced in the design in response to a call for tenders for the Florence sports arena, drawn up between 1955 and 1956 by Studio Nervi. The roof with its parabolic vault was intended to be a variant of the waved shell made out of closed ashlars prefabricated in ferrocement, made famous by *Salone B* in Turin, but no longer with diaphragms. It was intended that the light should filter from the saw-toothed walls to be erected in accordance with the system designed for the hangar in Buenos Aires.

Nervi had adapted the mushroom-shaped roof envisaged for E42 to his new system of prefabrication in ferrocement several times. Nervi also revisited the same motif, as usual made out of prefabricated elements, for the design of an exhibition centre in Caracas in 1956. Now the profile was altered to form a sequence of roofs in order to top off the centre of the building with a dome surrounded by a radial vault with waved arched beams and to obtain a ring of V-shaped pillars between both roofs. After extending the dimensions to create an 87-metre diameter dome, the structure reappeared with the same profiles in the design submitted for the competition to build a *Palazzo dello Sport* and *Fiera del Mare* building, in Genoa, drawn up with Daneri in 1958.

The certain arbitrariness of design and aggressive formal effect that Nervi suggested as being present in the asymmetrical profile of the Unesco Secretariat Piazza canopy also guided his hand in designing upward cantilevered claw flaps in the grand roof that he intended to build for the entrance pavilion to the *Torino Esposizioni* complex, designed with Mario Passanti, Paolo Perona, and Luigi Ravelli in 1956-57. In this case too, Nervi adopted a "large thin vault with sections of various shapes, stiffened by internal ribs."[22]

[22] Nervi, *Nuove strutture* cit., p. 31.

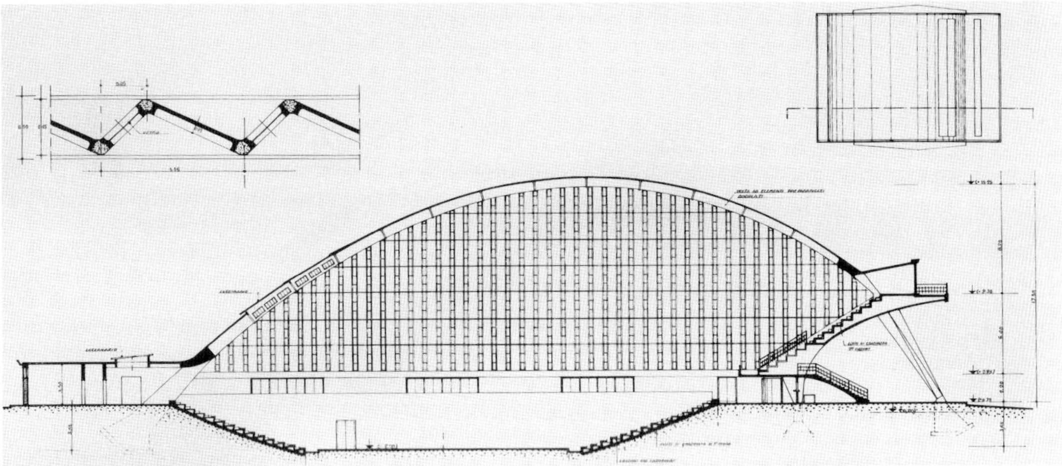

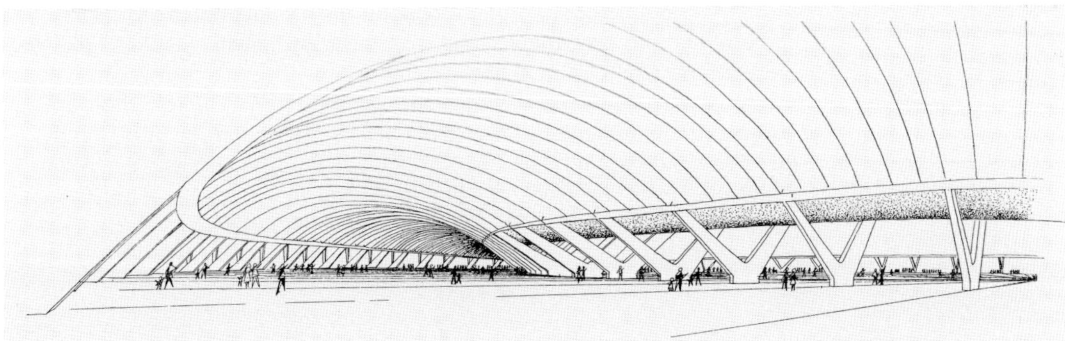

356. P. L. Nervi and A. Nervi, competition design for *Palazzo dello Sport,* Florence, 1955-56. Cross-section and detail of infill wall (P. L. Nervi, *Buildings, Projects, Structures. 1953-1963,* New York, Frederick A. Preager, Publisher, 1963, p. 20).

357. P. L. Nervi, design for Caracas exhibition centre, 1956. Perspective drawing (*The Works of Pier Luigi Nervi,* London, The Architectural Press, 1957, p. 141).

358. P. L. Nervi and L. C. Daneri, design for the *Fiera del Mare,* Genoa, 1958-61. Model (P. L. Nervi, *Buildings, Projects, Structures. 1953-1963,* New York, Frederick A. Preager, Publisher, 1963, p. 146).

359. P. L. Nervi, M. Passanti, P. Perona, L. Ravelli, design for entrance building at the *Torino Esposizioni* complex, Turin, 1956-57 (BSTA, 024-2079).

360. P. L. Nervi, structure of the Italian Institute of Culture in Stockholm, designed by G. Ponti, Gärdesgatan 14, Stockholm, 1955-58. "*Soluzione 'A',*" drawing dated 23 June 1955 (CSAC, n. inv. pra 14856, n. id. 1058, coll. 150/4).

Continuous shells were largely excluded from his repertoire because he preferred roof structures that he variously described as geodesic or corrugated, according to the type. For this reason, together with the Solvay storage shed in Milan, and the Common Salt storage shed for the *Manifattura Tabacchi* in Bologna, a version of the design of a roof for the small auditorium of the Italian Cultural Institute in Stockholm, constructed between 1953 and 1955 by Ponti, occupied pride of place in his production due to his use of a shell folded into great flaps, similar to those of Baroni, Candela, and Musmeci, almost like a pointed hat with tension members, a transcription of the images invoked as examples of strength through form. The resulting structure adopted the configuration of crossed beams interwoven on two slightly angled flaps and should be seen as a homage to Ponti's masterpiece in Milan, in which Nervi was involved – the Pirelli skyscraper.

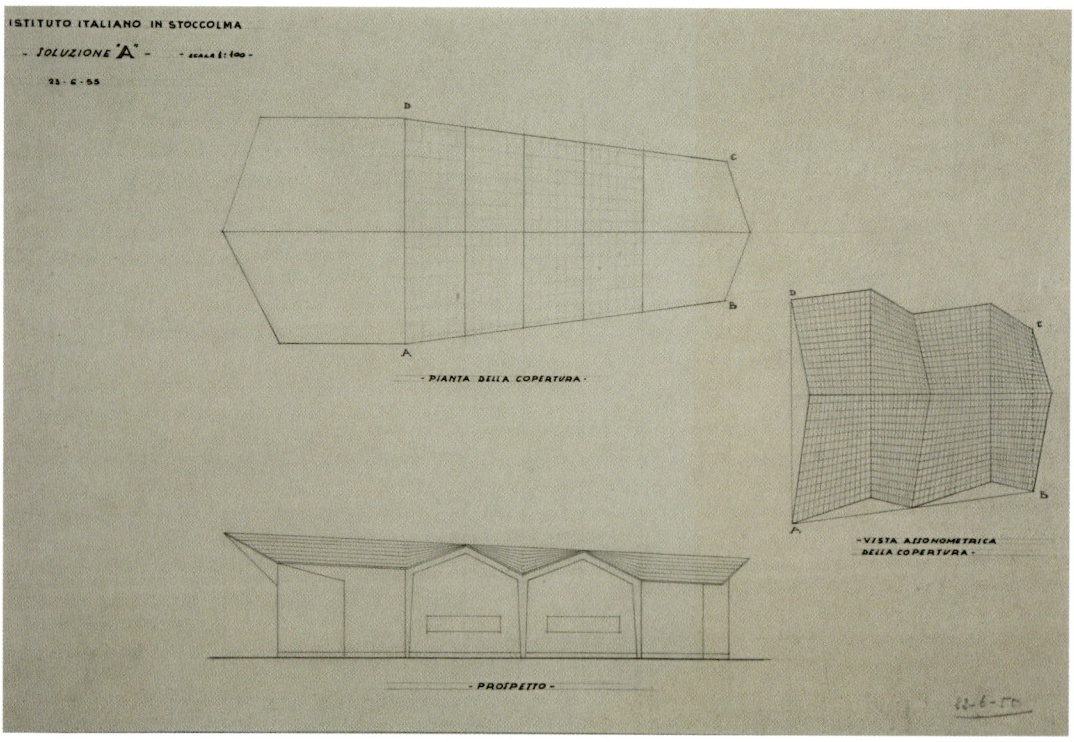

A sunburst structure: the *Palazzetto dello Sport*

After the roofs designed during the first half of the 1950s, Nervi was again able to erect, still using his own firm, another grand structure that followed on from his experiences in Turin and extended the system of structural prefabrication in ferrocement to a dome shape, as he had already suggested in the design for Vienna: the *Palazzetto dello Sport* in Rome, designed with Vitellozzi, was built by Nervi & Bartoli in 1956-57 and used for sporting events during the 1960 Olympic Games. The design dated back to 1954, when Vitellozzi asked for Nervi's assistance in designing the prototype for a stadium of "average size and economic type," that he had been commissioned to build by the Italian National Olympic Committee (CONI).[23]

Because the size was smaller than that of the sports arena in Vienna, Nervi aimed to use his system of prefabricated *tavelloni* in ferrocement that he had already used to build the ribbed shell of the half-dome of *Salone B* in Turin in order to cover with a segmental dome the space of a 60-metre diameter floor plan. The ribbed shell of the *Palazzetto dello Sport* was arranged in the form of a radial structure consisting of 36 reinforced concrete spurs with all their branches exposed in order to make the work look like a peripteral temple. It comes as no surprise to learn that the spurs and their connection to the ribbed shell were the aspects most studied by Nervi.

In the first versions of the design, the spur followed the curve of the dome but was broken down into a lower sector, clearly designed as an actual spur with curves emphasising the support on the ground, and an upper forked sector with V-shaped beams in contact with the dome edge. Nervi raised that edge with triangular pleating to create a connection with the V-shaped beams that is not merely formal but intended to render the static points – into which the spiral ribs of the shell flow after traversing the band of raised edges – more expressive. The aim of that particular band was to connect the number of V-shaped spur beams to those of the spiral ribbing, performing a role not unlike that of the pendentives.

361-362. P. L. Nervi and A. Vitellozzi, *Palazzetto dello Sport*, Piazza Apollodoro 10, Rome, 1954-57.

361. "Progetto esecutivo. Prospetto (Working project. Elevation)," 1954 (CONI).

362. "Sezione Trasversale (Cross-section)," 1954-55 (BSTA, 024-2081_2).

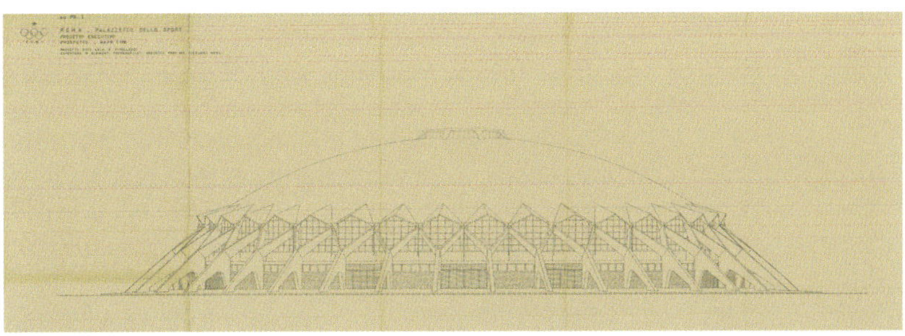

[23] *Relazione*, 19 September 1957, CONI, binder 36a.

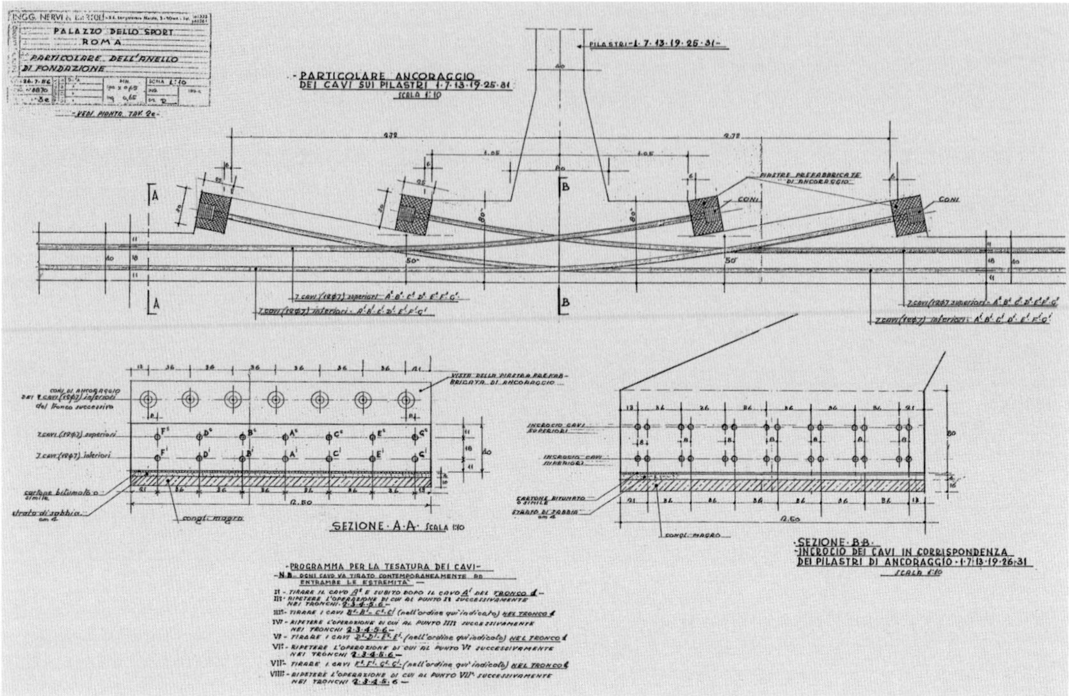

363. "Particolare dell'anello di fondazione (Detail of the foundation ring)," drawing dated 26 July 1956 (CONI).

Without altering the structural role and the number of structural components, in the final version Nervi changed the design of the spur to infuse it with a character that gives a greater sense of dynamic balance and loads descending from the dome in accordance with a taste that owes much to the flowing lines of a Gothic structure with rampant arches and buttress. In order to achieve this effect, he now used his knowledge of form to model a structure made out of continuous, inclined forks with three slender ends. The vertical support of the previous spur was reduced to a pillar that is as unobtrusive as possible, almost like a temporary strut of the fork. He defines this motif as a Y-shaped "trestle."[24] This term was intended to reflect the symbolic value intrinsic in the configuration he had now established for the structure, which was like a temporary work, almost hovering as though it would only be able to bear the load once the dome had been assembled and closed. This new hierarchy between the elements of the same configurations revealed the approach determining the juxtaposition between the mighty cantilever beams and the bracing arches in the cantilever beams of the *Stadio Berta* canopy.

Nervi's final design was what ensured that the spurs, now trestles, were able to take their rightful place in the graceful structural and construction device that was responsible for determining the thin dome of the *Palazzetto dello Sport*. Only by assuming that shape was it possible to ensure that the trestles no longer appeared to be resting on the ground as the abutments of a structure that is pushing down on them; instead, they became the extreme tentacles of that structure. The fact that Nervi intended to conceal the stresses of his cast-in-place reinforced concrete structure to make it more similar to a

[24] Nervi, *Nuove strutture* cit., p. 34.

364. P. L. Nervi and A. Vitellozzi, *Palazzetto dello Sport*, Piazza Apollodoro 10, Rome, 1954-57. Formwork for pouring trestles, 1957 (Maxxi - PLN, F00232).

ferrocement structure is also confirmed by the foundation system, consisting of a powerful and invisible ring of prestressed reinforced concrete encircling the trestles, like the "*sodo*" positioned by Brunelleschi at the base of his dome. The heads of the pre-stressing cables are located at the base of the trestles, which are thus, for all intents and purposes, stitched firmly to the ring.

Once he had introduced the trestle, Nervi gave up the pleated fold at the edge of the dome, which now became a wave to be created using special small vaults. The branches between the trestles and the ribs of the dome take place in the pendentive band in the same way as with the fans built in *Salone B* in Turin, using special prefabricated *tavelloni* made out of ferrocement. Compared to the patterns composed for Torre in Pietra and Turin, Nervi achieved a consistent design in the ribbed shell of the dome, with variations limited to the ring of long, narrow *tavelloni* around the oculus, whose dimensions were calculated to obtain spiral lines that flowed unbroken to the pendentive band. Because the *tavelloni* were prefabricated, Nervi was able to build the entire scaffolding of the dome. This included the whole pendentive band to be left on site once the reinforcement had been assembled and the concrete had been poured, in order to create decoratively interwoven invisible force lines. The result was another hyperstatic ribbed shell. Because the arena was located below the natural ground level and dug out only once the dome was complete, the relationship between the dome structure and the ground took into account the

365-367. P. L. Nervi and A. Vitellozzi, *Palazzetto dello Sport*, Piazza Apollodoro 10, Rome, 1954-57.

365. Construction of *forme madri*, 1957 (Maxxi - PLN, F00271).

366. Laying of dome *tavelloni*, 1957 (Acm-EPFL, 0172.02.0003.2137).

367. Positioning the steel reinforcement and pouring the calotte, 1957 (BSTA, 024-368).

construction at the design stage and was calculated to reduce the height of the temporary scaffolding required to assemble the prefabricated parts.

The construction work was carried out directly by Nervi & Bartoli without putting the work out to tender because it was believed to be the only firm capable of carrying out to perfection the process involved in what its own author had now identified as the "Nervi system." The building site was opened in July 1956; the construction manager was the engineer Giacomo Maccagno, and the plan was to finish the work in 420 days. All the cements were produced by the Unione Cementi Marchino cement plant in Civitavecchia. During construction work on the foundations, defects were identified in the concrete composition and quality, including the prestressed annular beam.

In order to produce the trestles, Nervi & Bartoli prepared formwork made out of wooden boards that could be reused several times, as had also been done by Fourré & Rhodes et Dumez for the pilotis of the Secretariat and *Bâtiment des Conférences*. The boards were not laid to create a pattern as they had been at the Parisian building site, and the concrete surface finish must have been imperfect because plaster cladding was required, as stated in the work specifications. Only when work was complete on the construction of the trestles did the Italian National Olympic Committee decide to leave the concrete exposed, to do away with the cost of plastering.

To achieve the exact configuration of the different folds of each set of *tavelloni* in ferrocement, Nervi changed the system he had used in Torre in Pietra and Turin and adopted layout criteria involving the use of special devices and tools indicative of his technical mastery, which was almost like that of a medieval builder. He built an actual-size segment of dome on the building site using a wooden board and had its surface plastered. In order to design the grid of lines to be followed by the geometrical pattern of the *tavelloni*, he used a stretched wire, as specified in the design, and prepared a special tool with a sliding bar made from a wooden truss. The grid of lines defined the axes of the dome ribs that were used as a basis for constructing a masonry form directly on the dome segment for each meridian. This form, always clad in plaster and with moulded sides, was used to manufacture what Nervi defined as the "*forma madre*" (mother form).[25] Once the *forma madre* was overturned, it generated a certain number of counter-forms that were used in turn to manufacture the *tavelloni*. This process ensured that the *tavellone* was identical to the *forma madre*. In the jargon used by the Nervi & Bartoli workers, the *forma madre* was referred to as "*nonna*" (grandmother), the counter-form as "*mamma*" (mother), and the *tavellone* as the "*figlia*" (daughter).[26] Unlike the *tavelloni* manufactured by Nervi in previous building sites, this ferrocement reinforcement was made up of a grid of bars enfolded by a single net. The small waved vaults of the pendentive band were made out of three special *tavelloni*; the ribs branching from the top of the trestles were the result of assembling other special pointed triangular *tavelloni*. Bars emerged from each *tavellone* to create links between the parts during the assembly of the reinforcements and concrete pouring in order to convert the part to a monolithic structure.

Production of the *tavelloni* proceeded at a rate of thirty units per day. By the end of December 1956, all the *tavelloni* were ready and stacked according to the various shapes. They were assembled in approximately thirty

[25] Pier Luigi Nervi, letter to Bruno Magrelli, 7 November 1957, CONI, binder 16.

[26] Tullia Iori, *Pier Luigi Nervi. Annibale Vitellozzi. Palazzetto dello sport a Roma. Un prototipo ripetibile e a buon mercato*, in, "Casabella," LXXIII, October 2009, No. 782, (pp. 51-109), p. 55.

368-371. P. L. Nervi and A. Vitellozzi, *Palazzetto dello Sport*, Piazza Apollodoro 10, Rome, 1954-57.

368. View (photograph by G. Gherardi and A Fiorelli; Unesco).

369. Detail of dome *tavelloni* (Acm-EPFL, 0172.02.0059.0356).

370. Aerial view (Unesco).

371. Detail of trestles (Unesco).

working days with the aid of a scaffolding of Innocenti pipes and a Wolf crane positioned near the oculus of the dome. For financial reasons, Nervi decided to install the *forme madri*, even though they were wrapped up in a thick oily layer that had been spread to make it easier to strip out the counter-forms. After positioning the reinforcement, the concrete was poured, again with the aid of the crane to lift the bucket through the oculus. Nervi & Bartoli was to perfect this method of carrying the concrete at their other building sites. One insulating layer containing vermiculite, one of Bitragar and sand, and alternating layers of Vetroflex and asphalt were laid on the calotte. Everything was coated with a bituminous paint known as Flicolor.

As soon as the dome was complete, in February 1957, the intrados was painted using limewash to increase the brightness of the *Palazzetto*, as well as to correct the *tavellone* production defects that were always present in Nervi & Bartoli's building sites. The scaffolding was then removed in order to dig out the games arena and then carry out the masonry and finishing work. Despite the limewash, this did not present the appearance of "numerous stains" left by the layer of "greasy substances" on some *tavelloni*.[27] However, these were only the special *tavelloni* that were used to form the *forme madri* and which were "therefore repeatedly soaked in oily solutions," as Nervi explained when he detailed a number of defective elements – one in every 108 in the same parallel, therefore 16 in total.[28]

The *Palazzetto dello Sport* was the result of the inexpensive assembly of universally recognisable elements designed to create different structural rhythms according to the kind of concrete, even though they were related through their design. Nervi portrayed it not in terms of any special aspect of its engineering but rather as an "honest work."[29] This definition underlies both the strengths and weaknesses of his idea of building correctly, which was dedicated to controlling the manufacturing process, his techniques, his costs, and the execution times. The adjective "honest" has nothing to do with the truth pursued by his contemporaries through the expression of exposed materials. Nervi himself was engaged during this very period in the construction project for the Unesco headquarters and in perfecting treatments for concrete intended to be left exposed, which was destined to become an international emblem of that type of truth. It comes as no surprise to learn that he saw the *Palazzetto dello Sport* as an "honest work," even though the main material was buried under layers of paint.

[27] Bruno Magrelli, letter to Nervi & Bartoli, 31 October 1957, CONI, binder 15bis.
[28] Pier Luigi Nervi, letter to Bruno Magrelli, 7 November 1957, CONI, binder 16.
[29] Pier Luigi Nervi, *Pier Luigi Nervi scrive*, in, "L'architettura. Cronache e storia," IV, January 1958, No. 27, p. 591.

chapter eleven
controlled brutalism: the *palazzo dello sport*

Study and construction of variable cross-section pillars

After the start of work at the building site in Paris for the construction of the Unesco headquarters, Nervi changed his opinion of the wooden formwork that he had continually criticised until the mid-1950s, up to the publication of the essay *Costruire Correttamente. Caratteristiche e possibilità delle strutture cementizie armate*, partly due to the creative limitations imposed on the production of complex figures. His designs for variable cross-section pillars signified his awareness of the creative potential inherent in the wooden formwork that until then he had ignored. The ruled surfaces and changing cross-sections were the clearest proof of his enthusiasm for imagining ever-varying pillar shapes that were the result of his static conception and building site virtuosity.

For its works after the experience with the Unesco construction project, Studio Nervi began to design the laying of the wooden boards using schematic but significant drawings. Nervi's new care over the quality of the concrete surfaces led him in many cases to make life-size models in order to check the quality of the surfaces, as had been done on the Unesco building site. Now Nervi's prototypes were no longer concerned only with checking the static performance of a structure or experimenting with a construction technique but belonged to the kind of prototypes that architects were increasingly having made on building sites to allow them to go beyond the roughness and imperfections of the first *béton brut*. Sandblasting, which was always found effective for correcting surfaces after stripping on the Unesco building site, was now included among Nervi & Bartoli's practices for achieving the goal of perfect concrete. The artistic refinements of Nervi & Bartoli were also crowned by the use of bush hammering for cast-in-place concrete and very sophisticated methods of treating the prefabricated parts. However, Nervi never belonged to the ranks of builders who decided to choose *brut* routinely for the manufacture of concrete, and he had no compunction about continuing to use painting, as in his early works.

372-373. P. L. Nervi (with the preliminary cooperation of M. Piacentini), *Palazzo dello Sport*, Piazzale dello Sport, Rome, 1954-59.

372. Preliminary project with flat roof, 1954-55 (P. L. Nervi, *Buildings, Projects, Structures. 1953-1963,* New York, Frederick A. Preager, Publisher, 1963, p. 82).

373. "Sezione. Vista Interna (Cross-section. Internal view)" (Acm-EPFL, 0172.02.0003/2140).

Nervi attempted his first significant exercise in controlling the quality of cast-in-place concrete in wooden formwork left exposed when he constructed the *Palazzo dello Sport* in Rome's EUR. This was the outcome of a joint-venture with Piacentini and Valle, commissioned by the Italian National Olympic Committee in 1954 and built by Nervi & Bartoli between 1957 and 1959 (the construction manager was the engineer Gustavo Baracchi; the building site was organised and subdivided into 48 sectors). The *Palazzo dello Sport* is colossal, with a dome measuring 100 metres in diameter, larger than the *Palazzetto*, and almost a physical expression of the limits glimpsed by Nervi in the processes of creating large structures. It constitutes a veritable catalogue of concrete elements, precast and cast in place, which had been tried and tested by Nervi & Bartoli and was taken here to a level that had never previously been achieved by the firm in terms of refined lines and perfected workmanship. This was also the first time that Nervi decided, right from the design stages, to leave all the skeletons of the cast-in-place reinforced concrete exposed and, in the wake of the experience he accrued in Paris, to state the necessary technical details for achieving perfect, unclad surfaces along with the work to be carried out after stripping, in the event of defects of execution.

The engineer who had cut his teeth building grand vaults set out to differentiate the structure of the *Palazzo dello Sport* from that of the *Palazzetto dello Sport* and hoped to surprise his audience with a bold flat roof that he designed in 1954–55, in the form of radial beams converging into a "rose window," stiffened by slabs aligned in accordance with the moments, transposing certain principles that informed the design for the *Bâtiment des Conférences* roof. His structural design saw the re-emergence of the interplay of forces that he had experimented with in his first flat roofs with central skylights, in the *Teatro Banchini* and the *Cinema Teatro Augusteo*. A system of opposing motifs guided the design of the arch structure of the tiers, dotted with inclined pillars with variable cross-sections loaded by the supports of the radial roof beams. "[…] it would be at the limits of current production capabilities and dangerous to attempt," he admitted.[1]

Between 1955 and 1957, Nervi drew up the design for the *Palazzo dello Sport* with the roof structure in the form of a dome, no longer with the *tavelloni* of the *Palazzetto* roof, but with waved beams in ferrocement so that the shell could be stretched over the vast circle measuring 100 metres in diameter. The floor plan radiated the aura of a Beaux-Arts composition with an orderly succession from the central playing field, the discs of the tiers, extensive areas for the outflow of spectators, narrow stairs and washrooms, and lastly the grand radial access stairs.

For the structure, Nervi included the device of curved tiers and posts used in the previous project for the lines through which the weight of the dome descended, with additional inclined posts that then branched out into the fans with three ribs supporting the dome ring. The study sketches for both the node onto which the variable cross-section inclined pillars and those of the fan motif or those of the "rose window" bars converged in the centre of the dome are eloquent demonstrations of the creative margins in the design

[1] Pier Luigi Nervi, *Architettura e strutturalismo*, in, "Casabella Continuità," XXIII, July 1959, No. 229, p. 5.

[2] *Relazione sull'andamento dei lavori per la costruzione del "Palazzo dello Sport," alla data del 31 ottobre 1958*, CONI, binder 4.

[3] Tullia Iori, *Pier Luigi Nervi, Marcello Piacentini e gli altri*, in, Pace, *op. cit.*, (pp. 97-103), p. 101, note 7.

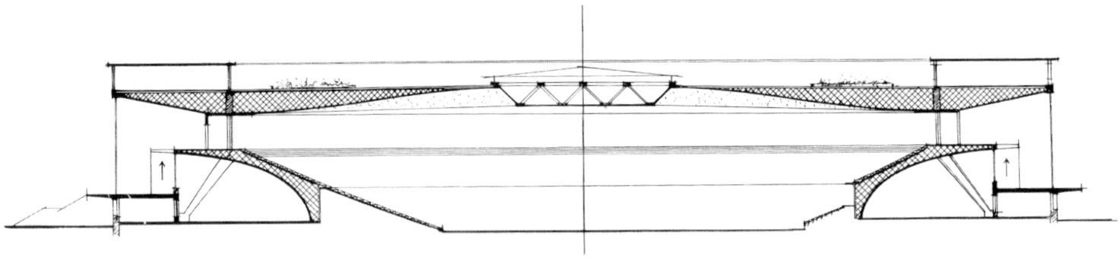

of ribs with unequivocally ornamental features that still existed in Nervi's structural system.

When developing the precast ferrocement ashlars of the arched roof beams, Nervi changed from the waved profile that he still employed in 1957 to the V-shaped cross-section profile designed in 1958, perforated stiffenings for the passage of the piping, and an infill slab, which was also prefabricated. The ashlars were fitted without any visible stiffening diaphragm and also without the subtle relief strip planned for the dome in Vienna, but simply juxtaposed in order to achieve a perfect continuity of lines from the abutment ring to the rose window. That joint detail, so simple and essential, revealed that Nervi's goal of a perfect structure, welded into a monolithic structure by hyperstatic traits that made it simultaneously shell and ribbing, involved the clear-headed renunciation of the aesthetic value that he had discovered in the diaphragms used in *Salone B* in Turin. The gallery stairway is another virtuous figure of Nervi's structure for the *Palazzo dello Sport*, shaped with lines expressing the plasticity of the reinforced concrete and characterised by the way the pairs of supports converge unusually on points that make them appear to be traversed by an animal-like tension, like paws drawn up ready to pounce – "dinosaur-like" stairways were to be defined during the course of the building site.[2] After the design for the sports arena in Vienna and after the *Palazzetto dello Sport*, the *Palazzo dello Sport* in Rome is a quintessential example of the way Nervi's whole structures "breathed" through the special cavities that could be found everywhere, not only in the pleated dome shell from which air emerged at the top, or those formed deliberately under the tiers and above the gallery, from where air was sucked and which carried pipes or housed the machinery for the air conditioning system developed by Società Dell'Orto-Chierigatti of Milan.

Piacentini's inability to manage the design of Nervi's complex structure meant that he was relegated at the outset to the role of the author of a Fascist era façade, an academic with a "Roman," "masculine" "character," as he himself pointed out,[3] until he was effectively ousted from the design of a work that was comprehensively stripped back to its structure. The choice of a continuous casing of glass along the gallery marked the final dissolution of the architecture of Piacentini's façade, and the triumph of Nervi's structural

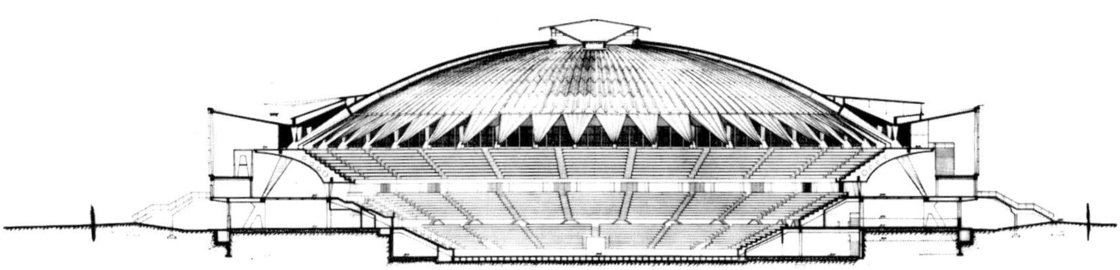

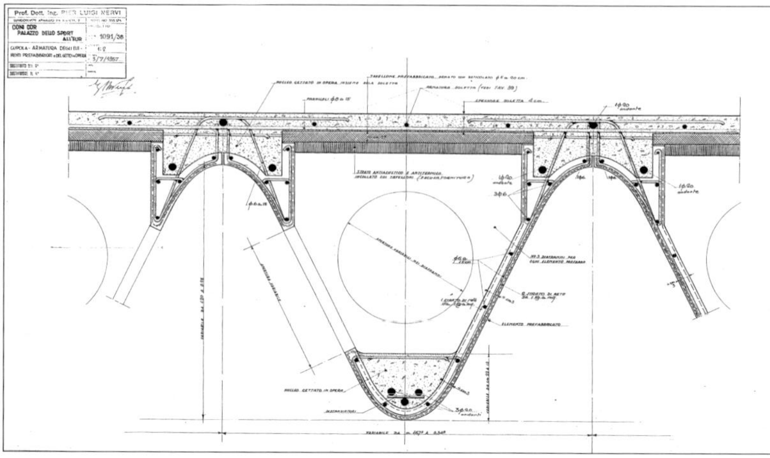

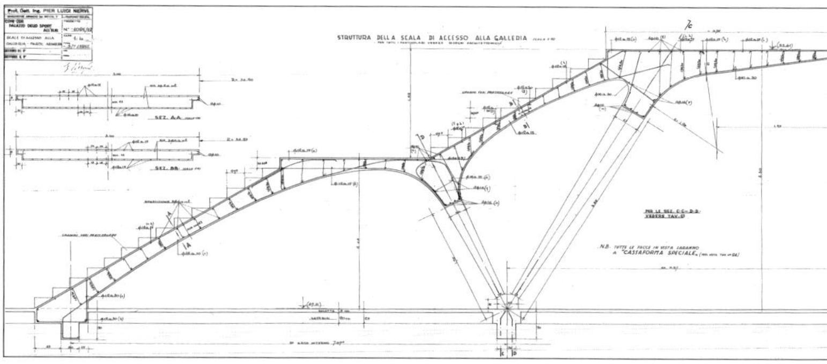

374-375. P. L. Nervi (with the preliminary cooperation of M. Piacentini), *Palazzo dello Sport*, Piazzale dello Sport, Rome, 1954-59.

374. "Cupola. Armatura degli elementi prefabbricati e del getto in opera (Dome. Reinforcement of prefabricated elements and in situ pouring)," drawing dated 3 July 1957 (CONI).

375. "Scale di accesso alla galleria. Partic.[olare] Armatura (Access stairs to gallery. Detail of reinforcement)," drawing dated 3 July 1957 (CONI).

architecture. Through that clear screen it was possible to see the animal-like stairs, the inclined variable cross-section pillars, the beams and arched shells of the tiers, and even the uncertain reverberation of the dome ribs. Unsurprisingly, the sequence of photographs assembled by Nervi to illustrate the *Palazzo dello Sport* opens with a night shot where, against the impenetrable black background, the artificial lighting emphasises the strength of the structural lights, as in an X-ray of the skeleton or a photograph of photoelastic examinations.

For the first time, Studio Nervi drew up working drawings, specifications, and design reports that went into details of the manufacture of all elements, even explaining, through a series of graphic diagrams, the specific processes used to prefabricate the ashlars – the ashlars described were still the waved type in ferrocement. The documents were used to announce a call for tenders for the award of a contract to build the rough structure to a firm that still had to be able to follow Nervi & Bartoli's technical methods (Nervi waived his rights to all the patents for the technical processes required for construction of the work). Nervi also drew up similar, yet less detailed, documents in approximately the same period for the construction of the *Stadio Flaminio* in Rome, which was also to be used for the Olympic Games.

The documents prepared for the call for tenders are the most powerful evidence of a radical cultural change by Nervi in this concept of reinforced concrete and his formal values. Against the backdrop of the *Palazzo dello Sport* building site, Nervi signed up fully to the aesthetics and principles of international brutalism and went beyond the statements he made in *Costru-*

ire Correttamente on wooden formwork. In his previous work, the *Palazzetto dello Sport*, he had not yet refined the technical solutions required to achieve an ideal level of finish for his exposed concrete. Not until the new work did he decide to try out the techniques that he had practised and seen on the Paris building site.

The "*Capitolato speciale di appalto*" (special tender specification) revealed the significance that Nervi attributed to the *Palazzo dello Sport* as one of the "most challenging and significant examples of present-day cement construction technique."[4] His intention was clearly to direct a work in Rome that would serve as a counterpart of the Unesco headquarters and become known for the formal quality of the finish of its reinforced concrete structures – this is what he was alluding to when he used the words "fundamental and essential architectural function." "[...] the aesthetic and architectural result is therefore very largely determined by the quality and accuracy of its execution."

The temporary works that he had always condemned in his writings since the 1940s, the wooden formwork, would have to be executed, as Nervi himself wrote, with "absolute accuracy," paying attention to the "formation of the edges," because the "pattern of the boards" will be "visible on the surface of the pourings" and will constitute an "interesting aesthetic enhancement." He was actually specifying the concrete pouring criteria that had been followed on the Paris building site. He therefore recommended a system of tongue and groove joints and "extremely accurate pourings to prevent concrete cobweb patterns ("*ragnature*") and areas with superficial defects that would not be acceptable under any circumstances." The cult of perfect cast-in-place concrete had reached the point where "if the Construction Manager irrevocably determines the pouring of an exposed structure to be superficially defective, its demolition and reconstruction shall be ordered." For the execution of parts in ferrocement, he recommended "perfect application of special executive procedures involving prefabricated elements." Nervi specified in his notes accompanying the working drawings the manufacturing methods of both types of formwork planned for concrete to be cast in place, which he described as "special formwork":[5] one version made out of ferrocement and another made out of wooden boards connected by means of tongue and groove joints. It is significant that Nervi mentioned some of his recent works, the dome in Chianciano and the Unesco buildings, in the specifications regarding the quality of the concrete.[6]

The variable cross-section pillars played a fundamental role among the various parts of the *Palazzo dello Sport* intended to emphasise the formal qualities of Nervi's structures and his new reinforced concrete. At a certain point of the design, the issues of the material and structure must have seemed to him to be unresolvable, or at least difficult to reconcile, in the shape and surface of the pillars, if the goal had to be a perfect and unique understanding of the nature of reinforced concrete and the static genesis of the figure. Even though his Parisian experience had revealed to him the expressive power of impression lines, he was now apparently unconvinced of this approach, and his insight about the essence of concrete may have been behind his momentary uncertainty.

[4] *Capitolato speciale di appalto delle opere in cemento armato*, p. 5, CONI, binder 4.

[5] Drawing No. 1091/24, CONI.

[6] *Capitolato speciale di appalto delle opere in cemento armato* cit., p. 9.

For the finish of the pillars, Nervi hesitated between shaping the surface using "special staved formwork," where the regular impressions left by the boards emphasised the geometry of the ruled surfaces, and a perfectly smooth surface to be obtained by using his ferrocement to create a special mould with two sections that were connected in situ. In both cases, Nervi nevertheless wanted the two different qualities to be perfect: the subtle impressions of the lines and the smooth surfaces without any ripples. The use of the term "stave" instead of "board" for the wooden formwork is evidence of his intention to build a device so perfect that it would be unusual for contemporary Italian building sites. One particular adjective clarified the level of polish he wished to obtain for the concrete made using the ferrocement mould: he specified it was to be executed in an "extremely accurate" manner.[7]

The opposition between wood and ferrocement that apparently should not have concerned him any longer cropped up again at this time in an isolated reflection, far from his experience with Breuer and Zehrfuss, and detached from his excited discovery of *béton brut* in Marseille. That opposition no longer concerned building site economy or technical limitations in the construction of a figure but instead led him to the heart of a dilemma over the formal values of the surface of his variable cross-section pillars, in other words the different reactions of that surface to the light that revealed – in the two different moulding processes – either the physical essence of the material or the geometrical and static constitution of the structure. There is no doubt that if it had not been for his experience in Paris, Nervi would never have questioned the values of the material surface. The option of a ferrocement mould appealed to him because the absolute level of polish that he had already managed to impress on the cast-in-place reinforced concrete for his floors with isostatic ribs could now be extended to other complex static figures, and it had thus become possible for concrete to be established everywhere due to its essence as an artificial product unlike any other. That alienation of the material, which is unrecognisable as the *béton brut* embraced by Marcel in the Fontenoy canopy of the Secretariat, was probably the effect that Nervi desired for his concrete if he were merely following his instinctive understanding of the nature of the compound. The fact that Nervi was aware he had stumbled on a question that was far from merely technical, but which embraced two views of the material, which could be implemented using either wooden formwork or ferrocement moulds, is borne out by the fact that he published a sheet of sketches for the purposes of comparison showing a "study of the inclined pillar formwork, ferrocement solution" alongside drawings and photographs of the pillars ruled by lines and the wooden formwork.[8] In his study sketches and working drawings of the pillars, Nervi had not yet opted for one or other of the options, although he depicted the generating lines of the ruled surfaces that coincided with the boards in the case of the wooden formwork: "surfaces produced using 'special formwork' (in ferrocement or wood boards connected by means of tongue and groove joints)."[9]

For Nervi, choosing the wooden formwork option did not mean slavish devotion to the aesthetic of *béton brut* and the decorative effects achieved by laying the boards but instead meant using the components of the formwork to convert the concrete surface into a geometrical site for the impressions generating the figure – marks indicating the construction of the ruled surface. In places where the weaving of the boards into a continuous ascending strip did not succeed in enfolding the figure, Nervi planned to use staves "arranged in a horizontal direction,"[10] as in the variable cross-section pilotis of the Unesco Secretariat.

The inclination of the pillar gave rise to a special formless base connection that Nervi isolated with a smooth surface, a kind of "base" recessed by a few centimetres in relation to the stem. He initially planned to clad this with a ring of metal, almost as if in reference to the steel hinge he had devised for the pillar in studies dating from 1956.

The trend of the inclined pillar cross-section was also subjected to different solutions that indicated the margins for formal creativity in the study of a figure, which Nervi however always justified by static reasoning alone. In the design of July 1957, the circular base gradually modulated up to a rectangular top, following a similar growth trajectory to that of the Unesco Secretariat pilotis. A marginal note to the drawings, stating that "the shape of the large pillars might end up being quite different from the working drawings, but would in any case feature ruled surfaces,"[11] indicating Nervi's desire to achieve a variable profile that could be distinguished from the one built for Unesco. The inclined pillar was built with a variable cross-section that started at the base with a rectangle extending in the direction of the gallery in order to stabilise the pillar (performing the same static function as the base ellipse of the Secretariat pilotis) and ended at the top in a rectangle oriented along the curved beams of the tiers, giving rise to a cruciform cross-section in some parts of the stem. Nervi proposed the same variable cross-section pillar profile for a design for Florence's municipal swimming pool in November 1957.[12] Nervi explained the significance of the spur, which had now become transformed into a variable cross-section inclined pillar, as a figure expressing the continuous quest for strength through form, in succinct terms: "The inclined pillars of the large gallery [...] exactly reflect the pressure line of the dome thrust arising as a result of vertical actions brought about by the dead weight of the roof perimeter ring and of the floor and overlying tiers."[13]

Nervi examined the arrangement of the formwork staves for the branching of the structures above the inclined pillar, made up of a base connection, the vertical pillars supporting the gallery roof, and the coupling of the dome abutment fan. Even though it was intended to be a monolithic structural node, Nervi broke down the parts with two grooves or "*scuretti*,"[14] achieved by positioning a slat in the formwork at the concrete casting, as in the walls of the *Bâtiment des Conférences*. A note about the choice of two alternative types of formwork is shown on the drawing: "The exposed faces of the main structures in cast-in-place reinforced concrete will remain as they were when stripped, without further plastering or cladding. The associated formwork will be constructed with particular care out of parallel boards measuring 8-10 centimetres in width and connected by means of tongue and groove joints: or the pouring surfaces will be perfectly smooth and the formwork in this case will be constructed out of ferrocement."[15]

In April 1958 or thereabouts, the decision to build the variable cross-section inclined pillars with rectangular base and top through the use of wooden formwork led Nervi & Bartoli to construct a prototype just as Fourré

[7] Drawing, 9 June 1957 in, Nervi, *Nuove strutture* cit., p. 74.
[8] *Ibid.*
[9] Drawing No. 1091/15, CONI.
[10] *Ibid.*
[11] *Ibid.*
[12] See drawings No. 3363-6bis, 3363-6ter, 3363-6e; 3363-12e, CSAC, n. inv. pra. 337, n. id 1385, coll. 41/3.
[13] Nervi, letter to Viganò, 20 November 1959 cit.
[14] Drawing No. 1091/24 cit.
[15] *Ibid.*

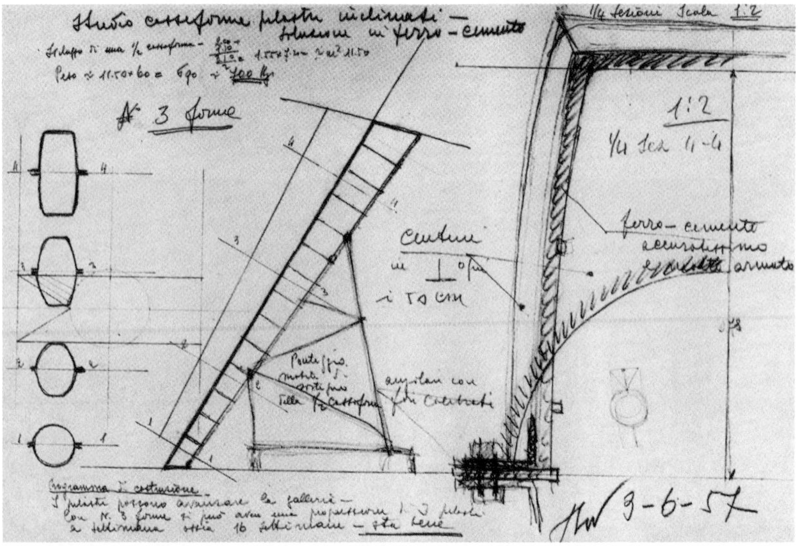

376-379. P. L. Nervi (with the preliminary cooperation of M. Piacentini), *Palazzo dello Sport*, Piazzale dello Sport, Rome, 1954-59.

376. "Studio casseforme pilastri inclinati. Soluzione in ferro-cemento (Study for inclined pillar formwork. Ferrocement option)," drawing dated 3 June 1957 (P. L. Nervi, *Buildings, Projects, Structures. 1953-1963*, New York, Frederick A. Preager, Publisher, 1963, p. 74).

377. Formwork of test pillar, April 1958 (photograph by G. Gherardi and A. Fiorelli; AFG, 1019).

378. Test pillar, April 1958 (photograph by G. Gherardi and A. Fiorelli; AFG, 928).

379. Execution of inclined pillars, May-September 1958 (photograph by G. Gherardi and A. Fiorelli; AFG, 1352).

EDILMAC

380. P. L. Nervi (with the preliminary cooperation of M. Piacentini), *Palazzo dello Sport*, Piazzale dello Sport, Rome, 1954-59.
Execution of inclined pillars and steps, May-September 1958 (photograph by G. Gherardi and A. Fiorelli; AFG, 1261).

& Rhodes et Dumez had done on the Paris building site. This was intended to check the results of the formwork system on the concrete finish – "to check the volumetric and architectural effect as well as the size of the formwork," as was written in one document.[16] Nervi & Bartoli built reusable formwork that could be opened, equipped with a steel frame that also included adjustable ties to reinforce the base sector where the pouring pressure was greatest. It also featured double planking: the layer in contact with the concrete was made out of planed slats that were connected by means of tongue and groove joints and fitted in accordance with the ruled surface geometry. The two parts of the formwork were joined by a dense set of bolts. A system of staves similarly reinforced by a metal frame was also used to produce the reusable opening formwork of the stumpy vertical pillars that flared toward the top, which were, as usual with a variable cross-section, located in front of the inclined pillars. At the end of May, the first four inclined pillars had been constructed as well as the corresponding flared vertical pillars, while 14 inclined pillars and the corresponding vertical flared pillars were already in place by the end of June. Both types of pillars were completed at the end of September.[17] Encouraged by the success of the work, Nervi proposed, still in September, a similar kind of variable cross-section pillar, now arranged vertically, for the construction by Nervi & Bartoli of a diving board for the swimming pool at the *Foro Italico* in Rome.

The system of formwork prepared by Nervi proved itself capable of producing pillars with almost perfect impressions, where even the joint between the two sectors was aligned with the joint between two staves – hence without its specificities being expressed. The building site photographs captured the quality of Nervi's *béton brut* immediately after stripping away the formwork and revealed the builder's anxiety to achieve a final result without imperfections, that was up to the level of international exposed concrete masterpieces.

Floors with isostatic ribs

In the same way as the variable cross-section pillars, other figures helped characterise Nervi's cast-in-place and exposed reinforced concrete structure. The gallery stairways became another significant part of his particular brand of *béton brut*. These were also achieved by carefully assembling the formwork staves, which were all laid horizontally, to facilitate construction in accordance with the curvature of the ramp structure. The fact that the legs of the stairways are also traversed by impressions of boards, again fitted horizontally, should not be overlooked in order to appreciate the "hardness," as Le Corbusier would have defined it,[18] that Nervi sought when setting up the formwork, without ever indulging in any decorative effect. While the vertically ruled surfaces of the inclined pillars express the geometrical genesis of the figure, the stairway surfaces, all horizontally striped, stand as an admonishment against the decorative style fashionable in international *béton brut*. The *béton brut* that Nervi produced for the *Palazzo dello Sport* clearly belonged to the radical international approach of which the works of Bo Bardi and the Brazilian architects were prominent examples, with the essential difference that Nervi always set out to assemble boards of the same size to perfection.

The reinforced concrete of the *Palazzo dello Sport* also saw the reappearance of the isostatic ribs and finish implemented by Nervi & Bartoli in the floors for the *Opificio Gatti* in Rome and the *Manifattura Tabacchi* in Bologna. Nervi had also proposed the use of Arcangeli's floor in the design for the Unesco headquarters to characterise the exposed reinforced concrete structure. When we consider the presence of ferrocement *tavelloni* in the arched shell under the tiers, the V-shaped ashlars, also in ferrocement, of the dome beams and the floor with isostatic ribs shaped using ferrocement, Nervi's uncertainties over the type of material to be used for the formwork of the variable cross-section pillars is a sign of his aspiration to achieve a single formal expression for all the cement structures in the *Palazzo dello Sport*, in other words one that most closely reflected the only possible truth Nervi could conceive of for that artificial material.

For the main floor of the *Palazzo dello Sport* gallery, Nervi firstly envisaged joists and slabs to be cast in place in "special formwork" to control the quality of the "exposed surface (intrados)." As with the variable cross-section pillars, as an alternative he envisaged using "perfectly smooth and regular" ferrocement formwork.[19] For the final version, he chose the deep beam solution of the floor with isostatic ribs that had been one of the options back in 1949 for the floors of the *Manifattura Tabacchi* in Bologna. For the floor of the *Palazzo dello Sport*, Nervi wished to attain the level of perfection that he had achieved in floors previously constructed by Nervi & Bartoli, which he indicated as a template in the tender specification: "The intrados surface finish must be of the type achieved in the *Manifattura Tabacchi* Bologna."[20] In order to correct certain imperfections that had arisen specifically in the Bologna floors where the reinforcement netting had been visible through the surface, Nervi advised either form spacers to be prefabricated in cement or spreading a "very dense cement grout"over the formwork.[21]

[16] Mario Saini, letter to Gustavo Baracchi ,10 July 1959, CONI, binder 8.

[17] *Relazione sull'andamento dei lavori per la costruzione del "Palazzo dello Sport," alla data del 30 settembre 1958*, CONI, binder 4.

[18] Gargiani, Rosellini, *op. cit.*, pp. 550-53.

[19] Drawing No. 1091/8, CONI.

[20] *Capitolato speciale di appalto delle opere in cemento armato* cit., p. 12.

[21] *Ibid.*, p. 9.

381. P. L. Nervi (with the preliminary cooperation of M. Piacentini), *Palazzo dello Sport*, Piazzale dello Sport, Rome, 1954-59.
Portion of floor with isostatic ribs (photograph by G. Gherardi and A. Fiorelli; P. L. Nervi, *Buildings, Projects, Structures. 1953-1963*, New York, Frederick A. Preager, Publisher, 1963, p. 70).

In April 1958, Nervi & Bartoli developed a design for their usual formwork that involved scaffolding sliding along rails in order to pour the floor with isostatic ribs. The forms were made out of ferrocement and joined together in the same way as the scaffolding used for the *Opificio Gatti*. Nervi & Bartoli began to build the first 12 sectors of the gallery floor in May and completed the work at the end of September. The building site documents do not specify any finish after stripping. The photographs circulated by Nervi immediately after completing the *Palazzo dello Sport* depict the usual isostatic floor sectors coated in white paint.

Prefabrication in ferrocement

At the design stage, when the contract for the work had not yet been awarded to any building firm, Nervi prepared a set of documents to explain the ferrocement prefabrication process typically used by Nervi & Bartoli and which he intended to use for the *Palazzo dello Sport*. The crucial element of the roof was represented by the beams, which were to be prefabricated out of waved ashlars. The stiffening diaphragms were inside each ashlar, as in the waved beams used for San Michele di Pagana, but contained a hole as a "passage" to allow the installation of system piping. As planned for the Vienna sports hall, the ashlars were closed by *tavelloni*, against which a sound absorbent panel was applied (Nervi also explained that the panel performed an insulating function). The fact that all the ashlars, except those of the annular section around the oculus, are perforated transforms the dome shell into an acoustic sounding box traversed by system piping and insulated against heat and sound. For the dome ashlars cross-section, in addition to the usual waved cross-section, Nervi also planned a V-shaped cross-section, again with internal stiffening diaphragms and perforated to allow the passage of piping. This V-shaped ashlar was also to be constructed for the dome of the *Palazzo*.

The transition of the shell profile from waved to pleated took place against the backdrop of his growing popularity in the field of contemporary reinforced concrete construction of jointed shells with a corrugated cross-section, of which the Unesco *Bâtiment des Conférences* was a prime example. From then on, Nervi preferred putting together beams that give rise to a pleated shell characterised by sharp edges, even though he continued to describe these structures as "waves." Only in his final prestigious masterpiece, intended to house no less a figure than the Pope, did he revisit the typical waved design of his first celebrated shells – not without a touch of nostalgia.

For prefabrication of the ferrocement ashlars, Nervi considered extending the system for creating several forms that he had already tried out for the *tavelloni* in order to produce many parts simultaneously. He therefore drew a "diagram of the successive stages of element fabrication"[22] in order to illustrate the construction process, involving a masonry matrix, a ferrocement form (referred to on the *Palazzetto* building site as the "*forma madre*"), and a counter-form, again in ferrocement. Nervi specified the terminology of his prefabrication process: "ground form," "prototype element," "forms," "prefabricated elements." In the jargon used by the workers, "prototype element," "forms," and "prefabricated elements" corresponded to "*nonna*," "*mamma*," and "*figlia*." The "ground forms" were made out of masonry with a "perfectly smooth" contoured surface.

In order to shape the curve of the beams, Nervi specified the construction of a portion of the dome on the floor ("form on ground"). On this he constructed the six "ground forms" of the elements making up a beam and the three "ground forms" of the abutment fan (in the drawing, the "ground form" is referred to as a "template"). As usual in his other building sites, at this stage of developing the structural prefabrication process, Nervi was still "lime washing all the elements making up the hall roofing vault,"[23] with the exclusion of the fan connections along the abutment. In order to achieve perfect parts in ferrocement, he had to use "forms" with "internal surfaces (those corresponding to the visible surfaces of the finished element) [...] that were absolutely regular and smooth." The polished surface was also a consequence of the consistency and manner of application of the cement mortar, to be mixed so that it was "doughy" and spread by hand "in order not to leave internal voids."

When the work was awarded to Nervi & Bartoli, Nervi went back to the system of pre-fabricating the beam segments and fan structure that he had used in the building site at San Michele di Pagana. In one sector of the building site, subsequently shielded by sliding roofs, his workers built nine "ground forms" corresponding to the types of V-shaped ashlars to be prefabricated. The forms were made out of brick masonry with the cavities lined with smoothed mortar and with a curved base and sides. Through the use of special templates, made out of a wooden truss girder, each positioned in the cavity of the ground form to be shaped, the workers succeeded in obtaining the required curved surfaces for each beam section. Still following Nervi & Bartoli's routine, the workers inserted the pre-shaped metal reinforcement in the various forms, applied prefabricated perforated diaphragms, and then spread the cement mortar. The parts were removed from the forms by means of a crane sliding along rails located alongside the forms.

Prefabrication of the V-shaped ashlars began in March 1958, when Nervi & Bartoli prepared "counter-forms," "moulds," and "rigs" (*banchi*) and had already completed 126 units.[24] In a departure from his other building sites, this time Nervi finished off his V-shaped ashlars using a trick learned from the Unesco building site. All the parts were taken from the store to be sandblasted and then re-stacked. Sandblasting was not included in the special tender specification, and Nervi & Bartoli had to issue an invoice in order to be paid for this work.[25]

[22] Drawing No. 1091/34, CONI.
[23] *Capitolato speciale di appalto delle opere in cemento armato* cit., p. 16.
[24] *Relazione sull'andamento dei lavori per la costruzione del "Palazzo dello Sport," alla data del 31 marzo 1958*, CONI, binder 4.
[25] Mario Saini, letter to Gustavo Baracchi, 10 July 1959, CONI, binder 8.

382-384. P. L. Nervi (with the preliminary cooperation of M. Piacentini), *Palazzo dello Sport*, Piazzale dello Sport, Rome, 1954-59.

382. Ground form for prefabrication of V-shaped ashlars, March 1958 (CONI).

383. Building site, 1958 (photograph by G. Gherardi and A. Fiorelli; AFG, 1041).

384. Mould for prefabrication of spindle-shaped pillars for the facade, 1958 (photograph by G. Gherardi and A. Fiorelli; AFG, 1033).

In order to check the installation of the ashlars and the installation of the piping for the systems, in May Nervi began planning the assembly of a special actual size model on the building site or an "experimental sector of the dome,"[26] which was to be built in approximately June–July in the centre of the *Palazzo*.

Building work began on the dome with the installation, in October, of the permanent ferrocement formwork for the fans, with the rising of the metal scaffolding. It continued with the installation of the ashlars, laying of the reinforcement, and pouring of the cement coating. By the end of November, all the ashlars had been prefabricated, although *tavelloni* in ferrocement still had to be constructed to be applied above the ashlars.

In areas of the building site to be used for prefabrication, near the "ground forms" for the production of the V-shaped ashlars, on the other side of the crane rails, the template of a vault sector known as a "plafond"[27] or "gallery false ceiling"[28] was prepared using wooden framework. This was produced using diamond-shaped *tavelloni* between the curved beams of the tiers. The *tavelloni* were prefabricated and put in place between May and October 1958. In the special tender specification, Nervi used yet another term to refer to what his

[26] See Nervi & Bartoli, letter to the Construction Management, 19 May 1958 and Bernardo Barotta, *Pro memoria per il Sig. Saini*, 11 June 1958, *ibid.*; *Settore sperimentale della cupola. Analisi del prezzo*, 11 June 1958 and Mario Saini, letter to Gustavo Baracchi, 11 June 1958, CONI, binder 4.

[27] *Capitolato speciale di appalto delle opere in cemento armato* cit., p. 10.

[28] *Relazione sull'andamento dei lavori per la costruzione del "Palazzo dello Sport," alla data del 31 luglio 1958*, CONI, binder 4.

385-388. P. L. Nervi (with the preliminary cooperation of M. Piacentini), *Palazzo dello Sport*, Piazzale dello Sport, Rome, 1954-59.

385. Actual-sized model of the dome, June-July 1958 (photograph by G. Gherardi and A. Fiorelli; AFG, 1264).

386. Assembling the permanent formwork for the fans, October 1958 (photograph by G. Gherardi and A. Fiorelli; AFG, 1472)

387. Assembling the permanent formwork for the fans, October 1958 (photograph by G. Gherardi and A. Fiorelli; AFG, 1554).

388. Prefabrication of diamond-shaped *tavelloni* for the gallery false ceiling, May-October 1958 (CONI).

workers called in their jargon "*mamme*," which were now defined as the "negative" of the parts: "On that template will be prepared the negative of the prototype elements from which the forms are to be obtained as in the case of the dome elements. Given that 48 units had to be produced for each element, fewer forms were required than those required for the dome element. In both of these cases, the number of forms depended on the prefabrication programme."[29]

The other prefabricated elements intended to support the continuous circular glazing of the *Palazzo dello Sport* were the spindle-shaped pillars in reinforced concrete with a three-arm cruciform cross-section constructed between May and September 1958. Drawings such as the one depicting the ironwork in the reinforcement of these pillars offer exemplary evidence of the control of technical accuracy pursued by Nervi in the construction of his various concrete parts. In order to prefabricate the spindle-shaped pillars, Nervi invented a special mould consisting of three shaped parts, one in masonry moulded on the ground and two in ferrocement used to produce sides that could be lifted once the pouring had set. All the surfaces of the three parts of the mould were "perfectly smooth and regular."[30] In order to accelerate the production of the pillars, Nervi used his usual assembly line-based principles – the two ferrocement parts were moved by means of a hoist running on rails so that they could immediately be reused in the adjacent masonry form. The tiers were also other parts prefabricated by Nervi, shaped using his usual system of ground forms.

Finishing works

Once all the cast-in-place concrete parts had been stripped, and the prefabricated parts had been fitted, the *Palazzo dello Sport* looked like a colossal structure with all its concrete surfaces exposed. Only the photographs remain to bear witness to the effect of the Nervi system in its rough form, stripped absolutely naked, because Nervi and his customer decided to take steps to

[29] *Capitolato speciale di appalto delle opere in cemento armato* cit., p. 8.

[30] Drawing No. 1091/25, CONI.

389-391. P. L. Nervi (with the preliminary cooperation of M. Piacentini), *Palazzo dello Sport*, Piazzale dello Sport, Rome, 1954-59.

389. Unfinished gallery false ceiling, 1958 (CONI).

390. Gallery false ceiling with painted *tavelloni* (CONI).

391. Structure in unfinished reinforced concrete, 1958 (CONI).

change the finish of some parts with the advice of Vitellozzi. Again with the aim of correcting the stains left on the diamond-shaped *tavelloni* during the prefabrication process, the ribbed, curved shells of the gallery were painted, accentuating the join between white infill parts and the grey structure in *béton brut*. The complex joint between pillar and fan, which had already proved problematic due to the polished surfaces and surfaces striped by impressions converging on it, was modified by painting the lower sector grey. Even though the sandblasting of the dome ashlars was intended to guarantee a finely worked surface to be left exposed, before the building was opened, the dome was nevertheless painted white, as were the curved shells, with the exception of the fans. As with the sandblasting work, it was not included in the specification, and Nervi & Bartoli had to issue a separate invoice for this work.[31] The reasons that led Nervi to paint his prefabricated elements were determined by the usual manufacturing defects affecting the ferrocement parts and also the need, in the case of the dome, to make the space brighter. The floor with its isostatic ribs was probably also painted white at this stage.

Once he had decided to sandblast the dome ashlars, Nervi decided to extend the same treatment to the most significant parts of the cast-in-place reinforced concrete structure, such as the stairways in the gallery and the variable cross-section inclined pillars. For this purpose, between February and March 1959, he carried out sandblasting tests on the "exposed reinforced concrete

[31] See Nervi & Bartoli, invoice No. 596 to CONI, 13 September 1959, CONI, binder 8.

392-394. P. L. Nervi (with the preliminary cooperation of M. Piacentini), *Palazzo dello Sport*, Piazzale dello Sport, Rome, 1954-59.

392. Building site, 1958 (photograph by G. Gherardi and A. Fiorelli; AFG, 1850).

393. View with parts of the painted structure, 1959 (photograph by G. Gherardi and A. Fiorelli; AFG, 2360).

394. Night shot, about 1960 (photograph by O. Savio; property of the authors).

structures."[32] Following the results, between June and July, Nervi demanded that surfaces of the cast-in-place concrete should be sandblasted, even though the CONI technicians believed "this finish to be 'pointless' in many parts of the structure"[33] (Vitellozzi was also consulted at this juncture to determine which parts should be sandblasted).

As a result of the technical devices implemented by Nervi & Bartoli for the *Palazzo dello Sport*, Nervi succeeded in establishing himself as the manufacturer of a form of *béton brut* in which the significance and effects of the impressions were modulated and had no further need of manipulation to achieve a perfect appearance in terms of line and surface finish, as has been the case for the *Stadio Berta*. This did not mean that Nervi shared the faith that prompted his contemporaries to leave all types of concrete exposed, even that blemished by *malfaçons*. "It is better to hide reinforced concrete if it is not well studied and executed,"[34] was his advice to students. As soon as he had completed the *Palazzo dello Sport*, in February 1960 Nervi spoke at his course in the Faculty of Architecture about formwork, the nature of concrete, and its manufacture. His words revealed the intensity of his participation in the cultural climate marked by the era of brutalism and some of his core values as well as, perhaps, his difficulty in deciding between expressing the truth of the concrete or that of the structure. This was an enigma he could never resolve that was apparent in his choices of formwork materials.

"At first formwork was thought of as something just to hold the concrete. When it was stripped, the concrete could be stuccoed over. Now there is a desire for sincerity. But the desire to express the natural structural element introduces a problem. [...] at first the form work ruled the form that concrete could take. The resulting structure was one of petrified wood. Everything was box-like and there was no ribbing. [...] the form has become very important through the desire to express the natural material [...]."[35]

[32] See Nervi & Bartoli, invoice No. 580 to CONI, 16 March 1959, *ibid.*

[33] Luciano Berti, *Pro memoria per il Dr. Saini*, 24 July 1959, *ibid.*

[34] Einaudi, *Pier Luigi Nervi: lezioni romane* cit., p. 116.

[35] *Ibid.*, p. 110.

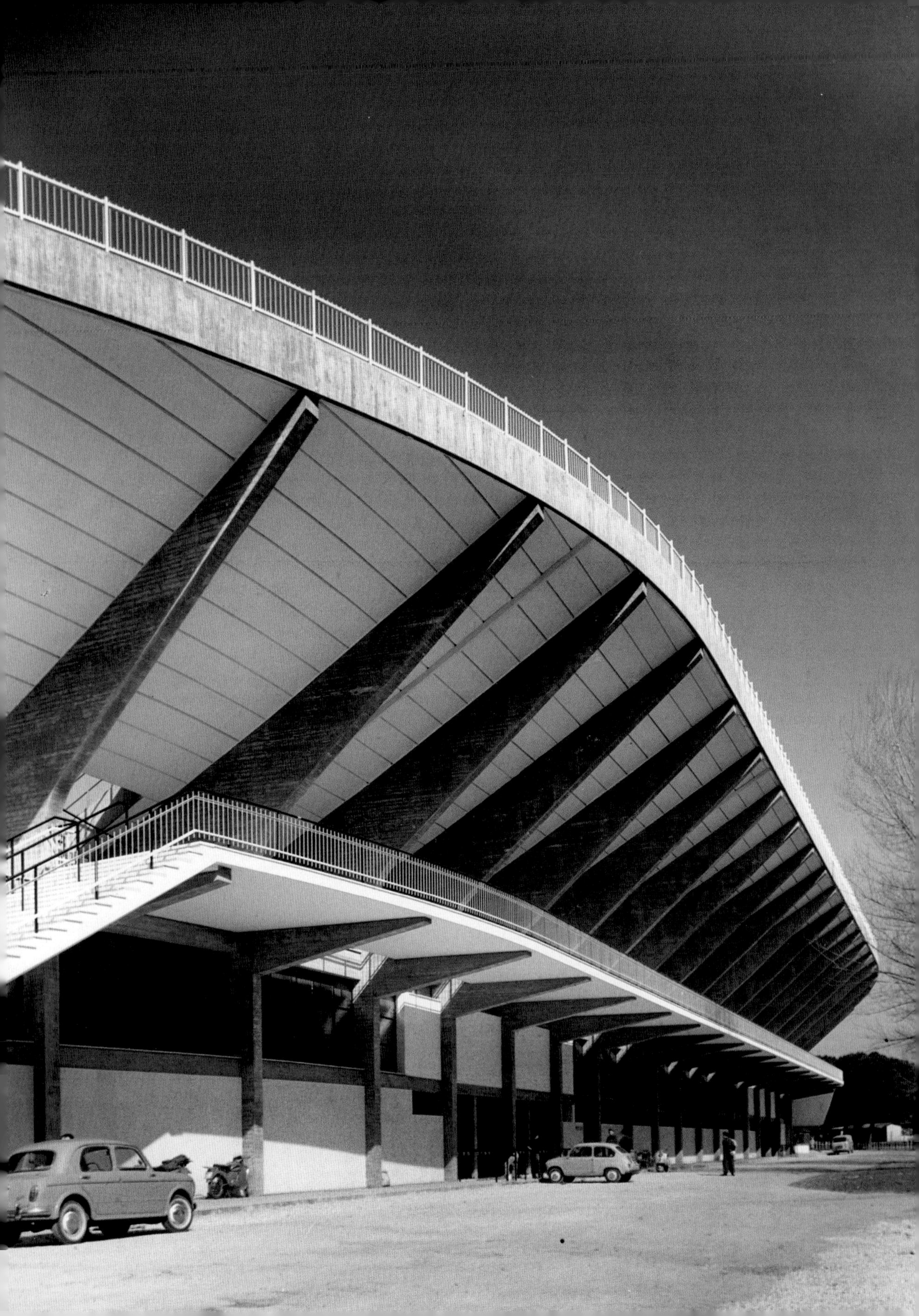

chapter twelve
the mirage of the matter's essence

Clad or exposed concrete

While the construction of the Unesco headquarters gave Nervi an opportunity to learn and try out some tricks of the trade when working on exposed concrete surfaces, the *Palazzo dello Sport* was the work that ultimately revealed his ability to offer a personal interpretation of *béton brut* stripped of any decorative aesthetic – even though certain contradictions remained unresolved relating to details of the clash between the two approaches to moulding the concrete, that is using wooden boards or ferrocement. There can be no doubt that for Nervi the concrete surface could not continue to be crucial to the truth of the material, though in his view the invisible reinforcements continued to perform this role. If we fail to consider that the essence of that material is embodied first and foremost by its mechanical make-up, we cannot fully understand the significance that Nervi attributed to the impressions left by the formwork and the special aesthetic value of his *béton brut*. It is clear that this view, clearly expressed when the technical drawings of the metal reinforcements are compared with the photograph of the *Stadio Berta* canopy, was still a powerful part of Nervi's vision, even after he had discovered concrete's surface qualities. This is borne out by the important role still played by the drawings of those reinforcements when he represented his structures. Without an understanding of those irons bent in accordance with certain lines, sized and interwoven to harness the forces, and even aesthetically representing the strength, it was as if it were not possible to grasp the reason for the "hardness" of his concrete surfaces enfolded by the reinforcements, protected and consolidated into a form that conferred strength. One is left with the basic impression that Nervi meant to transpose the intrinsic qualities of ferrocement into reinforced concrete, and not merely when he used smooth moulds in ferrocement. It was as though the very idea that he had formed of that cast-in-place material during the 1950s entailed a single possible interpretation of the concrete surface as an opaque film with its own invisible mechanics, and that the film would be all the more real if it were devoid of markings, as is the

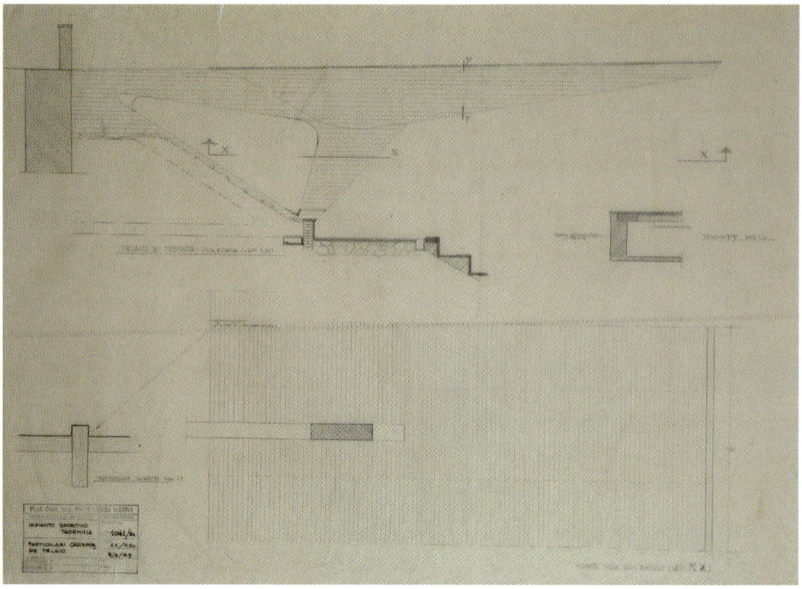

395. P. L. Nervi and A. Nervi, *Stadio Comunale*, Taormina, 1955-59. "Particolari casseforme telaio (Details of frame formwork)," drawing dated 3 April 1959 (CSAC, n. inv. pra 1016, n. id. 14794, coll. 158/4).

case with ferrocement. All this suggests that Nervi had seen a mirage of the secret essence or nature of reinforced concrete, that he had even inferred the way it could be represented, that he was already aware, as Kahn would subsequently become, of the symbolic power contained in the rich smoothness of the exposed concrete surface. Yet even Nervi was not always able to resist the appeal of *béton brut*.

Only in the wake of the results achieved with the construction of the *Palazzo dello Sport* did Nervi finally decide that in future structures, he would do away with the cement cladding of his reinforced concrete that, until then, he had always planned or used. One telling detail, which occurred around 1955, that illustrates the crucial transition in his aesthetic vision of that material, was that the execution of the *Palazzetto dello Sport* trestles marked the last time he used Nervi & Bartoli's usual irregular formwork to obtain a form of concrete whose surface finish was not yet controlled, because it was due to be plastered with cement mortar – as the various spurs or arches of Nervi's previous masterpieces in Turin had been. The layer of white paint applied in place of the cladding revealed the impressions of Nervi & Bartoli's last imperfectly manufactured wooden formwork.

In his other works dating from the late 1950s, Nervi opted for a form of exposed concrete whose manufacture was controlled in accordance with criteria implemented in the *Palazzo dello Sport*, when his intention was to establish himself as the exponent of his own personal view of *béton brut*. The Taormina *Stadio Comunale* (Municipal Stadium), built between 1955 and 1959; the Florence *Piscina Comunale* (Municipal Swimming Pool), designed between 1957 and 1958; and the roof of the Cuore Immacolato di Maria church in the Ina-casa neighbourhood of Borgo Panigale near Bologna, constructed between 1957 and 1962, are significant examples of Nervi's new approach to exposed concrete structures whose surfaces were increasingly accurate and well-designed.

The fact that 1955 was a crucial year in Nervi's development is confirmed by the fact that the design for the Taormina *Stadio Comunale* drawn up that year described a structure entirely in reinforced concrete, cast in place but not yet left exposed. The perforated supports with inclined lines, similar to the usual spurs and referred to as "frames,"[1] and the canopy, also with inclined lines to improve the view, were initially supposed to be clad: the former structures with "cement plaster," and the latter with "civil plaster." When the building site was opened in 1959, with the contract awarded to the Ditta Domenico Maiolino of Messina and managed by the engineer Aurelio Caruso, Nervi made substantial changes to the finish of the work that now reflected the experience he had gained in the *Palazzo dello Sport* and his current project to build the *Stadio Flaminio*.

Nervi designed the details of the formwork construction and the method of laying the boards in horizontal layers in the frames. He ordered the use of boards without joints in the inclined post in order not to break up the integrity of the structure, aligned the vertical joints in the cantilever beam section to perfection, and designed the intrados surface of the canopy with a regular pattern of boards laid at right angles to the frames. Each portion of the intrados, corresponding to a single stage of the pouring, was delimited by a groove in order to mitigate any changes in the colour of the concrete between one pouring and the next. The design of the Taormina stadium is one of the first to emerge from the Studio Nervi engineering department that specified the pattern of the formwork boards, right down to the details of the joints.

The fact that Nervi now intended to set himself up as an authoritative exponent of the exposed reinforced concrete aesthetic, perhaps even in competition with the version popularised by Le Corbusier, is confirmed by the justification that he gave to the Mayor of Taormina for doing without any cladding: "to achieve visible surfaces that are so good-looking that they do not require plastering, by paying particular attention to the construction of the formwork and concrete mixture pouring."[2] Nervi warned that if there were "areas of the pouring that were porous or poorly executed or deformations of the formwork, it would no longer be possible to correct them,"[3] and the contractor Maiolino therefore had to be well aware of the "greater executive accuracy necessary for the reinforced concrete structure, which must remain exposed without plastering"[4] in view of the fact that, after stripping, the surfaces would simply be "cleaned by light sandblasting with compressed air."[5] The construction of the *Stadio Comunale* in Taormina was intended to become an emblematic work of the "technique of finishing exposed reinforced concrete structures without plaster" that "had recently been refined and is spreading rapidly," Nervi went on to confirm, with the authority due to him as one of the parties responsible for the Unesco headquarters and the creator of the *Palazzo dello Sport*.[6] Despite the working drawings of the formwork and the specifications for the pourings, the concrete produced for the *Stadio Comunale* in Taormina was not of the same quality as that obtained during the same period by Nervi & Bartoli.

The same concern over the construction of formwork able to perfectly mould the concrete lay behind the design for the *Piscina Comunale* in Florence, for which Nervi used some of the motifs emblematic of his structures, including variable cross-section pillars – the same used in the *Palazzo dello Sport* – and

[1] Drawing No. 1041-1e, CSAC, n. inv. pra 1016, n. id. 14794, coll. 158/4.
[2] Pier Luigi Nervi, letter to the Mayor of Taormina, 6 May 1959, quoted in, Antonio Marino, Laura Marino, *Taormina. Lo stadio e il progetto della piscina*, in, Bianchino, Costi, *op. cit.*, (pp. 213-221), p. 221.
[3] *Ibid.*
[4] *Ibid.*
[5] *Ibid.*
[6] *Ibid.*

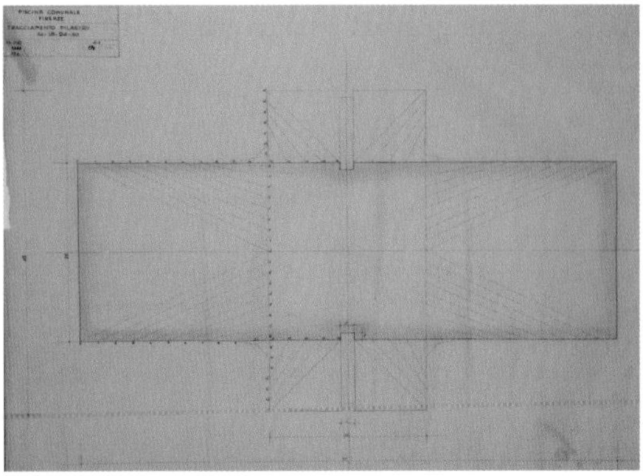

396. P. L. Nervi, Municipal swimming pool, Lungarno A. Moro, Florence, 1957-58.
"Tracciamento pilastri (pillar layout)," drawing dated 13 November 1957 (CSAC, n. inv. pra 337, n. id. 13685, coll. 41/3).

beams with a V-shaped cross-section to be prefabricated using his usual "ground forms." In this case too, the variable cross-section pillar was dependent on installing boards with different sizes and shapes in order to reflect the geometry of the structure more effectively, usually revealing the generating figure through the impressions. Nervi departed from the approach he had used for the *Palazzo dello Sport* pillars by introducing a groove along the installation lines of the two sections making up the pillar formwork in order to overcome any imperfections. This was a practical way of avoiding the perfect mechanical device used to join the formwork of the Roman building site.

Again with the aim of obtaining perfect concrete, in the church in Borgo Panigale, Nervi used his usual system of permanent ferrocement formwork, this time to build not the usual ribbed domes but a special type of roof based on Arcangeli's floor with isostatic lines (the construction contract was awarded to the firm Fratelli Donati under the management of Glauco Gresleri and with Gogliardo Tubertini as foreman). In order to roof the circular hall designed by Vaccaro, Nervi used a structure of the kind made famous by Wright, which was new to his repertoire and consisted of a variable cross-section cruciform pillar ending in a mushroom-shaped slab (there was no formal connection between the pillar and floor). In this case, the choice of the cross was in tune with the sacred nature of the space and also stiffened the base of the structure.

The cast-in-place reinforced concrete pillars were constructed using formwork made out of wooden slats. The combination of four units produced a roof with lines that were no longer isostatic but sometimes even symbolic, considering the central cross formed by the joints between the ribs. The ribs of each unit were concentric and controlled by a compositional and constructive geometry that was no longer connected in any way with expressing the static forces. These ribs were then a manifestation of the manufacturing process, based on a kind of coffering produced by assembling the *tavelloni*.

On the Borgo Panigale building site, as previously in the *Palazzo dello Sport*, Nervi adopted a process of prefabricating the *tavelloni* in which he removed the intermediate stage – the element sometimes referred to as the "prototype" and sometimes as the "*mamma*." Perfect working of the *tavelloni* is demanded in the tender specification because the "stripped surfaces" must "imperatively" remain "exposed."[7] These ferrocement *tavelloni* were among the few produced in accordance with Nervi's system to be left unpainted.

[7] ASG, *Capitolato*, item No. 18.

397-399. P. L. Nervi, structure of the *Chiesa del Cuore Immacolato di Maria*, designed by G. Vaccaro, Via G. Mameli 5, Borgo Panigale, Bologna, 1957-62.

397. Prefabrication of *tavelloni*, 1961 (ASG).

398. Building site, photograph dated 30 September 1962 (ASG).

399. View of roof (ASG).

A pillar with a variable cruciform cross-section, a roof with rhetorical ribs of static inspiration, cast-in-place concrete, and precast ferrocement are the fundamental ingredients that turned Vaccaro's design into an apotheosis of Nervi's structure, now bent to create a mystical space, where the crosses and weaves multiplied beyond the needs of the static structure.

The *Stadio Flaminio* in Rome

It is almost a paradox that even by the end of the 1950's, Nervi almost never succeeded in controlling the formal outcomes when he used the system of prefabricating parts in ferrocement, which he had always seen as a potential method of obtaining surfaces without imperfections. Quite the contrary, compared to the results that Nervi & Bartoli could achieve with cast-in-place concrete, the advantages seemed to be outweighed by the need to control the design of the formwork and carry out sandblasting after stripping. In this way, the wooden formwork that had been literally banned by Nervi ended up becoming an intrinsic part of the building of his structures. The formwork impressions became technical and aesthetic factors that Nervi attempted to control from the stage of drawing up his detailed designs, ever sensitive to changing attitudes in the international architectural debate. Nervi never signed up completely to the principles of brutalism, even in works that were intended to experiment with the formal qualities of this material. He continued to strive for the unachievable goal of executive perfection, which led him to use masking.

The design and construction of the *Stadio Flaminio*, between 1956 and 1959, probably revealed Nervi's aspiration to achieve a work that represented exposed concrete and the usual remedies implemented to overcome execution defects.

The stadium structure appears as a personal critical review of the complex design stages of *Stadio Berta* in Florence, in which Nervi, due to the cultural climate imposed by international brutalism, was freed of his concern to hide the material and even the structures, and exposed the entire system, which was made perfect by the clever way he broke it down into cast-in-place parts and prefabricated parts.

Due to a condition in the competition documents requiring the dimensions of the old stadium to be respected, Nervi exploited the inclination of the cantilever beams of the tier to attach a canopy at the top, while he reduced the support point that made the extension possible to a slender, inclined steel pole. The line of poles also distinguished the cast-in-place reinforced concrete from the precast concrete in the cantilevered section, made possible by 15-metre long waved beams made out of ferrocement and lightened by holes. The set of structural parts outlined a combination of taut lines in space through the plastic expression of the forces and resistances.

The frames with two hinges and inclined beams to support the tiers were made out of reinforced concrete poured in wooden formwork that was designed in detail because the concrete was now characterised by the impressions. After Studio Nervi had already designed, in April 1957, the boards of the formwork for the *Stadio Comunale* of Taormina, and in July those for the variable cross-section pillars of the *Palazzo dello Sport*, in October 1957, Antonio took some time to depict the effect of the impressions left by the wooden boards with their veins in his drawing of the cross-section – one of the rare documents in which Nervi's specific brand of *béton brut* risked being identified with the expressive qualities of wood. The drawings for the *Stadio Flaminio* differed from those for Taormina in representing an irregularity in

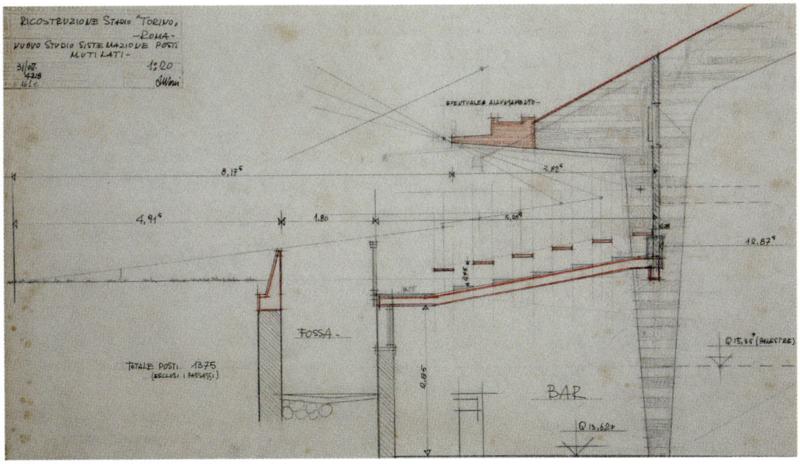

400. P. L. Nervi and A. Nervi, *Stadio Flaminio*, Viale Tiziano, corner of Via D. Pietri, Rome, 1956-59. "Nuovo studio sistemazione posti mutilati (New study for arrangement of seating for disabled)," drawing dated 31 October [1957] (CSAC, n. inv. pra 1119, n. id. 15050, coll. 17/4-5-6).

the working of the boards that ultimately also distinguished the finished work, in accordance with a taste for a random *béton brut* that probably had more to do with Antonio than with Pier Luigi. The ironic appearance of a vault in the study drawings of the wooden impressions could potentially be more than an allusion to the possibility of moulding images in concrete – the outcome of a discussion about the uncontrollable figurative effects of the marks left by the materials. It is likely that Nervi rejected any decorative effect left by the formwork in order not to distract the viewer's attention from the significance of the material as an instrument for the creation of the only figures that appealed to him: structural figures. From this perspective, Antonio's design represented a real betrayal of Nervi's idea of concrete in decorative terms, even if it was performed in accordance with criteria that were apparently practical and economic. Now the drawings of the *Stadio Flaminio* depicted the woven lines of the formwork, marking the advent of a kind of technical drawing that recorded the effects of the exposed concrete. For the construction, Nervi opted for formwork with boards that were "planed and joined by tongue and groove joints" and cleaned the surfaces by means of sandblasting that was not mentioned in the tender specifications, "in order to remove the surface stains while leaving the impressions of the boards visible."[8]

Confirmation of the almost primeval significance of concrete to Nervi is evident in his choice of the type of cement for the stadium. During the construction of the stand frames, Nervi & Bartoli used cements supplied by the CEMENTIR cement plant. However, the first trials resulted in a pinkish concrete mixture that Nervi rejected, clearly wanting only grey for the colour of his exposed concrete. He then turned to the Unione Cementi Marchino, which also provided the cement for the *Palazzo dello Sport* and was willing to meet Nervi's requirements. The concrete of the *Stadio Flaminio* was manufactured using cement produced in Guidonia by the Unione Cementi Marchino.

For the tiers, after those prefabricated for the *Palazzo dello Sport*, Nervi invented another structure made out of prefabricated elements that, once assembled, produced seats that could conceal the feet and cavities that could

[8] Pier Luigi Nervi, Antonio Nervi, *Stadio Flaminio a Roma*, in, "Casabella Continuità," XXIV, February 1960, No. 236, (pp. 23-31), p. 28.

be used for disposing of rainwater and the thermal insulation of the rooms formed under the tiers. Nervi patented the system in January 1957, specifying the treatment of the seats "in reinforced concrete with the upper face in crushed marble or hammered travertine in order to allow effective finishing of the exterior surface."[9]

For the steps of the *Stadio Flaminio*, Nervi proposed a kind of prefabrication that incorporated the cladding in the very procedure for prefabricating the parts, "arranging a layer of crushed marble on the form of the prefabricated element and then pouring the reinforced concrete mixture of the element."[10] Assembly of the parts gave rise to V-shaped joint lines designed by Nervi to create an ornament so that the surface appeared "regular" but, as he himself explained, "enlivened by the joints which were made particularly evident for this purpose."[11]

Like the entire stadium, the canopy of the covered stand was broken down into parts made out of cast-in-place reinforced concrete – the corrugated roof slab extending outward – prefabricated parts in ferrocement, and the V-shaped beams that formed the canopy extending toward the sports field and perforated for lightness and to install the artificial lighting system. The combination of beams gave rise to a pleated shell similar to the geometry of the beams in the dome of the *Palazzo dello Sport*, which was designed as a variant of the waved shell used in previous masterpieces by Nervi. In the canopy, Nervi positioned inclined supports along the seam between both types

[9] Pier Luigi Nervi, patent IT 564484, *Sistema di costruzione di gradinate di cemento armato ad elementi prefabbricati per stadi o tribune per spettaccoli, e relativi elementi prefabbricati*, 12 January 1957, p. 7, ACS.
[10] *Telai principali e gradinate*, p. 3, CONI, binder COR 1960.
[11] Nervi, letter to Viganò, 20 November 1959 cit.

401-402. P. L. Nervi and A. Nervi, *Stadio Flaminio*, Viale Tiziano, corner of Via D. Pietri, Rome, 1956-59.

401. Installation of prefabricated steps, 1957 (photograph by G. Gherardi and A. Fiorelli; AFG, 675).

402. View of painted tiers (photograph by G. Gherardi and A. Fiorelli; Acm-EPFL, 0172.02.0003.2146).

of concrete. They were made out of Mannesmann pipes filled with concrete in order to make them as slender as possible.

Colour was also used to remove defects in the *Stadio Flaminio*. The imperfect execution of the elements was noted during the installation stage of the first prefabricated steps. This was always caused by the same defect: scarring by "numerous stains due to the greasy substances used to facilitate the stripping."[12] In order to conceal the stains, the construction manager, engineer Bruno Magrelli, asked Nervi & Bartoli to carry out painting tests.[13] The ferrocement of the canopy beams was similarly stained, and the possibility of a "limewash painting" was mooted in October 1958, opposed by CONI,[14] and proposed by Antonio and Pier Luigi Nervi for a "good aesthetic outcome for such an important and demanding work."[15] It was only through the efforts of CONI that the concrete of the canopy was left in its natural colour, and it was decided to improve the surface quality using sandblasting alone.

[12] Magrelli, letter to Nervi & Bartoli, 31 October 1957 cit.
[13] *Ibid.*
[14] Mario Nervi, letter to COR, 1 October 1958 and Mario Saini, letter to Bruno Magrelli, 7 October 1958, CONI, binder 16.
[15] Pier Luigi and Antonio Nervi, letter to COR, 10 October 1958, *ibid.*

403. P. L. Nervi and A. Nervi, *Stadio Flaminio*, Viale Tiziano, corner of Via D. Pietri, Rome, 1956-59. View of covered stand (photograph by G. Gherardi and A. Fiorelli; Acm-EPFL, 0172.02.0003.2150).

The slender frames of the uncovered stands and those of the covered stand soar powerfully with an almost Brazilian structural taste, and the slab of the canopy is patterned with the thick ruling of lines that was envisaged at the design stage. It appears perfect only when viewed at a distance from the stadium. The other surfaces are smooth in grey cement, stained, or painted white.

The aesthetics of exposed concrete

When Nervi & Bartoli was engaged in completing the *Stadio Flaminio* and the *Palazzo dello Sport* in Rome, Studio Nervi designed a series of works in which the main structure continued to be designed as a combination of load-bearing elements, usually in the form of mighty pillars with a variable cross-section to be built in cast-in-place concrete and left exposed, and light roofing elements, to be prefabricated using the ferrocement technique and painted white. The railway station of Savona Mongrifone, built between 1958 and 1960, the *Palazzo del Lavoro* in Turin, built between 1959 and 1961, and the George Washington Bridge Bus Station in New York, built between 1958 and 1963, were Nervi's main works in the years around 1960. During the stages when he was developing his designs, however, Nervi often renounced his preferred technique after the 1940s, relegating this to the construction of secondary parts and continuing to promote it only abroad, in the United States, in Africa, or in Australia. All the great structures of works built by Nervi & Bartoli until Nervi's death were now characterised by the use of reinforced concrete, either cast-in-place or prefabricated. Nervi's last ferrocement

404-406. P. L. Nervi and J. M. Kyle, George Washington Bridge Bus Station, Broadway, West 178th Street, New York, 1958-63.

404. View of building site, photographed in 1962 (BSTA, 024-7044).

405. View of variable cross-section pillar, photographed in 1961 (BSTA, 024-7042).

406. Intrados of roof (BSTA, 024-7090).

was laid by US, African, and Australian companies that actually changed its constitution – from the thicknesses to the kind of reinforcement – to make it almost more similar to reinforced concrete. Nervi & Bartoli's Roman works, for the 1960 Olympic Games, from the *Palazzetto dello Sport* to the *Stadio Flaminio*, were destined to be the last works for which Nervi manufactured parts in ferrocement for his largest structures. His decision to give up ferrocement was influenced by various factors. These depended on manufacturing times and costs, the static capacity of individual parts, and also the superlative aesthetic quality of the surface that had become an essential feature of Nervi's large structures.

The development of Nervi & Bartoli's construction technique merely confirmed the new role that Nervi was carving out for himself as a builder in Italy at the end of the 1950s, who was the most experienced expert in exposed reinforced concrete. Le Corbusier's visits to Nervi's Roman building sites took on particular significance in this context, confirming a friendship and reciprocal regard between the two stars in *béton brut*,[16] even though each attributed a different role and importance to the material.

The design and construction of the Savona Mongrifone railway station provided evidence that Nervi had given up his system of ferrocement roofing beams. Because the building site area was so small, a system was devised to install the prefabricated elements by means of a crane with a modified arm to lift the heavy beams and which could slide along rails that could even pass over a floor. A rotary platform was provided to turn the beams from their position of manufacture in the ground form to their position in the roof, given that the arm of the crane could not turn due to the small size of the building site.

The station was, however, built out of cast-in-place reinforced concrete. The shape of the pillars became a formal device for distinguishing and char-

[16] Piero Ostilio Rossi, *"Mon cher ami..." Le Corbusier e Pier Luigi Nervi*, in, Francesca Romana Castelli, Anna Irene Del Monaco, eds., *Pier Luigi Nervi e l'architettura strutturale*, Roma, Edilstampa, 2011, pp. 40-65.

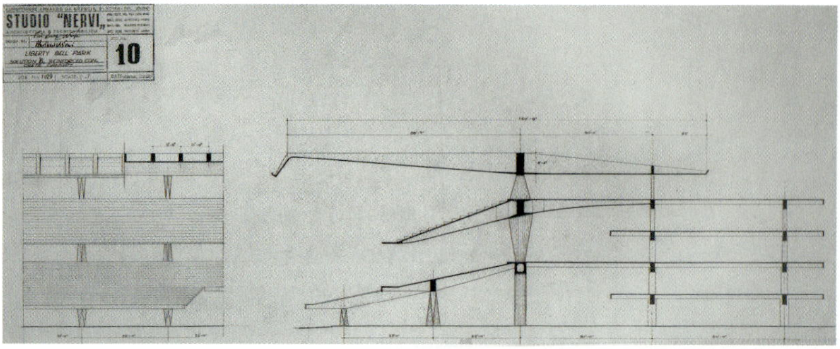

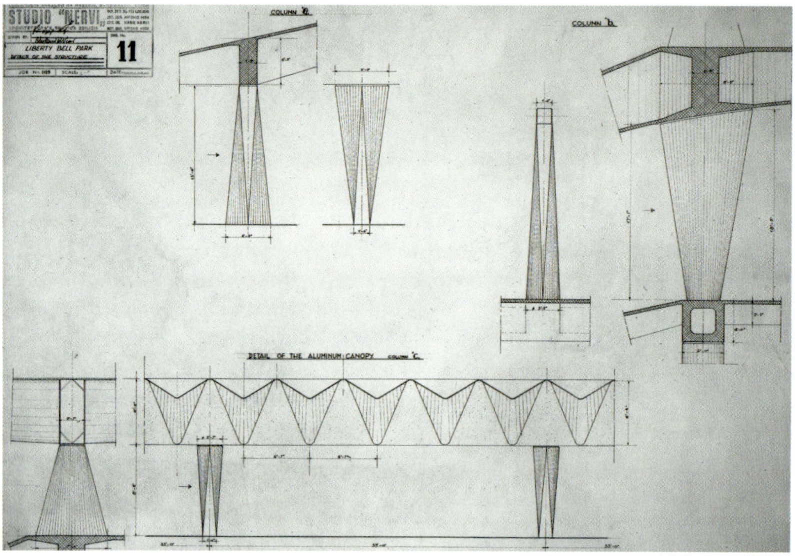

407-408. P. L. Nervi and A. Nervi, design for stands of the Liberty Bell Park hippodrome, Philadelphia, Pennsylvania, 1961.

407. "Solution 'B'. Reinforced concrete canopy," drawing dated 20 February 1961 (BSTA, 024-7598).

408. "Details of the structure," drawing dated 20 February 1961 (BSTA, 024-7600).

409-411. P. L. Nervi and A. Nervi, design-and-tender competition for passenger terminal of the new Savona Mongrifone railway station, Piazza A. Moro, Savona, 1958-60.

409. Perspective cross-section, about July 1958 (CSAC, n. inv. pra 1192, n. id. 15128, coll. 45/2).

410. Variable cross-section pillar of main facade (CSAC, n. inv. pra 1192, n. id. 15128, coll. 45/2).

411. "Particolare sezione pensilina (Detail of canopy cross-section)," drawing dated 10 February 1960 (CSAC, n. inv. pra 1192, n. id. 15128, coll. 45/2).

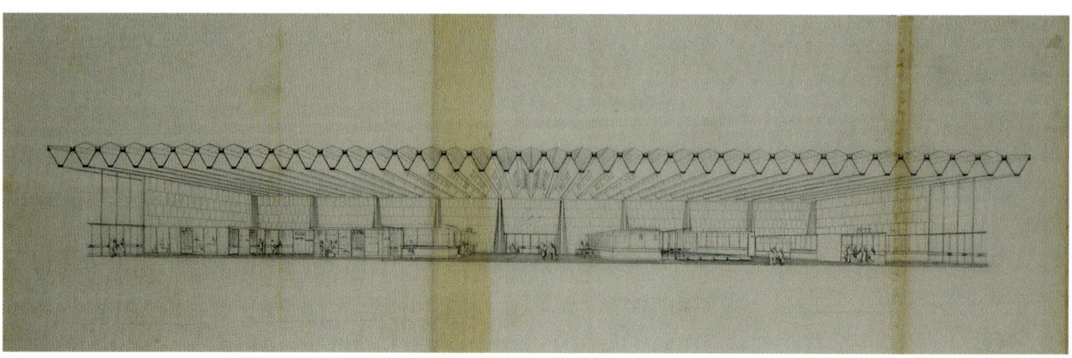

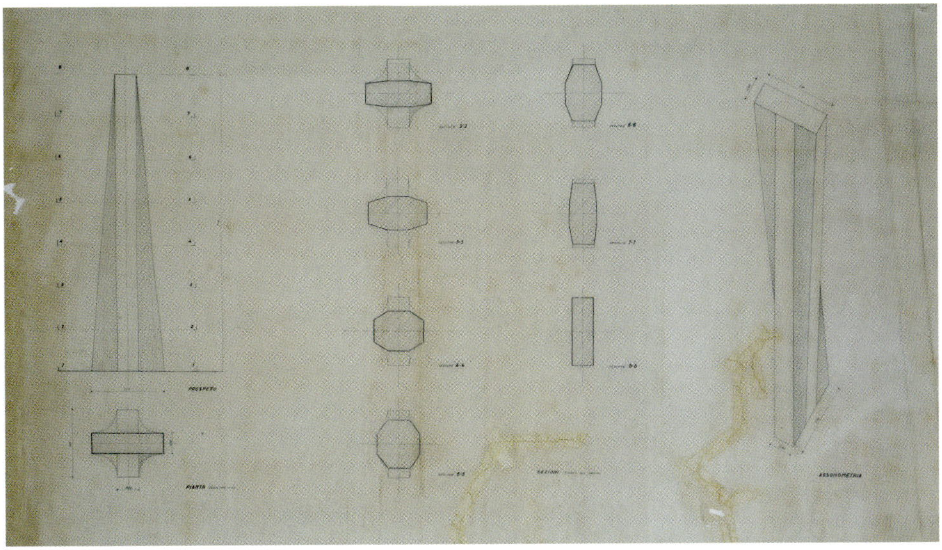
PROSPETTO
PIANTA
SEZIONI
ASSONOMETRIA

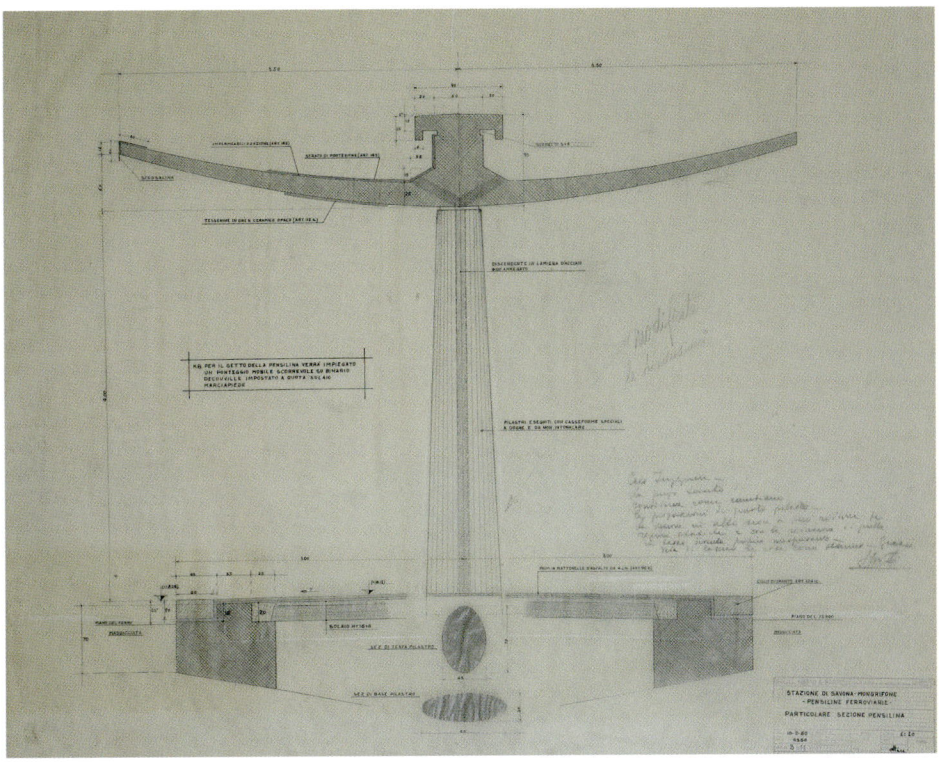
N.B. PER IL GETTO DELLA PENSILINA VERRÀ IMPIEGATO UN PONTEGGIO MOBILE SCORREVOLE SU BINARIO DECOUVILLE IMPOSTATO A QUOTA SOLAIO MARCIAPIEDE
STAZIONE DI SAVONA-MONGRIFONE
- PENSILINE FERROVIARIE -
PARTICOLARE SEZIONE PENSILINA

acterising the various parts of the complex. The variable cross-section pillars were for the monumental colonnade of the main building with the atrium. The cylindrical pillars supported the canopy connecting the atrium to the building housing the offices, waiting rooms, and washrooms. Lastly, the canopy was supported along the tracks by pillars that were again variable, but with an elliptical cross-section. In the drawings, Nervi specified a system of formwork with staves for the various pillars. The formwork of the variable cross-section pillars of the atrium were always made in two sections to be joined together and reused.

During the design of the canopies along the tracks, an engineer of the railway construction service proposed changing the sizes of the elliptical variable cross-section pillars in February 1960. Nervi's response demonstrated the extent to which those profiles had become for him expressions of the characteristic formal identity of his structure: "The upper cross-section cannot be

412-415. P. L. Nervi and A. Nervi, design-and-tender competition for passenger terminal of the new Savona Mongrifone railway station, Piazza A. Moro, Savona, 1958-60.

412. Formwork of variable cross-section pillars of main facade, photographed in 1960 (BSTA, 024-7051).

413. Variable cross-section pillars of main facade (BSTA, 024-7063).

414. Positioning the roof reinforcement, photographed in 1960 (BSTA, 024-7048).

415. Retaining wall for the railway platform (photograph by Studio Cresta; P. L. Nervi, *Buildings, Projects, Structures. 1953-1963*, New York, Frederick A. Preager, Publisher, 1963, p. 70).

reduced for static reasons, and reducing the lower cross-section would make it expressionless. See that you leave things as they are."[17]

The atrium roof was a pleated shall in cast-in-place reinforced concrete in formwork with carefully made wooden boards. Deep grooves along the intrados of the roof broke down the continuous pleated surface into segments that only apparently alluded to the parts assembly envisaged at the initial stage of the project. Nervi had begun to use the grooves in the *Palazzo dello Sport*, but only introduced them at the pouring stages in the pillars. Now those grooves assumed complex values and, in some cases, identified the pouring stages. The two grooves that run transversely to the pleating allude to the invisible stiffening added between the folds of the shell, while those that run parallel to the deep hollows in the pleating reflect a decorative intent, to scan the rhythm of the structure and act as a counterpoint to the set of cylindrical holes left by the form spacers on the most exposed faces of the pleating. The set of traces left by the grooves and holes assume a shaded effect that was able to camouflage the perception of execution defects or technical details such as the expansion joint. The presence of that pattern of grooves and holes was solid evidence of Nervi's adherence to the sophisticated technical and formal criteria that had become current in the manufacture of exposed concrete.

In order to supplement work for the 1960 Olympic Games, Nervi & Bartoli were commissioned, in 1958, to design and build the viaduct in Corso Francia that runs alongside the *Palazzetto dello Sport*, the *Stadio Flaminio*, and the Olympic Village.

This was the first time that Nervi had systematically used prestressed concrete. As in other works for the Olympic Games, the concrete was cast in place as well as prefabricated: in both cases it was left exposed. The viaduct was further confirmation of Nervi's work on the shape of the variable cross-section pillar for which he also considered a hyperbolic paraboloid with an elliptical cross-section in his initial sketches. The pillar and beam became a single sculptural structural element to be cast in place and in which the variable thicknesses of the edges, bevels, grooves, and formwork

[17] Drawing No. 4350-03 OFF, CSAC, n. inv. pra 1192, n. id. 15128, coll. 45/2

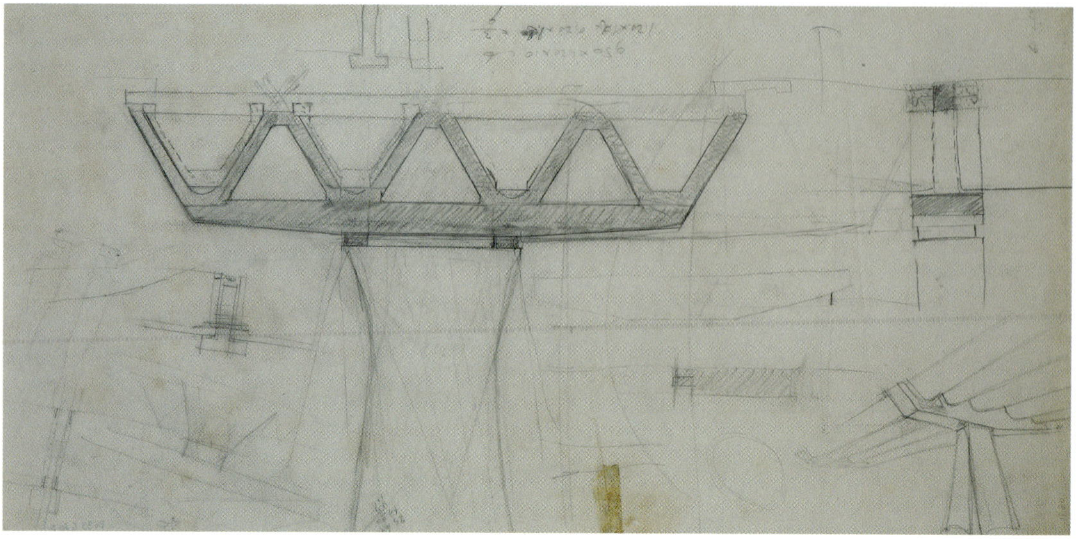

416-420. P. L. Nervi, viaduct in Corso Francia, Rome, 1958-60.

416. Preliminary study, 1958 (CSAC, n. inv. pra 1252, n. id. 15254, coll. 31/6).

417. View of building site with prefabrication of prestressed reinforced concrete beams, 1959 (photograph by G. Gherardi and A. Fiorelli; AFG, 2914).

418. Formwork of variable cross-section pillar, photographed in 1959 (photograph by G. Gherardi and A. Fiorelli; AFG, 4067).

419. Variable cross-section pillar, photographed in 1959 (photograph by G. Gherardi and A. Fiorelli; AFG, 4068).

420. View showing construction of cantilever beam formwork, 1959 (photograph by G. Gherardi and A. Fiorelli; AFG, 4069).

421-422. P. L. Nervi, viaduct in Corso Francia, Rome, 1958-60.

421. View of building site, 1959 (photograph by G. Gherardi and A. Fiorelli; AFG, 3273).

422. View (Acm-EPFL, 0172.02.0003.2098).

423. P. L. Nervi, design for viaduct for the Kaiser Steel Corporation, 1962 (CSAC, n. inv. pra. 510, n. id. 13971, coll. 281/2).

424. P. L. Nervi, design for viaduct, Ancona, 1964 (BSTA, 024-4473).

board impressions emphasised the continuity and the process of construction. Nervi fitted the precast reinforced concrete beams of the roadway onto that structure.

For the construction of the pillar with ruled surfaces, Nervi & Bartoli produced formwork with double planking using wooden boards, as for the pillars of the *Palazzo dello Sport*. The sides of the upper rectangle and those of the cross at the base were broken down into segments of equal length, each corresponding to one formwork board. The boards were joined by means of "*maschiettatura*" (tongue and groove joints) so that they could be bent slightly to follow the contour of the ruled surface.

Nervi & Bartoli produced three formworks for the pillars, reusing each one approximately forty times: one formwork for the isolated pillar with cantilever beam and two for the pair supporting the double cantilever beam forming the portal frame. The ruled surface configuration of the pillar contour meant that the structure of the formwork could be cut into sections at various levels, determined by the height of the roadway.

Nervi & Bartoli built sixteen V-shaped moulds in clad masonry in order to manufacture the roadway beams on site. The beams were installed by means of small trolleys on rollers. Two Ormig self-propelled mobile cranes, with three rubber wheels, dragged the beams into the correct position. The crane then lifted the beam to allow the workers to remove the roller trolleys.

Following the international success of his Roman works for the Olympic Games, Nervi proposed various replicas of the Corso Francia viaduct in Italy and abroad. These included his viaduct design for Ancona, dating from 1964, with large variable cross-section inclined pillars and a set of variants in steel designed on behalf of the Kaiser Steel Corporation based in Oakland, California, where the large pillars and deck reflected the structural motifs of his works in reinforced concrete. For the same company, he also designed a structure made entirely out of reinforced concrete for the Fort Sutter Viaduct in 1963.

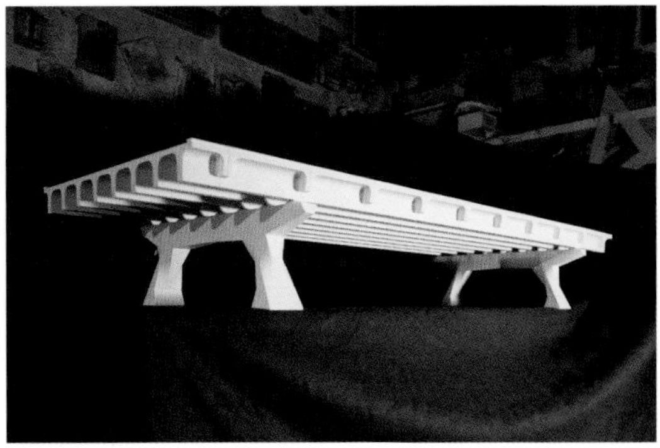

chapter thirteen monumental structural architecture

The Turin colossus: *Palazzo del Lavoro*

"[...] a large structure is always a victory over conflicting physical facts and that victory is all the sweeter when it can be won by achieving the exact proportions between the available resources and the forces to be overcome – or, in other words, the more technically perfect it is," was Nervi's personal dogma on contemporary structure.[1] It was again possible for that "large structure" to take shape and be erected in opposition to any "extreme form of static functionalism" due to a design submitted in response to a competition notice launched in 1959 to build the *Palazzo del Lavoro* in Turin. The work, subsidised by Fiat, was completed in only 11 months, in December 1960, and was intended for use as the headquarters of the Centro Nazionale per l'Istruzione Professionale when the event was over. Due to the limited time available between design and construction, Pier Luigi and Antonio Nervi gave up the idea of any kind of vault or dome; and even though the competition notice called for a surface free of support points, they raised a forest of sixteen colossal pillars in cast-in-place reinforced concrete, each crowned by a roof with a square plan and radial cantilever beams. This was initially planned in reinforced concrete and subsequently constructed in steel (the engineer Gino Covre and the contractor Antonio Badoni of Lecco were responsible for designing and constructing the steel structure).

The pillar, colossal as a pylon, again had a variable cross-section, with a cruciform base and round top, outlined by the lines of the ruled surfaces and featuring an oval cavity generated by the housing of a drainpipe and ladder to allow inspection of the roof. "The tripod or cross," explained Nervi, "is the best form for stability from overturning, while the circle was the best form to take the spoke-like cantilevers."[2] In order to construct the pillar, Nervi & Bar-

[1] Nervi, *Architettura e strutturalismo* cit.

[2] Einaudi, *Pier Luigi Nervi: lezioni romane* cit., p. 116.

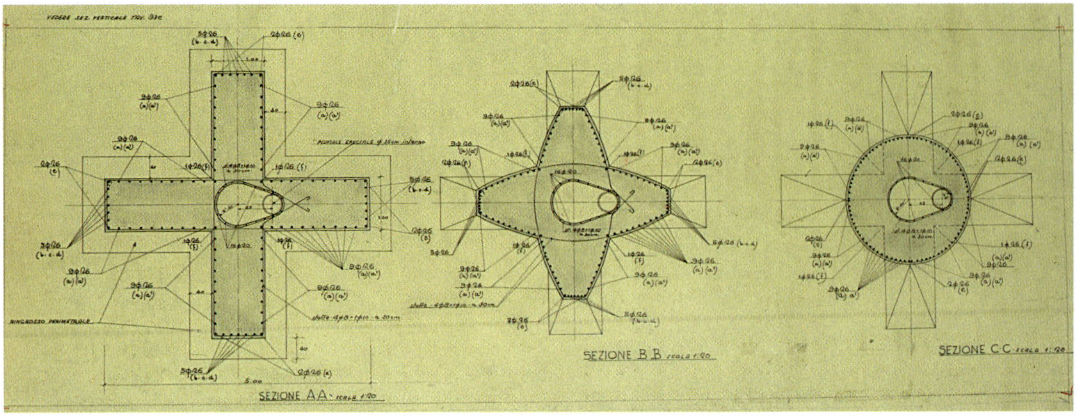

425-427. P. L. Nervi, A. Nervi and G. Covre, *Palazzo del Lavoro*, Via Ventimiglia, Corso P, Maroncelli, Corso Unità d'Italia, Turin, 1959-61.

425. "Pilastroni. Armatura sulle sezioni orizzontali (Pillars. Reinforcement on horizontal sections)," drawing dated 24 March 1960 (CSAC, coll. 257/2).

426. View showing installation of formwork for pouring of pillars, 1960 (photograph by A. Moisio; BSTA, 024-7568).

427. View of completed pillars, 1960 (photograph by A. Moisio; Acm-EPFL, 0172.02.0059.0345).

toli developed formwork divided into sectors. Although this was derived from the version used for the *Palazzo dello Sport*, it had now evolved into a complex machine made up of six sections, usually reinforced by metal hoops, each identified by a horizontal groove pressed into the concrete to avoid irregular joint marks. One formwork alone allowed the construction of sixteen pillars. The formwork had now become a powerful device, a work of building site engineering of which Nervi & Bartoli designed every technical detail. Nervi offered an accurate description when he explained the technical originality of his *Palazzo del Lavoro* ("this formwork was perhaps one of the most technical aspects of the entire work"[3]).

The action of pouring the concrete into the colossal formwork took place in six stages, corresponding to the grooves, and was performed in a single day. The work schedule involved leaving the formwork in situ for nine

[3] Pier Luigi Nervi, *Architettura strutturale con riferimento al Palazzo del Lavoro*, in, "Atti e Rassegna Tecnica della Società degli Ingegneri e degli Architetti in Torino," XV, June 1961, No. 6, (pp. 165-178), p. 170.

428-429. P. L. Nervi, A. Nervi and G. Covre, *Palazzo del Lavoro*, Via Ventimiglia, Corso P, Maroncelli, Corso Unità d'Italia, Turin, 1959-61.

428. Positioning of sliding formwork for pouring of perimeter gallery, photographed in 1960 (photograph by A. Moisio; BSTA, 024-7257).

429. Intrados of perimeter gallery floor, photographed in 1961 (Acm-EPFL, 0172.02.0003.2110).

days; on the tenth day it was removed and carefully cleaned so that it could immediately be reused. Once the pillar was stripped, it was sandblasted. The precautions taken by Nervi did not prevent the formation of some imperfections in the surface quality of the pillar concrete. Six months after their execution, some stains were still visible – even though Nervi gave a reassurance that these would disappear "over time" – as well as colour differences in places where the various pourings had been mended. Nervi admitted that he had "not been able to fully establish the reason for this."[4] The "dark" grooves impressed into the pillars served as optical corrections to the perception of the different colour gradations, and "they ensure a certain regularity and even the colour changes are not annoying," he explained.[5]

The other reinforced concrete parts of the *Palazzo del Lavoro* were the structure of the perimeter gallery, formed from cylindrical pillars patterned by the impressions of the formwork boards, and a floor, based on the customary isostatic line geometry. In the construction of the floor, Nervi & Bartoli proceeded as usual to install a scaffolding on rails fitted with ferrocement moulds. In the same way as Nervi had introduced pouring grooves in his pillars, he shaped the formwork in such a way that he was able to fit a wooden slat to impress a groove into the pouring to emphasise the pouring stages between one span and the next.[6] These multiple grooves may be seen as a device for controlling the concrete surface finish quality. Instead of the special

form he had used in the *Opificio Gatti*, this time Nervi decided to highlight the transition between these spans using the shadowy depths of the groove to eliminate any imperfect joint impression. The final result, which was of a floor crisscrossed by isostatic ribs and shaded grooves, undermined the monolithic aspect of the structure, revealing the individual components assembled to construct it. For the stripping, which had always been the critical step in the construction of floors with isostatic ribs or coffering, Nervi invented another device that he described in a conference held when the work was newly completed: "in order to facilitate the stripping, holes had to be left at the ends of the formwork segments and then struck with a hammer."[7]

Despite the sophisticated devices for controlling the pouring quality, the concrete floor surfaces of the *Palazzo del Lavoro* building site were also stained with the oil used to facilitate the stripping, and Nervi applied limewash as usual.

In the construction of the *Palazzo del Lavoro*, prefabrication was limited to the execution of long permanent formwork in ferrocement, which Nervi & Bartoli used to build the basement floor. In order to speed up manufacturing times, Nervi used a vacuum technique to remove water from the concrete and placed the permanent formwork into a special steam oven he had installed to further reduce the setting time.

[4] *Ibid.*, pp. 175-176.
[5] *Ibid.*, p. 176.
[6] See drawing No. 4558-60e, CSAC, n. inv. pra 409, n. id. 13836, coll. 275/2.
[7] Nervi, *Architettura strutturale con riferimento al Palazzo del Lavoro* cit., p. 177.

430-431. P. L. Nervi, A. Nervi and G. Covre, *Palazzo del Lavoro*, Via Ventimiglia, Corso P, Maroncelli, Corso Unità d'Italia, Turin, 1959-61.

430. Building site, photographed in 1960 (AST).

431. View, 1961 (BSTA, 024-2069).

The hoist placed against the *Palazzo del Lavoro* became another part of the sculptural structure, as with the stairways in the gallery of the *Palazzo dello Sport*. It was carefully designed in every construction detail of the formwork and also grooved by shaded lines corresponding to the pourings.

The sixteen colossal variable cross-section pillars, the cylindrical pillars, and the stairway were the main parts of Nervi & Bartoli's structure where exposed material took the limelight. The formwork impressions demonstrated the careful workmanship and justified Nervi's words about his Palazzo, spoken with the authority of someone who could produce his own *béton brut*. "The material must therefore be seen: stone must be stone, bricks must be bricks, iron must be iron, and reinforced concrete must be reinforced concrete. And reinforced concrete must not be clad in plaster."[8]

The pillars of the *Palazzo del Lavoro* were monumental to the point of being deliberately oversized, disturbing presences worthy of a latter-day hypostyle hall. Shaped with a variable cross-section to control the impressive mass of matter, they became obelisks erected to celebrate the formal power of reinforced concrete, contributing to make this particular work an example of the true "structural architecture" that Nervi intended to set against "excessive static functionalism." Yet those pillars cannot have removed the doubts that must have assailed Pier Luigi when Antonio suggested, during a discussion at the preliminary stages of the project, the use of "isolated elements": how was it possible to control the "unsightly appearance" of a pillar with colossal dimensions?[9]

The driving quest to express the political and social values of the Fascist regime through structural configuration was countered, after the tragedy of the war, by an inbuilt aspiration to the eternal values of an increasingly definitive and absolute technique, supported by the canonic models handed down through history, which became persistent references in Nervi's work. Nervi's personal creative trajectory, striving for purification with bursts of rhetoric, coincided with the new political and social climate established in Italy by the Democrazia Cristiana (Christian Democrats) and industrial capitalism. Even in the period that opened with reconstruction and resulted in an economic boom, Nervi therefore confirmed that he was still a creator of structures that always had symbolic connotations and to which the Turin colossus became a

[8] *Ibid.*, p. 165.
[9] *Ibid.*, p. 170.
[10] Giovanni Michelucci, letter to members of the competition panel, 23 December 1959, FM.
[11] Nervi, *Aesthetics and Technology in Building* cit., p. 23.

monument. The critique against the lack of "human scale" voiced by Michelucci, a member of the competition jury, is perceptive and lucid: "I believe that no one seeing that 'interesting' shape would imagine that any person, young or old, would linger 'in there' to study, converse or enjoy themselves; instead they would think of a temple dedicated, a few thousand millennia late, to unknown gods, or they would think of a museum for dinosaurs or other prehistoric animals. (Or they would think that at last the twenty years of fascism had found its architectural form, its spiritual measure)."[10]

The reinforced concrete order

With the erection of the *Palazzo del Lavoro*, the series of variable cross-section pillars had achieved the expression of a colossal order that through its size alone ultimately personified the aspirations of an industrial civilisation, because those images are the sometimes mechanistic reflections of building site economies, manufacturing times, and the entrepreneurial efficiency of Nervi & Bartoli. When the *Palazzo del Lavoro* was complete, Nervi drew up a table of drawings summarising the different variable cross-section pillar shapes he had built since the Unesco headquarters. Like Mies van der Rohe or Perret, Nervi revealed that he was engaged in a quest to seek a column order worthy of the technical properties of the new materials. In this way, his series of pillars became established as the original expression of Nervi's specific form of brutalism, definitively replacing his cartilaginous structures in ferrocement. "[...] I can assure you that the variation in form was inspired in each case by objective static and construction considerations. The transition between each section is obtained by joining, with straight lines, the corresponding points of the two extreme sections; these lines are easily realized with the planks used for formwork. It is obvious that statics offers, in each and every case, a definite direction, but that the detailing of forms and their interrelationship is a personal choice. To the wealth of form possibilities inspired by static factors are added those furnished by the building process."[11]

When he had reached the point of putting together an already extensive repertoire of exposed and variable cross-section reinforced concrete pillars,

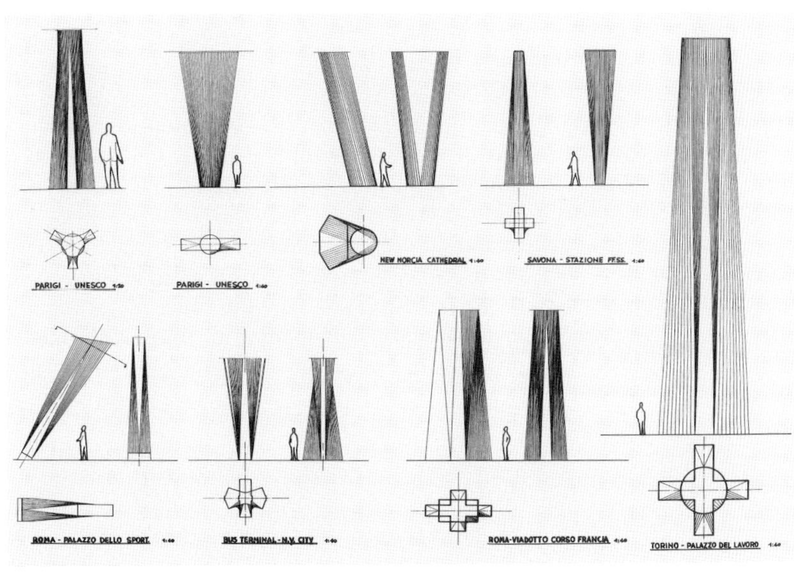

432. P. L. Nervi, plate showing variable cross-section pillars (P. L. Nervi, *Aesthetics and Technology in Building*, Cambridge, Massachusetts, Harvard University Press, 1965, p. 37).

433. P. L. Nervi and A. Nervi, competition design for a multipurpose hall, Kassel, 1961. Model (P. L. Nervi, *Buildings, Projects, Structures. 1953-1963,* New York, Frederick A. Preager, Publisher, 1963, p. 138).

all "capable of expressing a pure structural logic,"[12] Nervi resumed his reflection on the "empty rhetoric" that had been widespread during Fascism and Nazism, and on the use of that kind of "useless columns" or a "profusion of marble" to produce the "appearance and signs of an illusory power." "Looking back today on such works, at a distance of not even twenty years, we feel disgust because their empty rhetoric fossilizes the moral poverty and falseness of the ideas which inspired them."[13] Seen from this perspective, the material bareness and static reason of his pillars was intended to express, in the contemporary cultural climate, the concern for truth that had already assumed in Nervi himself the connotations of "truthful style," and that he reaffirmed in erecting the *Palazzo del Lavoro* to distinguish it from a form of truth reduced to the language of material alone, to the visible and controlled traces of its manufacture and to the ornaments of construction. Nervi's truth continued to be that of the structure and its genesis in the principle of "strength through form." Even though the substance of his thoughts on the style of truth remained unchanged, even in the examples he gave, compared to his previous pronouncements, the term "truth" in the words spoken by Nervi in London in 1961 took on new force through insistent repetition. At times this appears to have been dictated by the concern of someone who believed he had discovered the foundations of style and was eager to explain it at a particular time in history typified by an approach that can only have seemed to Nervi to be a frenzy of structural shapes dictated by will:

"The style whose beginnings we see can be called an authentic *style of truth*, because the ideal forms which inspire it are *true* in the fullest meaning of the word. The shapes of least resistance to penetration are *true* and immutable. A great arch whose average stress is the curve of its permanent load is *true*; the catenary of big suspension bridges is *true* and immutable; a great iron girder whose profile follows the line of bending moments is *true*. And besides, how can we deny the aesthetic emotion which these ideal forms give us and the sense of sublime serenity which we derive from them? How can we fail to rejoice at the fact that they can become the guide of mankind's aesthetic taste? How can we deny the value of an immutable guide, the same for all men, an efficacious method of promoting universal brotherhood, across all divisions of race, color or religion?"[14]

[12] Pier Luigi Nervi, *The Influence of Reinforced Concrete and Technical and Scientific Progress on the Architecture of Today and Tomorrow*, in, "Arts and Architecture," vol. LXXVIII, December 1961, No. 11, (pp. 16-17, 28-30), p. 28.
[13] *Ibid*, p. 17.
[14] *Ibid*., p. 30.
[15] Pier Luigi Nervi, *Chiesa per il quartiere ANIC di Ravenna*, in, "Fede e arte: rivista internazionale di arte sacra," XIII, 1965, No. 2, (pp. 165-171), p. 167.

434-435. P. L. Nervi and A. Nervi (with the cooperation of A. Arcangeli), design of a church for the Azienda Nazionale Idrogenerazione Combustibili village, Ravenna, 1960-64.

434. Sketch by A. Arcangeli, about 1960 (AA, folder No. 237 "Esposizione Parigi. Solai a nervature isostatiche e varie").

435. "Sezione trasversale (cross-section)," drawing dated 5 April 1963 (CSAC, n. inv. pra 469, n. id. 13921, coll. 275/5).

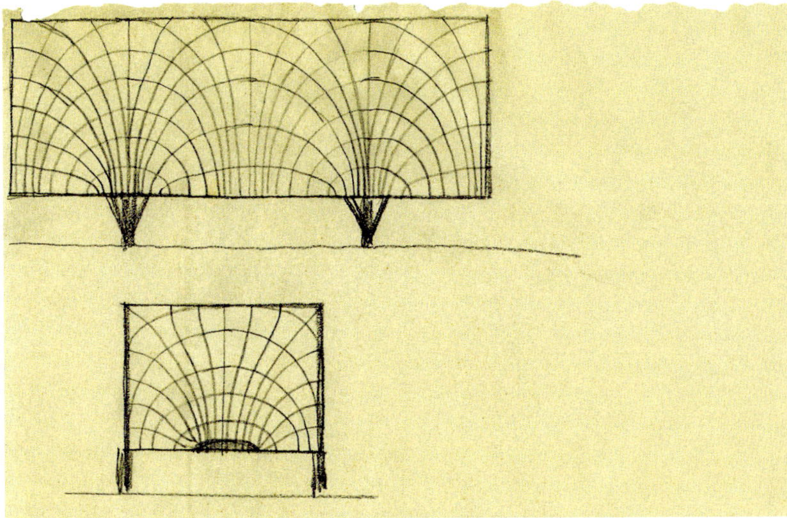

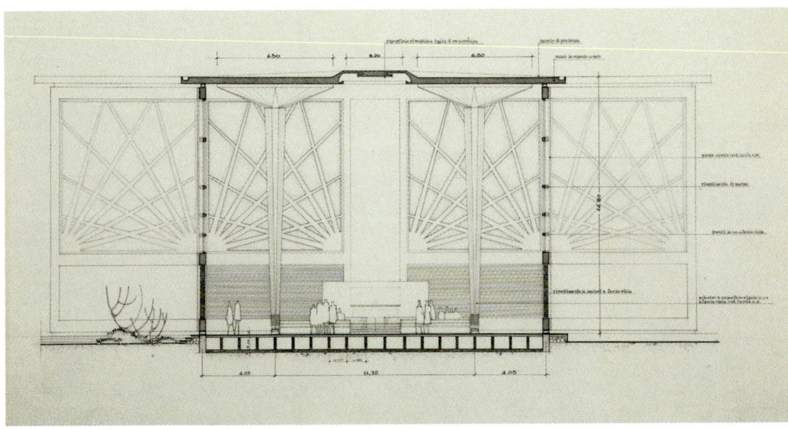

Yet it is impossible to escape the impression that Nervi had returned to the "empty rhetoric" that he had criticised in the "entrance halls of gigantic proportions" of the 1920s and 1930s, when the colossal pillars intended to represent the style of truth erected in the *Palazzo del Lavoro*, were observed through Michelucci's eyes.

The kind of structural unit built for the *Palazzo del Lavoro*, that had already been tried out in works such as the building at the Kursaal in Castelfusano or the church in Borgo Panigale, resurfaced in the structural configuration, also highly symbolic, of a church with a basilica floor plan for the parish centre of the Azienda Nazionale Idrogenerazione Combustibili village in Ravenna. The design was drawn up between 1960 and 1963. The church hall was punctuated by two rows of "mushroom-shaped structures,"[15] independent from one another according to the system designed for the *Palazzo del Lavoro*. They were made of pillars, as usual tapering toward the base, with their cross-section varying from a circle to a cross, and with their arms branched to give rise to a roof with vaguely Gothic ribbing. All parts of the mushroom-shaped structure were designed in exposed concrete. The envelope was a transposition of Gothic cathedral windows to a geometry based on iso-

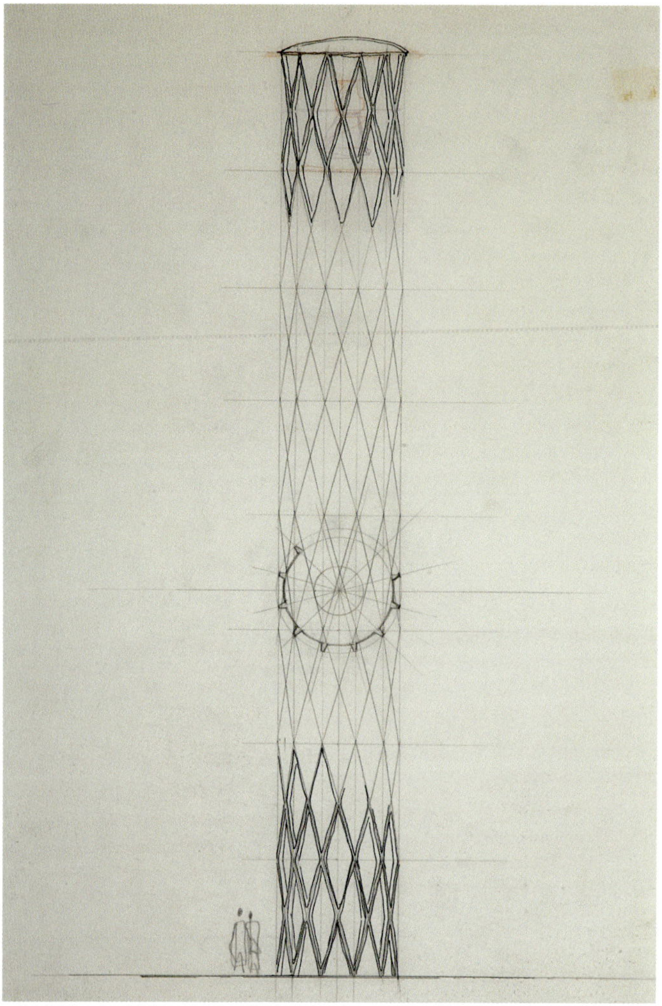

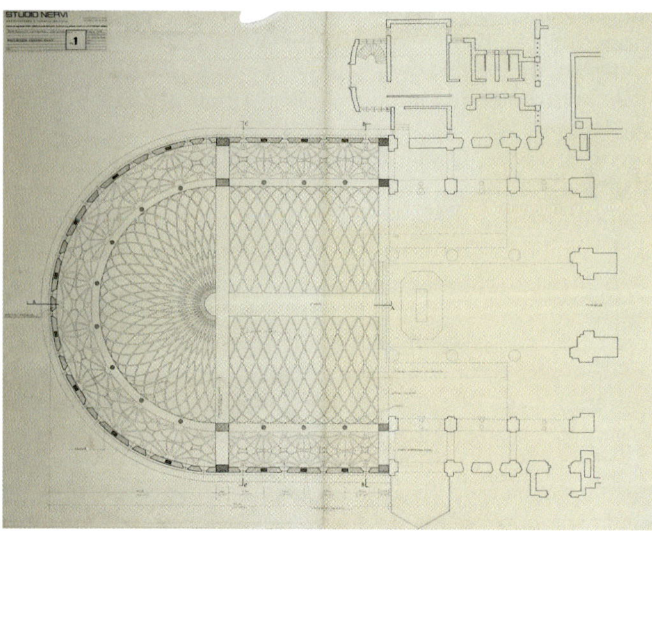

436. P. L. Nervi and A. Nervi (with the cooperation of A. Arcangeli), design of a church for the Azienda Nazionale Idrogenerazione Combustibili village, Ravenna, 1960-64.
Bell tower (CSAC, n. inv. pra 469, n. id. 13921, coll. 275/5).

437. Studio Nervi with Seely & Paget Architects, design for extension to Portsmouth Cathedral, United Kingdom, 1964-68. "Reflected ceiling plan," drawing dated 8 March 1966 (CSAC, n. inv. pra 1305, coll. 137/2).

static ribs, to be constructed in precast reinforced concrete. All the structural lines generated by the flow of forces and static insight that Nervi was so fond of citing, some already exploited for their ornamental potential in the floors, tended to lapse into a decorative effect, rendered even more refined by the surface treatments, such as the protective marble cladding of the window lines or the symbolic gold leaf of the cross moulded on the roof. The bell tower also became a structure with complex lines, in the form of steles or columns: a cylindrical version was also planned, to be built using diamond-shaped *tavelloni* with a stretched geometry, prefabricated in ferrocement. It comes as no surprise that during a trip to England in January 1963, Nervi was impressed by a visit to Basil Spence's Cathedral in Coventry, in which the reinforced concrete was modelled in accordance with English Gothic geometries extraordinarily similar to the ribbed grid patterns of his own works.[16] He proposed a design of his own that would demonstrate the historical value of his isostatic lines when he took on the enlargement of Portsmouth Cathedral in England between 1964 and 1968.

The Burgo paper mill in Mantua and the Fiat water tower in Turin

Nervi's critique of the structural virtuosity of Niemeyer or Saarinen was not limited to his articles published in the *Critica delle strutture* series in "Casabella Continuità," but also found an outlet in his works.[17] The colossal order of the *Palazzo del Lavoro* was one of Nervi's first attempts to demonstrate his own view of a "structural architecture" that he tried to develop using a combination of motifs typical of his system, including variable cross-section pillars and ribbed shells, as well as, to a more limited extent, by experimenting with

[16] Pier Luigi Nervi, letter to Basil Spence, 21 January 1963, Maxxi, P54.
[17] Pier Luigi Nervi, *Critica delle strutture*, in, "Casabella Continuità," XXIII, January 1959, No. 223, p. 57; Pier Luigi Nervi, *Critica delle strutture. Cinque Ponti*, in, "Casabella Continuità," XXIII, February 1959, No. 224, pp. 53-54; Pier Luigi Nervi, *Critica delle strutture. Rapporti tra ingegneria e architettura*, in, "Casabella Continuità," XXIII, March 1959, No. 225, p. 50; Pier Luigi Nervi, *Critica delle strutture. Modello e imitazione*, in, "Casabella Continuità," XXIII, May 1959, No. 227, pp. 50-51; Pier Luigi Nervi, *Architettura e strutturalismo* cit.

438. Studio Nervi with G. Covre, Burgo paper mill, Viale di Poggio Reale 9, Mantua, 1961-62.
Option with arched structure, drawing dated 17 March 1961
(Maxxi - PLN, F08956).

439-441. Studio Nervi with G. Covre, Burgo paper mill, Viale di Poggio Reale 9, Mantua, 1961-62.

439. Positioning of permanent formwork for trestles (AUTCB).

440. Building site, photographed in August 1962 (AUTCB).

441. View of a trestle (BSTA, 024-2930).

motifs that were completely foreign to his creative career. In the wake of the *Palazzo del Lavoro*, Nervi also began to combine cast-in-place reinforced concrete with steel structures. Nervi continued to control the quality of the cast-in-place concrete by taking great care over the construction of the wooden formwork, although he was seeking other solutions that would guarantee a finish without the unpredictable aspects of the traditional system.

The work that confirmed Nervi's decision to configure new roof structures that marked a departure, like the *Palazzo del Lavoro*, from the variety featuring domes and vaults, was the large building produced for the Burgo paper mill in Mantua in 1961–62, which was intended to house a long Beloit Italia machine used for paper production (Nervi & Bartoli made the concrete works, while Badoni produced the metal parts). After planning his usual arched structures, Nervi opted to cover the 250-metre long and 30-metre wide area by using two colossal trestles and cables to suspend the roof in accordance with a device inspired by the structural system of suspension bridges and intended as a critical response to the structural acrobatics evident in the spectacular Brazilian architecture of Niemeyer and Bo Bardi. The positioning of the trestles was designed to make them asymmetrical so that they had to be tilted to anchor the cables. If Nervi's report that the "roof floor and chains were initially supposed to be made out of reinforced concrete"[18] is true, then the first version of the structure for his paper mill should be considered a variant of those reinforced concrete prestressing systems implemented by Morandi for bridge construction. A comparison with the similar prestressed structures of Morandi's strictly symmetrical bridges reveals the expressive tension that animates the trestles in Nervi's system. The inclination converts a large static pillar into a structure that is closer to Nervi's static insight, always dependent upon motifs determined by the need to come up with vaults, domes, or arches

[18] Pier Luigi Nervi, letter to Ulrico Hoepli, 3 July 1962, Maxxi, folder *Costruire Correttamente*.

and turn the large pillar almost into a flying buttress or a secret version of his variable cross-section inclined pillars. The chains suspending the roof are motifs that draw essential lines of force in space – a pure expression of the style of truth. They give substance to the catenary curve that is one of the ideal forms in Nervi's repertoire of strength through form.

It is clear that, as in the preliminary stages of his design for the *Palazzo del Lavoro*, Nervi also intended the paper mill to be a work that would demonstrate the creative potential of reinforced concrete, taking its place in the scenario of international structural architecture and even assuming the connotations of a reply to the illogical yet appealing structure of the Museu de Arte de São Paulo. The introduction of steel for the roof and cables (the calculations were carried out by Covre), radically altered the image of the paper mill and the role of the reinforced concrete, which was limited to two trestles in the finished work.

The construction of the trestles became a pretext for implementing one of Nervi & Bartoli's most carefully executed versions of exposed concrete. Nervi invented a kind of permanent form for these trestles that was destined to be repeated, with changes, in some of his other works: rectangular slabs of reinforced concrete to be left in situ, moulded so that small reinforcement pillars could be constructed once they had been assembled. Due to the contours of the trestles, Nervi & Bartoli prepared an abacus composed of 152 slabs of different sizes, all seven centimetres thick and produced using moulds made out of wooden boards.

The building site started in May 1961; four Wolff cranes on rails were installed to carry out the work. The feet of the trestles were constructed out of reinforced concrete poured in wooden formwork, but constructed using boards the same height as those of the moulds so that the differences were imperceptible. The holes for the wooden form spacers were nevertheless still visible in the feet. Wooden formwork was also used for the connection beams between the trestles – the beams at the top were actually box beams made out of steel clad in concrete. Once the slabs had been installed to a height corresponding to one stage of the pouring, with the aid of Innocenti scaffolding pipes, the sides of the formwork were consolidated using metal ties anchored to irons protruding from the slabs. The reinforcement pillars were then poured to obtain what Nervi described as a "self-supporting hollow structure."[19] The drawings produced by Nervi & Bartoli depict all the various stages of the pouring, confirming the extent to which Nervi had incorporated the concrete manufacturing processes into the design of the structure to perfection, in a manner that went well beyond the processes of *béton brut*.

The four trestles were completed in January 1962. In April, the floor on which the paper production machine would stand was built out of prefabricated slabs in reinforced concrete. Between June and September, the metal roof and aluminium facades were installed. During 1962, Nervi & Bartoli also built a warehouse for storing timber in accordance with a system they had already used for the *Manifattura Tabacchi* in Bologna as well as two water filtration tanks out of the usual triangular *tavelloni* prefabricated in ferrocement.

With its grand trestles extending toward the sky, carefully arranged so that they were tilted and therefore able to express the dynamic, almost Gothic tension between supports and cables, and with the beam at the top aligned in the direction of the cables, the Burgo paper mill became a manifesto for an architecture based on the image of a structure generated not by calculation but by the interplay of forces. These are embodied as motifs that are sometimes almost anthropomorphic, along the lines of Nervi's usual shapes but still quite unusual among his repertoire of structures, a demonstration of that "formal expression"[20] of "structural schemes"[21] that Nervi wished to restore to the level of primordial understanding so that, as he himself wrote, the "spontaneous static scheme" is "immediately apparent"[22] – a "style born out of adherence to natural laws."[23]

[19] Nervi, *Nuove strutture* cit., p. 166.
[20] Pier Luigi Nervi, *Le mouvement architectural actuel*, in, "L'Architecture d'aujourd'hui," vol. XXXI, September 1960, No. 91-92, p. 140.
[21] *Ibid.*
[22] *Ibid.*
[23] Pier Luigi Nervi, *Uno stile comune per l'architettura*, in, "Civiltà delle macchine," IX, October 1961, supplementary edition, (pp. 51-52), p. 52.

442. Studio Nervi with G. Covre, Burgo paper mill, Viale di Poggio Reale 9, Mantua, 1961-62.
Water filtration tank (AUTCB).

443-444. P. L. Nervi, water tower, Fiat Mirafiori complex, Via A. Faccioli 33, Turin, 1961-63.

443. Matrices for constructing the curved rectangular *tavelloni* of the stem (Maxxi - PLN, F08915).

444. View (photograph by T. and G. Vasari; Maxxi - PLN, F08931).

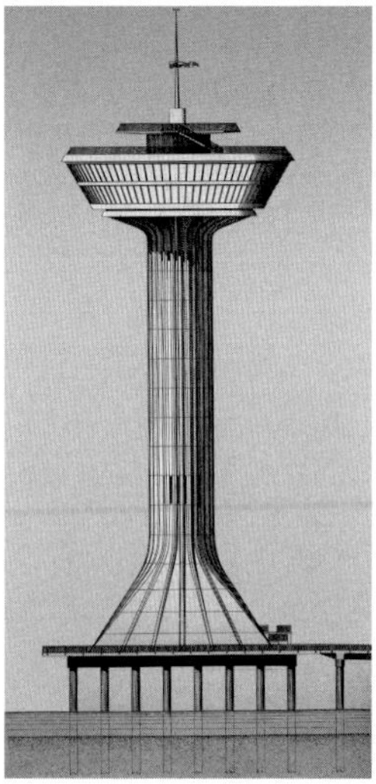

Again with the aim of obtaining a reinforced concrete structure with perfect finish, Nervi adopted the same system of permanent form made up of precast reinforced concrete slabs to erect a 68-metre-high water tower in the Mirafiori industrial complex between 1961 and 1963. Its purpose was to pressurise the firefighting system, and its height was to be exploited to create a belvedere from which to survey the city of Turin and the extent of the Fiat industrial plant.[24] The cylindrical stem of the water tower was built out of formwork consisting of one face to be left in situ, to form a perfect hardpan for the support, and one face to be stripped away after pouring. In the first case, Nervi used three-centimetre-thick curved rectangular *tavelloni* made out of ferrocement and fitted with perforated tabs so that they could be bolted together. These *tavelloni* were prefabricated on-site using masonry matrices finished with a layer of plaster. The face of each *tavellone* was covered by six holes left by the form spacers to create a decorative pattern of the type discussed with Breuer during the Unesco headquarters construction. Nervi used concrete cylinders for the form spacers, again as used in Paris. He planned to use the usual wooden board system for the inner face of the formwork. In order to construct the tank in the shape of a hyperbolic paraboloid, Nervi used another kind of permanent form that was perfectly finished and made out of triangular *tavelloni* in ferrocement, as usual prefabricated using the same system of *tavelloni*. After arranging four layers of metal netting, concrete was applied to build the tank structure. The inclined pillars and round beams of the belvedere were made out of reinforced concrete that was poured, in the case of the pillars, into permanent forms with a variable U-shaped cross-section, while a system of *tavelloni* prefabricated in ferrocement was used for the roof.

During the building site stages and when the work was complete, Nervi commissioned the Vasari studio in Rome to undertake the photo shoot of a work to which he attached particular significance. He considered it to be a monument to the process of constructing reinforced concrete in the modern era, implemented by means of a design and building site comparable to the industrial system used to manufacture Fiat cars. This was the first time that Nervi had left the *tavelloni* made in his ferrocement exposed. The entire water tower, however, was painted in the 1980s to conceal what had come to symbolise the greyness of factory life.

Expressions of structural architecture

A new structure with its lines modelled in accordance with static function and made out of finely worked exposed concrete again characterised the works of Nervi and his firm that were designed and built during the 1960s as further manifestations of his structural architecture. The invention of a structure that pretends to be cantilevered in relation to a central backbone, but nevertheless still belongs to the genealogy of the structures previously developed by Nervi, characterises the 1963 design for the *Palazzetto dello Sport* in Vicenza. The central beam supported by Nervi's variable cross-section pillars is a hollow structure designed to house the air-conditioning system. A special system of sliding formwork was planned to construct the variable cross-section beams

[24] Pier Luigi Nervi, *Serbatoio sopraelevato della capacità di m³ 2000*, in, "L'Industria Italiana del Cemento," XXXIII, March 1963, No. 3, pp. 193-200; A. B., *Un château d'eau de 68 m de hauteur et de 2000 m³ de capacité près de Turin*, in, "La Technique des Travaux," XL, March-April 1964, No. 3-4, pp. 85-88.

445. P. L. Nervi, A. Nervi, F. M. Imbert, design for a tower with restaurant at the Lido di Castelfusano, 1962-66 (BSTA, 024-2905).

446. P. L. Nervi, A. Nervi and P. Maltauro, design for *Palazzetto dello Sport*, Vicenza, 1963 (BSTA, 024-2188).

that branch from the hollow beam to create the roofs of the swimming pool and playing field. The technique of prefabrication was used to create the spindle-shaped cruciform pillars along the ends of the beams, derived from those in the *Palazzo dello Sport*, with arched roofing *tavelloni* to be fitted between the beams. All the exposed concrete was to be finely finished by sandblasting using compressed air, "to remove the superficial stains due to the oily substance that has to be spread on the formwork for clean and easy stripping" as Nervi explained.[25]

In the case of the design for the Risorgimento bridge in Verona, drawn up in 1963, the static intuition that Nervi frequently referred to was assured by a synthetic diagram representing the bending moments. This was designed on the basis of a drawing by the municipal technical department of a profile of the three-arched bridge. As with the static function of the Burgo paper mill, Nervi subdivided the length of the bridge into four segments in order to obtain two cantilevered segments at the end and reduce the number of pylons to two (the heads rested on retaining walls made out of reinforced concrete). Nervi did not, however, limit himself to modelling the cross-section of the deck box structure variable profile in order to adapt it to the moment diagram and the shear forces. Instead, he controlled the flow of water in the event of floods by using a ruled surface, the hyperbolic paraboloid, to configure the sides of the deck in order to direct the current alternately under the arches and over the pylons, never by opposing it with a flat surface. In this way, the surface almost appears to have been generated by the flow of the river's current when in flood, like the ice models created to compute the profile of the aerodynamic structures so admired by Nervi, which were shaped by a current of air.[26]

For the construction of the bridge, which was carried out between 1966 and 1968 by Edilbeton S.p.A. of Verona under the direction of the engineer Rocco Nicolò, Nervi drew up a detailed tender specification in 1966 to determine the kind of exposed reinforced concrete, all cast in place, that would be used to characterise the bridge. The surfaces, obtained using planed boards no longer than 8 centimetres in width, would be "left in their raw state as when stripped," but worked on by means of "light sanding using compressed air" to remove any stains left by the "oily substances" spread on the formwork to facilitate stripping, without altering the visibility of the "impressions left by the boards and joints."[27] In the technical documents, Nervi underscored the role of the woodwork for manufacturing the concrete, the "pattern of boards,"

[25] *Relazione architettonica e tecnica*, p. 5, Maxxi, P54.
[26] Pier Luigi Nervi, *Is Architecture Moving Toward Unchangeable Forms?*, in, Gyorgy Kepes, ed., *Structure in Art and Science*, New York, George Braziler, 1965, (pp. 96-104), p. 96.
[27] *Capitolato speciale di appalto*, 9 August 1966, p. 35, Maxxi, P26M.

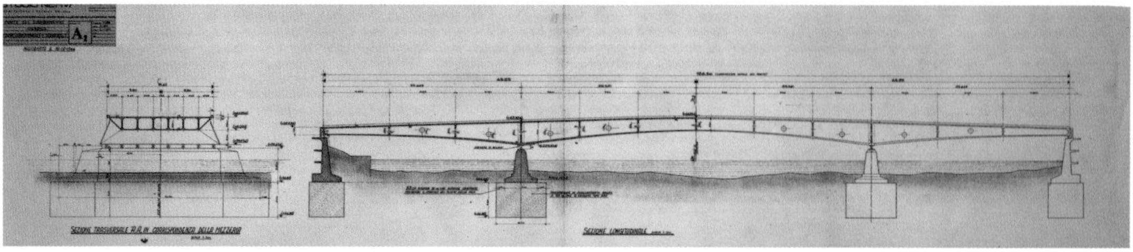

447-448. P. L. Nervi, *Risorgimento* bridge over the River Adige, Verona, 1963-68.

447. "Sezioni longitudinale e trasversale (Longitudinal and cross-sections)," drawing dated 10 December 1966 (BSTA, 024-4623).

448. View of deck box structure, about 1968 (BSTA, 024-6760).

and their valuable role of "aesthetic enhancement." "[...] the cement structures with exposed faces must be perfect, without finishing, because any concrete cobweb patterns or irregularities can no longer be removed," stated the Italian disciple of *béton brut*.[28] In order to control the pouring stages and remove any imperfections, Nervi included in the drawings what had become a distinguishing mark of his concrete: a groove, or *scuretto,* positioned at the point where he planned to extend the continuous pouring of one structural segment. At this point, Nervi's *scuretti* took on the significance of Kahn's pour joints: "The presence and the position of the *scuretti* will avoid highlighting the inevitable irregularities resulting from concrete casting, and herein lies their importance."[29] The colour of the concrete also became a concern to Nervi, who gave instructions for ensuring its "perfect uniformity" and for conducting an "in-depth experimental study of the aggregate grading." Lastly, he advised that the pourings should be "appropriately vibrated at depth, taking care not to cause the mixture elements to separate."

Another technical detail that was destined to further characterise Nervi's concrete was the cylindrical hole left by the screw used to fasten a clamp produced by Tempo of Milan. In the swimming pool in Via Mincio, in Milan, built between 1963 and 1964, Nervi & Bartoli used this device, which pro-

449. Studio Nervi, *Stadio Comunale* at Parco Agogna, Viale Marmo 8, Novara, 1963-76. View of covered stand, about 1976 (BSTA, 024-7380).

duced a pattern similar to that left on the Unesco headquarters or the Fiat Mirafiori water tower. "The vertical position of holes for 'Tempo' clamps is always halfway between the floor intrados and the base of the large beams,"[30] was written on a working drawing.

For the 1963 design of a stadium in Swindon in the United Kingdom and that of the *Stadio Comunale* in Parco Agogna in Novara, built by Nervi & Bartoli between 1964 and 1976, Nervi returned to the structural system that he had used in the *Palazzo dello Sport*, with inclined variable cross-section pillars and infill shells between the stand beams, in cast-in-place reinforced concrete for the stadium in Swindon, and in triangular *tavelloni* prefabricated in ferrocement for the stadium in Novara. For the latter (construction work began in 1971 under the direction of Desideri and the engineer Giuseppe Rossi, who was construction manager), Nervi supplemented the reinforced concrete structure with a steel canopy, proposing a composite structure like structures of the *Palazzo del Lavoro* and Burgo mill paper – the roof covering, which was initially supposed to be made out of ferrocement *tavelloni*, was made out of cement asbestos. In order to "achieve a certain ease of dismantling and stripping with a view to future reuse,"[31] the bolted joint system designed by Nervi for the *Palazzo dello Sport* pillar formwork was changed and replaced by wedges.

As part of the project to construct buildings for use as restaurants owned by the confectionery business Motta on the Italian motorways, Nervi designed various types between 1962 and 1966, in the form of bridges for the towns of Greggio near Novara and Limena near Padua, and in the form of towers for Serravalle Scrivia, in the province of Alessandria. In accordance with a model introduced by Melchiorre Bega, the bridge restaurants became a pretext for envisaging other sculptural pylons shapes – his variable cross-section pillars were also present. He additionally developed a complex profile for the platform in order to achieve patterned effects in the finished work as a result of the interplay of formwork boards. His aim was to show travellers a structure with sophisticated, moulded lines that had the stature of a work of art. For

[28] *Ibid.*, p. 3.
[29] *Ibid.*, p. 5.
[30] Drawing No. 4903-14, CSAC, n. inv. pra 1257, n. id. 10259, coll. 36/2.
[31] Mario Desideri, letter to Giuseppe Rossi, 8 October 1971, Maxxi, P96.

450. Studio Nervi, design for bridge restaurant, Greggio. "Studio prospettico (perspective study)," drawing dated 29 January 1961 (CSAC, coll. 17/1-2-3).

the tower restaurants, that were configured as colossal pillars with a variable hollow cross-section, Nervi went back to consider his designs and patents of the 1930s for revolving buildings.

Nervi & Bartoli was awarded the contract to build a bridge restaurant in Limena, designed in conjunction with Bega between 1964 and 1966. The restaurant took the form of a colossal Vierendeel girder. Thinking back to the Burgo paper mill trestles, Nervi was clearly looking for figures in his engineering repertoire that could be transformed into an expression of his structural architecture. He again configured the position and size by means of the moment diagram generated by the support on the four variable cross-section pillars. All of these were different from the octagonal openings of the Vierendeel girder, initially fitted with contracted flanges to form a frame. Nervi's aim was to use those openings to reveal the flow of forces between pillars and girder. In one study, he actually envisaged adding a reinforcement for each pillar. The model of the Vierendeel girder, executed in 1965 with a slab of perforated tiolon, underwent load tests at the Bergamo Istituto Sperimentale Modelli e Strutture (the reinforcements and contracted flanges were removed). Nervi also used the usual ribbed floor to create the roofs of the buildings at the end of the bridge restaurant. The fact that the secondary ribs create ring sectors in each compartment of the floor revealed Nervi's decorative intent in resolving some practical problems involved in the installation and stripping of the large moulds for the square spans.

In the working drawings, Nervi planned every element of the formwork manufacture, including the position of the Tempo clamps, the vertical position of the boards, and the pouring horizontal groove, which was aligned with the longest axis of the octagonal openings. When deprived of the frame

451-452. P. L. Nervi, A. Nervi and M. Bega, bridge restaurant over the A4 Brescia – Padua motorway, E70, km 355.2, Limena, 1961-65.

451. "Vista della dogatura del pilastro (View of pillar staves)," drawing dated 27 October 1965 (CSAC, n. inv. pra 1167, n. id. 15103, coll. 17/1-2-3).

452. View, about 1966 (BSTA, 024-2898).

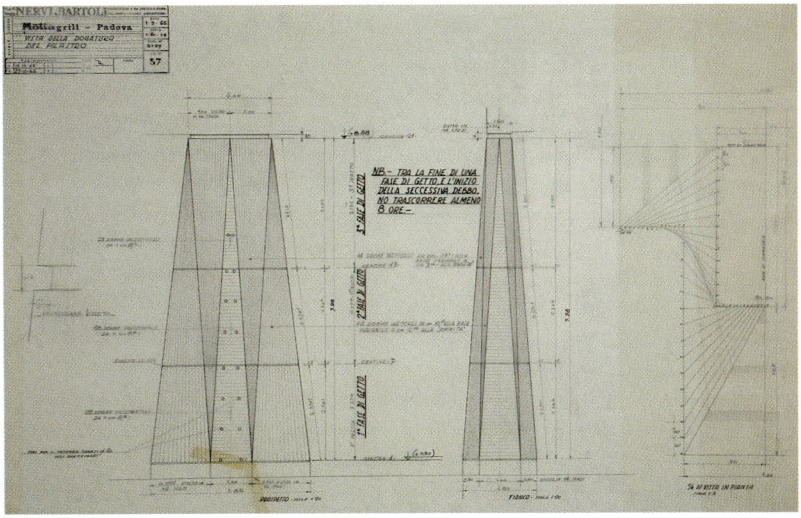

and reinforcements, the Vierendeel girder looked like a monolith resting on the pillars. Prefabricated beams in reinforced concrete were used to create the exposed weave of the floor between the pair of Vierendeel girders (the roof consisted of a brick-cement floor). For the variable cross-section pillars, Nervi, as usual, examined the position of the boards and divided their height into three parts, identified by grooves corresponding to the pouring stages.

Even though the structure was organised in accordance with conventional figures of pillars and floors, Nervi cited it as the crucial instrument in determining the formal and symbolic quality of the headquarter of the *Officina Carte Valori* (Banknote Printing Works) of the Banca d'Italia, in Via Tuscolana in Rome, designed between the end of 1961 and 1965, and built by the firm Astaldi S.p.A. of Rome between 1966 and 1975 (the engineer Vittorio Morucci was appointed to manage the work). Nervi distinguished the shape of the pillars according to whether they belonged to the internal pattern of the building (in this case, they were cylindrical and mushroom-shaped, to facilitate foot traffic and the organisation of the rooms), or defined the perimeter of the building (in this case, they became cruciform with a spindle-shaped profile). For the floors, he also initially designed a beam geometry able to create a decorative grid, but this did not allow for the need to house the system piping; therefore he adopted a reinforced concrete floor without ribs. Variable

453-454. P. L. Nervi and A. Nervi, headquarters of the Banca d'Italia *Officina Carte Valori*, Via Tuscolana 417, Rome, 1961-75.

453. View, about 1975 (BSTA, 024-3415).

454. View, about 1975 (BSTA, 024-3423).

cross-section figures were proposed for the entrance canopy and the chimney stack. The entire building was made out of exposed reinforced concrete, as usual manufactured in accordance with the practices that were now routine for Studio Nervi, described in reports and specifications drawn up using the same wording. This was a sign that the Nervi style no longer took the form of prefabrication in ferrocement but in cast-in-place reinforced concrete and carefully worked in accordance with certain criteria, including Tempo clamps, formwork with planed, hinged boards, pouring grooves, and sandblasting of the surfaces after stripping. "[...] the expressiveness of the facades is largely determined by the cement structure left exposed without plaster or cladding of any kind," specified Nervi.[32]

Prefabrication through the use of ferrocement in addition to reinforced concrete and using moulds made out of coated masonry or ferrocement, still with the vacuum technique, again became part of the company's routine production practices when it constructed the grand *Stabilimento Cromo* for Fiat in Venaria Reale, near Turin, between 1962 and 1963. The prefabrication concerned all the main and secondary parts, with the exception of the pillars; and included the colossal roof trusses, perforated contoured beams and guttering,

[32] Pier Luigi Nervi, letter to Tullio Riccio, 12 December 1961, Maxxi, P66.

455-458. Società Nervi & Bartoli, Fiat Cromo plant, Viale Carlo Emanuele II 150, Venaria Reale, 1962-63.

455. Production of prefabricated elements using the *Vacuum* system, 1963 (BSTA, 024-2379).

456. Precast reinforced concrete roof trusses, photographed in 1963 (BSTA, 024-5350).

457. System for transporting precast reinforced concrete beams, photographed in 1963 (Maxxi - PLN, F06612-3).

458. General view of building site, 1963 (BSTA, 024-2388).

459. Studio Nervi, design of "studio di serbatoi di grezzo sommersi da 22.000 m^3. Pianta (Study for submerged crude oil tanks of 22,000 m capacity. Plan)," drawing dated 6 March 1970 (CSAC, n. inv. pra 1351, n. id. 15331, coll. 169/5).

460. Nervi & Bartoli, design for nuclear vessel. "Soluzione A con anello esterno in C°.A°. Sezione orizzontale a metà contenitore con indicazione armatura delle pareti verticali (Option A with outer ring in reinforced concrete. Horizontal section halfway down vessel showing vertical wall reinforcement)," drawing dated 18 December 1965 (BSTA, 024-4662).

which were made out of reinforced concrete; and the infill panels, large roof tiles, and joint covers, which were made out of ferrocement. Nervi & Bartoli ran their building site using a pneumatic system to transport the concrete and special machinery to build the concrete substrates (a paver mounted on rails), and a Fiorentini mobile crane to handle the prefabricated parts. The roof trusses and panels were installed between August and October 1963, and the complex was almost complete by December.

Between 1962 and 1970, Nervi designed experimental construction systems, always involving the use of reinforced concrete, including prestressed concrete, for the manufacture of underwater tanks and nuclear reactor vessels, the former intended for companies such as Ente Nazionale Idrocarburi (ENI), Società Nazionale Metanodotti (SNAM), and Shell Italiana. For the tanks, which were to be constructed while floating and then sunk, in accordance

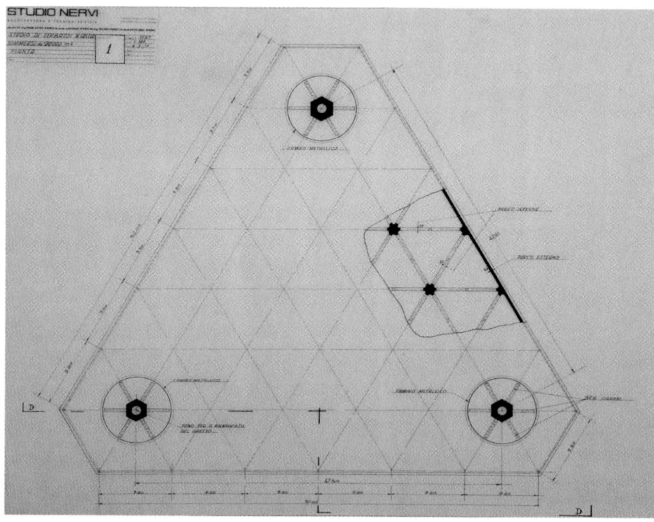

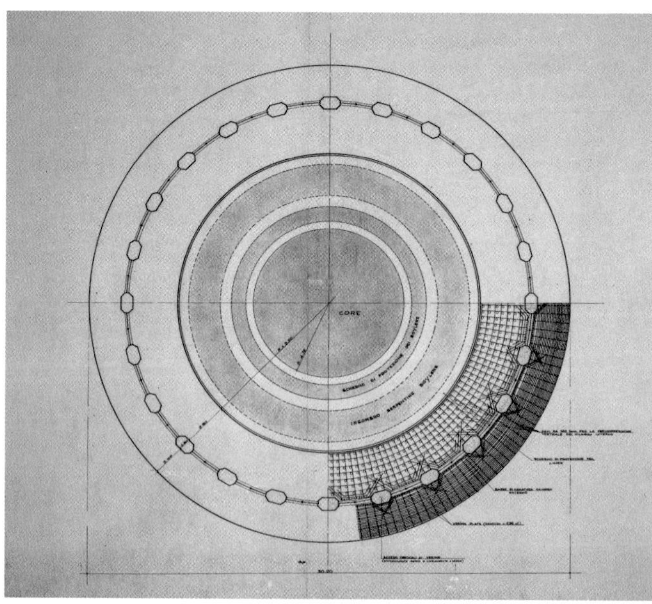

with traditional hydraulic engineering criteria, Nervi planned prefabricated elements measuring up to 20 centimetres in thickness, sometimes involving geometries with a triangular base for the foundation, roof, and tank. For the nuclear reactor vessels, he designed a construction system in prestressed reinforced concrete through the use of a device patented by Freyssinet that he attempted to popularise in the United States.

Despite the effectiveness of their implementation and the ensuing construction speed, the construction techniques used to build the Venaria complex were mainly based on patents filed between 1943 and 1949. This revealed that from the 1960s, Nervi & Bartoli was no longer able to introduce substantial technical innovations or construction breakthroughs in its building sites. The production of prefabricated elements that won the firm commissions due to the speedy construction they guaranteed was still subject to the massive use of skilled labour, which was increasingly costly due to a gradual improvement in the economic conditions of post-war Italy.[33]

International spread of the Nervi style

While Studio Nervi and Nervi & Bartoli gradually abandoned using prefabrication in reinforced concrete or ferrocement for the construction of large roofs in Italy (one typical example was a design for a flat roof in steel for the *Palazzo dello Sport* in Milan in 1969), the structural figures of Nervi's work based on prefabricated, assembled elements, made famous by the *Palazzetto dello Sport*, began to be exported throughout the world. This led to the design and construction of large roofs in the United States, Africa, the United Kingdom, and Brazil. The building work was carried out by local firms under the management of engineers from large architecture and engineering firms. Examples of the spread of the Nervi style, in its most traditional form, were roofs built for the field house and the ice rink at Dartmouth College, in Hanover, New Hampshire; for the Cathedral of Saint Mary of Assumption,

[33] See Sergio Poretti, *Struttura e architettura nel modernismo italiano*, in, "Rassegna di Architettura e Urbanistica," XLI, January-August 2007, No. 121/122, pp. 30-31.

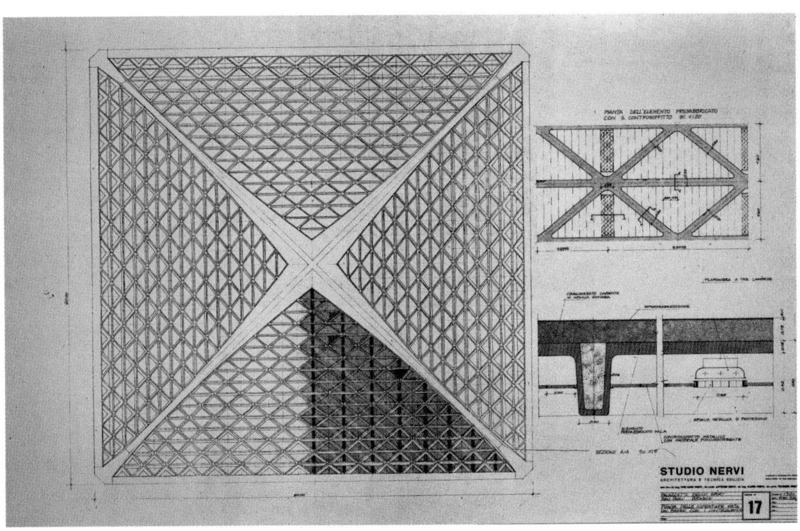

461. Studio Nervi, design for *Palazzetto dello Sport*, São Paulo, 1971-75. "Pianta delle coperture vista dal basso con i controsoffitti (Plan of roofs seen from below with false ceilings)" (BSTA, 024-5215).

462-463. P. L. Nervi and Campbell & Aldrich Architects (with the cooperation of J. Minnich), Field House, Dartmouth College, South Park Street, Hanover, New Hampshire, 1960-62.

462. Building site, photographed in October 1961 (BSTA, 024-3255).

463. Construction of prefabricated *tavelloni*, photographed in 1962 (BSTA, 024-3258).

at Cathedral Hill in San Francisco, designed by Pietro Belluschi; for the Congress Centre in Norfolk, Virginia;[34] for the Good Hope Center in Cape Town, South Africa; and for the funeral chapel for the President of the Republic, Félix Houphouet Boigny, in Yamoussoukro, Ivory Coast. Nervi's work was generally limited to providing designs, sometimes after having their structures checked using models at ISMES. He rarely followed developments on the building sites, except in cases such as the site in Cape Town, where he set up an office under the management of his son Mario. He often encountered so many technical and labour-related difficulties of various kinds when there was a need to manufacture his special *tavelloni* on site that, in the case of the Cathedral of Saint Mary, he proposed to produce all 1600 parts of the roof at the Nervi & Bartoli site in Via della Magliana then send them by boat to San Francisco.[35] All the prefabricated *tavelloni* used for the roofs of Dartmouth College, the Cathedral of Saint Mary, or the Congress Centre in Norfolk were made out of reinforced concrete and not ferrocement. This was a reflection of the technical development of Nervi's system but also of difficulties in finding wire netting and small-diameter steel bars in the United States.[36] His attempts to export his structures to the United States clearly revealed the special and now outdated personality of Nervi the engineer and builder. He had become a relic of an Italian production and building site system that was still at the level of a craft industry.

A significant variant of "structural architecture," aptly inspired by the monuments of Mies van der Rohe, was developed in 1964 for the design of a Memorial dedicated to Enrico Fermi to be erected on the Chicago University campus. Confirming the rhetorical drive that continued to inspire his inventions, Nervi decided to use four colossal columns, again with ruled surfaces and tapering towards the top, but to be built out of colossal funereal pieces of black Nuraghe granite that were 50 centimetres high and connected by stainless steel foils visible in the joins between the parts. There is no doubt that this kind of stone drums column reflected the inventions he envisaged for designs such as the *Monumento alla Bandiera*. For the cantilever roof of the Memorial to Fermi, he envisaged a kind of floor with isostatic ribs, to be produced out of prefabricated elements in ferrocement, in this case executed with sophistication using a paste made out of white crushed marble, to be poured into moulds treated with Rugasol, so that it would not become stained. "[...] the white will therefore be absolute," wrote Nervi.[37]

[34] For Nervi's activities in the United States, see Alberto Bologna, *Pier Luigi Nervi negli Stati Uniti. 1952-1979. 'Master Builder of the Modern Age'*, Firenze, Firenze University Press, 2013.
[35] Nervi entered into agreements with the Fratelli Delfino company, ship brokers from Rome, to send the entire roof structure to San Francisco via Panama, see correspondence exchanged between 19 July and 8 October 1965, Maxxi, P70A.
[36] Bologna, *op. cit.*, p. 181.
[37] *Progetto Preliminare per il "Memorial Enrico Fermi." Chicago, Relazione Tecnica*, Maxxi, P56A.

464-466. P. L. Nervi, A. Nervi and Campbell, Aldrich & Nulty Architects, ice rink, Dartmouth College, South Park Street, Hanover, New Hampshire, 1967-76.

464. Preparatory model for inclined perimeter columns, June 1971 (BSTA, 024-4575).

465. Inclined perimeter columns, photographed in 1976 (BSTA, 024-7008).

466. View, photographed in 1976 (BSTA, 024-7030).

467-468. P. L. Nervi, consultancy for structure of Saint Mary of Assumption Cathedral, designed by P. Belluschi, McSweeney, Ryan & Lee Architects and L. Robinson, 1111 Gough Street, San Francisco, 1963-71.

467. Internal view of model, 1964 (BSTA, 024-2174).

468. Raising and positioning formwork for pillars, photographed in 1968 (BSTA, 024-4486).

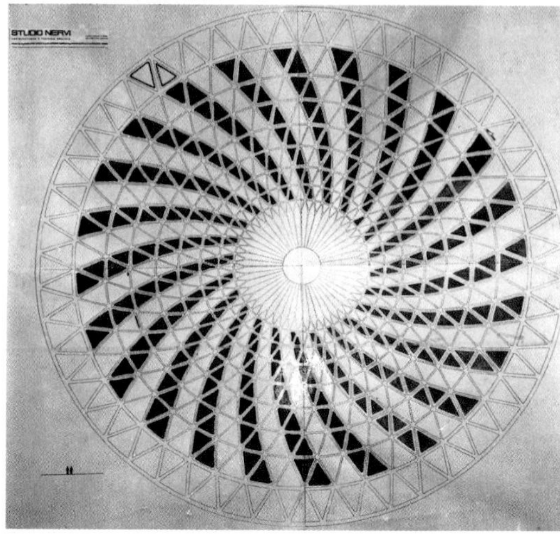

469-471. Studio Nervi with William and Tazewell & Associates Architects, Cultural and Convention Center, 201 East Brambleton Avenue, Norfolk, Virginia, 1965-71.

469. Intrados of covered arena dome, drawing from 1966 (BSTA, 024-2829).

470. Intrados of covered arena dome, drawing from 1966 (BSTA, 024-2834).

471. Assembling the prefabricated *tavelloni* of the covered arena dome, photograph dated 1 September 1970 (BSTA, 024-4479).

472. Studio Nervi with Powell and Moya Architects, design for Pitts Rivers Museum and Centre for Anthropological and related studies, Oxford, 1966-68 (BSTA, 024-4954).

473-474. Studio Nervi with Colyn & Meiring, Good Hope Centre, Sir Lowry Road, Foreshore, Cape Town, 1969-79.

473. Building site (BSTA, 024-6550).

474. Preparatory model of floor (BSTA, 024-4227).

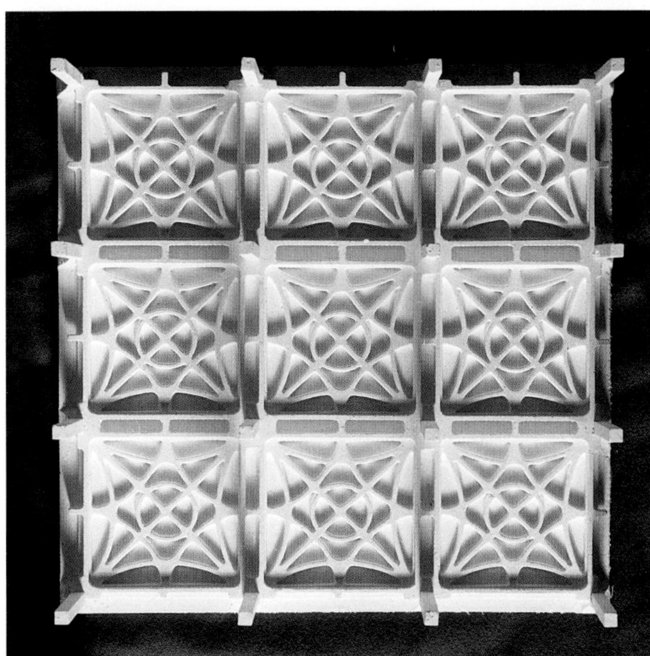

475-477. Studio Nervi with Colyn & Meiring, Good Hope Centre, Sir Lowry Road, Foreshore, Cape Town, 1969-79.

475. Assembling prefabricated *tavelloni* of the covered arena dome (BSTA, 024-6967-21).

476. View of the covered arena 1979 (BSTA, 024-6961-14).

477. Bush hammered concrete at the base of the covered arena dome (BSTA, 024-6969-16).

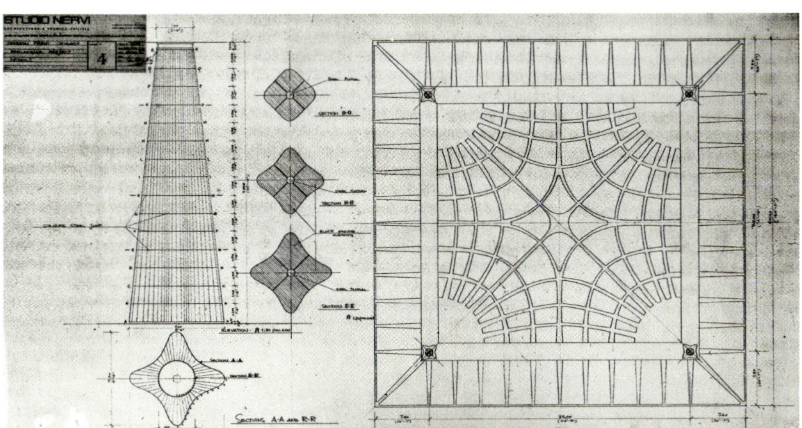

478. P. L. Nervi, design for the Enrico Fermi Memorial, Chicago University campus, 1963-66. "Preliminary Project. Details," drawing dated 28 December 1964 (BSTA, 024-4895).

479-480. Studio Nervi with A. Scattolin, headquarters of Cassa di Risparmio di Venezia, Campo Manin, Venice, 1963-72.

479. Matrix for construction of formwork, photograph dated 28 November 1968 (photograph by G. Giacomelli; Maxxi - PLN, F08450).

480. Matrix for construction of formwork, photograph dated 17 December 1969 (photograph by G. Giacomelli; Maxxi - PLN, F08481).

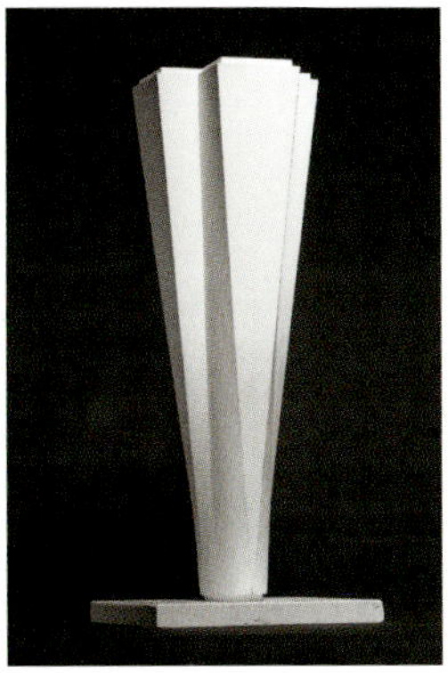

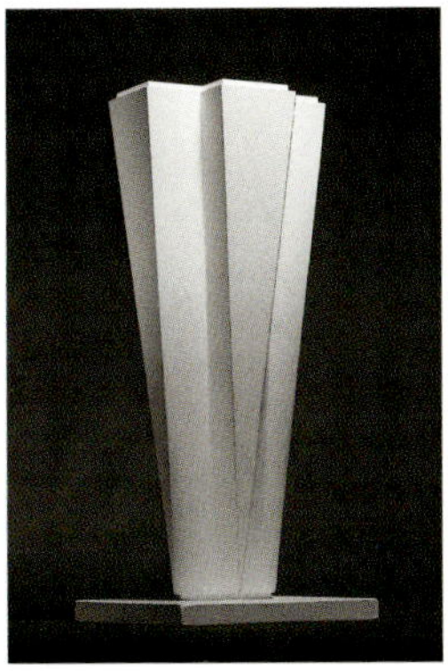

481-482. E. Beaudouin, A. Camenzind, and P. L. Nervi, headquarters of *Bureau International du Travail,* Route des Morillons 4, Geneva, 1966-75.

481. Preparatory model for entrance pillars (BSTA, 024-4220).

482. Preparatory model for entrance pillars (BSTA, 024-4221).

483-484. Studio Nervi with Cabinet J. Leon, funeral chapel for President of the Ivory Coast Republic F. Houphouët-Boigny, Yamoussoukro, 1975-78.

483. View, 1978 (BSTA, 024-6839).

484. View, 1978 (BSTA, 024-6740).

chapter fourteen
the last works by studio nervi

The *Aula per le Udienze* in the Vatican, the apotheosis of Baroque

Nervi's creative path, starting from his debut during fascism with works symbolising a political revolution, up to the masterpieces he erected to celebrate the productive forces of the democratic state built at the end of the war, and including the monuments he erected against the backdrop of the economic boom, all demonstrate the extent to which the structures of one of the twentieth century's greatest engineers are rhetorical devices and not simply systems of static forces or construction processes. Nervi's last celebrated work, the *Aula per le Udienze Pontificie* (Papal Audience Hall) in the Vatican, commissioned by Pope Paul VI for the Second Vatican Council, did not escape the rhetorical urge that breathed life into his other masterpieces. The candour that pervaded the structure and its material, the use of artificial and natural light, the design of the eye-shaped openings, the interplay of lines traced by his signature ribs all display that Nervi was now folding, modelling, and manipulating the figures in his repertoire to provide a setting for the Pope exactly as Bernini had done. Just as Bernini deployed fine materials to accentuate the theatrical effects of his spaces, Nervi adorned or clad his concrete to the point that its nature was transfigured in ways that were unprecedented in his work. No other contemporary engineer dared to design his structures in a way that went beyond the boundaries of static and construction rules, which improvements in mathematical modelling and technology were making increasingly dependable and objective. Silence from Italian and international critics eloquently reflected the way Nervi became lost when he strayed from the basic values of his structural architecture.

In the preliminary plans for the Hall with its trapezoid floor plan, from 1963–64, apart from some attempts to use the triangular geometry of his *tavelloni*, Nervi decided to use waved beams, to be prefabricated in reinforced concrete, to draw the gaze of the faithful toward the papal throne. He planned to achieve this by arranging them along the long axis in a fan arrangement. He explored the profile of the beams not due to their static or structural char-

485-488. P. L. Nervi (with the cooperation of A. Nervi), *Aula per le Udienze*, Vatican City, 1963-71.

485. "Pianta dei soffitti (plan of ceilings)," drawing dated June 1968 (SCV).

486. "Volta: armatura e getto longit.[udinale] onde prefabbr.[icate] (Vault: reinforcement and longitudinal pouring of prefabricated waves)," drawing dated 19 June 1968 (SCV).

487. Preparatory model for pillar with ruled surface (BSTA, 024-3249).

488. "Studio colonnato atrio (Study for atrium with columns)," drawing dated 13 July 1967 (BSTA, 024-2878).

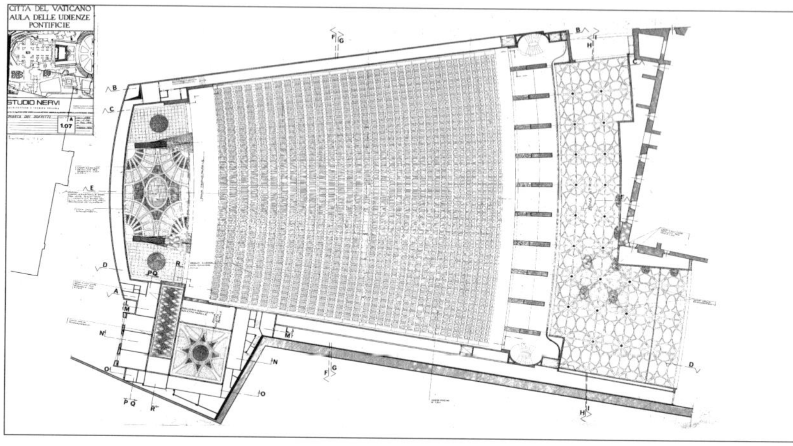

acteristics but for the effect that they had on the perspective of the figure of the Pope. When he still planned to contain the volume of the Hall within the contours of the Leonine Wall, Nervi designed his beams to follow an asymmetrical path, swelling at the base of the Hall and directed toward the throne, almost as though to confer authority and distance on the figure of the Pope. For the beam cross-section, he planned waves, then pleats, and then incorporated perforations for the installation of the lighting and ventilation systems. When the Hall was constructed, an air conditioning system with grilles was installed in the floor.

The abutment structure of the vault, to be made out of cast-in-place concrete, was specially modelled to create an appropriate stage set for contemplation of the Pope. Nervi also examined different variants of this structure, some powerfully sculptural with variable cross-section pillars distorted to create disproportionately amplified ruled surfaces that connected to the ribs of the vault. It was almost as if he were aiming to find a new configuration, in reinforced concrete, for the bronze canopy in St. Peter's. It is significant in this regard that in one version of the design documented by a model, Nervi designed pillars with shaded bands, to be achieved by a particular arrangement of the boards in the formwork, reflecting the grooves of the Renaissance pilasters that framed the papal throne in the same model. Nervi took a long time to find a solution to the ratio between the vault and the infill walls, undecided between a desire to identify a symbolic figure that expressed the sacred value of the Hall and an urge to declare the static function of its parts. In some studies of the facades, the roof structure is therefore completely detached from the infill by means of a continuous glazed slot. In others, the masonry bends to create an opening in the shape of a divine eye, with an arched profile opposing that of the vault. In other studies, which led to the final solution, Nervi designed a window in the shape of an actual eye.

The flat floors, one along the stand used by the Pope, the floor of the entrance, the floors of the room for private audiences and of the gallery linking the room to the Hall (known as the *Galleria di Gala*) provided him as ever with opportunities to show off his trademark structural geometries – a grid of ribs based on the isostatic ribs is a recurrent theme. Although Nervi had long been accustomed to converting the rules of Arcangeli's floor into ribs that were no longer isostatic, in the case of the *Aula per le Udienze*, the ribs now became grids suspended in space that were self-supporting, with their

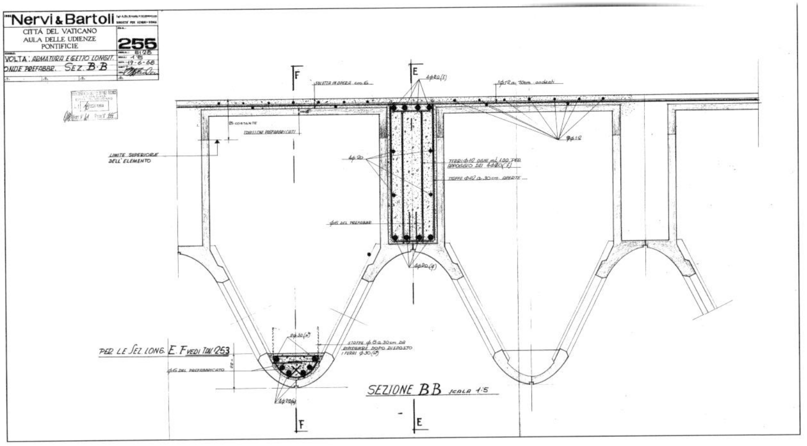
Nervi & Bartoli
CITTÀ DEL VATICANO
AULA DELLE UDIENZE
PONTIFICIE
255
SEZIONE BB scala 1:5

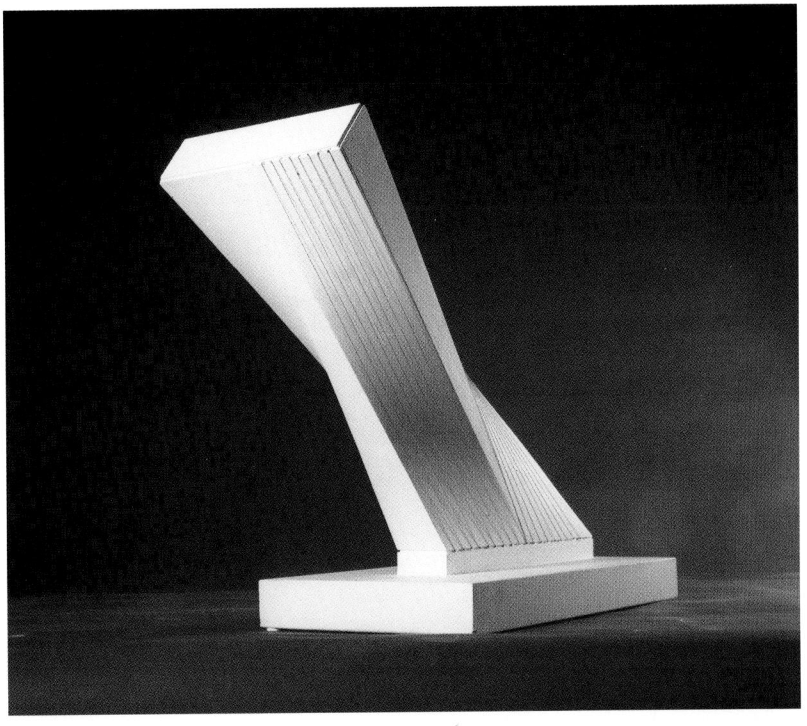

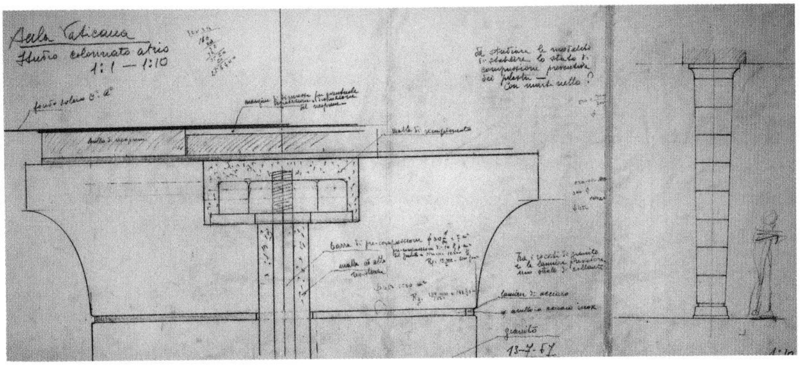
Aula Vaticana
1:1 — 1:10

489-491. P. L. Nervi (with the cooperation of A. Nervi), *Aula per le Udienze*, Vatican City, 1963-71.

489. Building site, photograph from October 1967 (BSTA, 024-2358).

490. Formwork of one variable cross-section throne pillar, photograph dated 3 October 1967 (BSTA, 024-2148).

491. Building site, photograph from October 1967 (BSTA, 024-2356).

compartments completely open, like a Gothic lattice. Sound-absorbent panels or gilded grilles were then rested above this to filter the artificial or natural light in order to produce Baroque lighting effects.

At the preliminary stages, the presence of exposed concrete had already been carefully scrutinised. Within the landscape of the Vatican monuments, the Hall generally appears to be enveloped in fine materials such as the bronze-coated brass slabs designed to clad the heads of the vault structure to form the eaves, the travertine marble used for the infill walls, or the marble blocks used to line the rose window. Nervi planned to use white cement for the exposed concrete.

The contract for the *Aula per le Udienze* was awarded to Nervi & Bartoli, and work began in May 1966, after the firm Castelli S.p.A. of Rome had finished demolishing the existing buildings. To reabsorb the stresses created by the extensive thermal expansions in the vault with its 70-metre span, Nervi had a special foundation built using Lorenz wells, bored piles, and ten chains in reinforced concrete, with a cross-section of 50 x 50 centimetres. These ran beneath the pavement along the length of the Hall in order to connect the ten vault abutment spurs to a special coffered beam located beneath the variable cross-section pillars of the papal throne. In July 1966, to check the chain device, he had models constructed in the ISMES laboratory. Nervi & Bartoli did not begin to build the chains until one year later, in July 1967. These were completed in October and held in a curved position throughout the building site, until the vault was finished. Between 1966 and 1967, Nervi continued to develop his design for the various structures (he did not decide to produce the atrium false ceilings using ribs inspired by Arcangeli's floor until December 1966).

For the roof, from August 1964 Nervi constructed prototypes at the Via della Magliana laboratory in order to check the sizes of the parts, the way they were detached from the moulds, the surface treatment methods and, last but not least, the colour of the concrete. The tests prompted him to reject the usual wave ashlar with slender sections like those made out of ferrocement in favour of two-piece segments made out of reinforced concrete with an S-shaped cross-section that could be joined together to produce a single ashlar. By October, "actual size models" of the roof ashlars were ready, "complete with a lighting system."[1] Nervi carefully studied the profile of the curves to en-

[1] Mario Nervi, letter to Renzo Ferrante, 14 October 1964, F34-SCV.

492-495. P. L. Nervi (with the cooperation of A. Nervi), *Aula per le Udienze*, Vatican City, 1963-71. Prefabrication building site, Santa Maria di Galeria, 1967.

492. Preparing reinforcement of waved vault ashlars, photograph dated 4 October 1967 (BSTA, 024-2153).

493. Pouring of concrete, photograph dated 4 October 1967 (BSTA, 024-2164).

494. Surface finishing of waved vault ashlars (BSTA, 024-2325).

495. Construction of formwork for papal tribune false ceiling, photographed in 1967 (BSTA, 024-3227).

able him to apply the concrete without having to resort to a counter-form. The material was spread with ladles, as for the beams of *Salone C*, and remained in place before setting, partly due to the abundant metal reinforcement.

The colour and the finish of the waved ashlars surfaces were also tested. Nervi decided back in 1964 that the colour of the precast reinforced concrete used for the vault would have to be white. At this stage, however, it was still planned to make the structure in cast-in-place reinforced concrete at the vault abutment in grey cement and to leave it as it was following stripping. In this building site, as in those of his previous masterpieces, from *Salone B* in Turin to the *Palazzo dello Sport*, Nervi therefore continued to use the device of juxtaposing the two kinds of concrete with their different colours.

The type of cement that Nervi chose for the prefabricated parts was Italbianco 1000 made by Italcementi in Bergamo, one of the most expensive and refined "supercements." For the aggregate, he planned to use white Apuan marble quarried near Lucca. The stripping operations were planned to achieve a given surface treatment. The moulds were spread with a product known as Rugasol, which delayed the setting of the superficial layer of concrete and made it easier to extract the part. While he was still carrying out tests at the Via della Magliana laboratory, Nervi explained how to work the parts when the surface layer was still fresh: "simple cleaning by means of a wire brush removes the layer of surface cement grout that has not set, displaying the underlying grain, which is perfectly smooth and free of stains."[2] This

[2] *Relazioni*, January 1965, p. 16, F6-SCV.
[3] Alberto Bologna, interview with Vittorio Nervi, 10 October 2011.
[4] Pier Luigi Nervi, letter to Carlo Pesenti, 14 March 1967, Maxxi, P83.

was the way the parts were to be worked – cleaned by a pressurised water jet while sorghum brooms were used for the brushing.[3]

Due to the small size of the building site and in order to ensure perfect workmanship, the parts were prefabricated in a special shed set up in 1966 on land owned by the Vatican in the town of Santa Maria di Galeria. Here Nervi continued the tests he had started in the Via della Magliana laboratory to develop all the technical details for prefabrication of the parts. He intended to achieve a "purity and beauty"[4] by executing the parts in white concrete. In order to transport the parts, railway tracks were installed linking the shed to the railway station at Anguillara Sabazia, less than one kilometre away. The parts were then sent to the railway station in the Vatican and taken by truck to the building site for immediate assembly.

496-498. P. L. Nervi (with the cooperation of A. Nervi), *Aula per le Udienze*, Vatican City, 1963-71.

496. Formwork for pouring ceiling of the *Galleria di Gala* (BSTA, 024-2758).

497. Unfinished intrados of *Galleria di Gala* ceiling (BSTA, 024-3286).

498. Intrados of *Galleria di Gala* ceiling (BSTA, 024-5114).

In January 1967, at Santa Maria di Galeria, Nervi & Bartoli began to prefabricate the parts of the vault using the usual procedure of masonry ground forms, but this time adding a removable wooden side to make it easier to remove the part. By the beginning of October, work on the prefabrication was in full swing. The organisation of the building site was similar to that of *Salone B*, but measures had been expressly taken to ensure perfect execution of the parts; they were produced under the shelter of a shed, equipped with a overhead crane and concrete batching plant with automatic mixers. Masonry moulds were positioned at man height to make it easier to fit the reinforcement and apply the concrete mixture to a flat surface. Thirty-six ground forms were prepared of two types of S-shaped cross-section units with flanges that were the mirror images of each other. Every 48 hours the parts were stripped. Surface treatments were then applied, and the pairs of S-shaped parts with flanges were then assembled to produce the waved ashlars. Ferrocement formwork was also prefabricated in the shed. This was constructed at all times in accordance with Nervi & Bartoli's routine procedures, but now without using moulds for large-scale production, because the formwork was to be used in accordance with traditional criteria. In other words, it had to be stripped after fitting the metal reinforcement and pouring the concrete to build the ceilings of the atrium, the throne, the Private Audience Room, and the *Galleria di Gala*.

From October 1967, Nervi began to build the structure out of cast-in-place reinforced concrete on the Vatican building site. He hesitated for a long time over the choice of colour and surface finish for the concrete. When he had not yet begun to carry out tests on the prefabricated parts, he planned to use grey cement and leave the surfaces intact after stripping, merely sandblasting them. After Nervi's initial recommendations for a structure made out of exposed grey concrete, in 1966 he examined various alternative solutions, for which he also planned to construct actual-size prototypes. In his desire to improve the quality of the pillars, he lapsed into a Baroque style of cladding. The same applied when he examined the panels to be prefabricated in concrete. These were to be grooved in an old-fashioned style with variable cross-section cavities, the flat faces of grooves coated with strips of gold-plated brass, and could potentially be used as permanent forms. Another simpler plan was to apply fine marble cladding to the concrete.

By the time tests were under way for the prefabricated roofing beams, in September 1966 Nervi had already given up the idea of turning the pillars into old-fashioned grooved columns and planned to treat the exposed concrete surfaces by hammering after stripping. He envisaged characterising the cast-in-place concrete using the same white cement that he was trying out for the prefabricated parts of the vault. He therefore consulted Italcementi to obtain information on the "interesting finishes" that could be achieved by using the cement Italbianco 1000.[5] At the same time he created, with the cooperation of ISMES, a "programme of research into concrete made with white cement to define the mechanical properties of the concrete mixture."[6]

In May 1967, by agreement with the Vatican Commission, Nervi decided to use a "concrete mixture of white and ultra-white marble"[7] for all his pillars, to be worked with hand-held or pneumatic chisels after stripping to give them the appearance of stone and match them with the precast concrete of the waved beams. Nervi predicted the surfaces would become dirtied by the hands of visitors: "when they pass by such an unusual material, they will instinctively want to touch it, and it will soon become dirty."[8] To avoid this

outcome, Nervi asked Italcementi to design a "surface waterproofing that will prevent the dirt from being absorbed"[9] or, in any case, a solution that would make it possible to achieve a washable surface. Before building the structure, Nervi also had tests carried out on the building site to "experimentally determine the shrinkage of the white concrete mixture."[10]

Nervi was now so utterly persuaded of the effect of the white concrete surfaces, whether bush hammered or brushed, that he continued to use the same concrete with the same treatment for the four large pillars and floor of the central hall of the Cassa di Risparmio di Venezia headquarters, designed with Angelo Scattolin and built by Nervi & Bartoli, or for the "large colonnade"[11] of the Bureau International du Travail in Geneva, designed in conjunction with Alberto Camenzind and Eugène Beaudouin.

By July 1968, all the formwork and metal reinforcements for the floor of the papal throne had been assembled, and the wooden formwork for pour-

[5] Pier Luigi Nervi, letter to Antonio Pesenti, 26 November 1966, Maxxi, P83.
[6] Emanuele Fumagalli, letter to Pier Luigi Nervi, 15 March 1967, *ibid.*
[7] Pier Luigi Nervi, letter to Carlo Pesenti, 23 May 1967, *ibid.*
[8] *Ibid.*
[9] *Ibid.*
[10] Drawing *Determinazione sperimentale del ritiro del conglomerato bianco,* 18 January 1968, CSAC, coll. 152/6.
[11] Nervi, letter to Pesenti, 23 May 1967 cit.

ing the side parts was under construction. By September, the scaffolding of Innocenti poles was complete and the first prefabricated element of the vault was positioned. The pourings connected the troughs, and crests of the waves were carried out using the sliding cranes. All the various components had been assembled by May 1969.

Despite the many prototypes constructed in the Via della Magliana laboratory and the Santa Maria di Galeria shed and his careful handling of the various ashlars, Nervi noted that the prefabricated parts had lost their original "brightness" once assembly was complete.[12] After considering the possibility of outsourcing the work to specialist stone working companies, the Nervi & Bartoli workers finished the concrete after stripping using hand-held tools for the inclined pillars of the papal throne and pneumatic hammers for the beam and other pillars. In a little over one month, by approximately May 1971, all the concrete finishes were complete, including the sandblasting and hammering of the isostatic ribbed floor surfaces.

Both the use of Apuan marble chips in the mixture and the surface working of the precast and cast-in-place concrete indicates that Nervi was imbuing the material he had used in a rudimentary manner since his debut as an engineer and builder with the quality of stone. It is therefore not surprising that stone also made an appearance in the design for the *Aula per le Udienze*, although treated in a special way and dressed in accordance with an approach typical of a builder specialising in reinforced concrete. In the first versions of

499-501. P. L. Nervi (with the cooperation of A. Nervi), *Aula per le Udienze*, Vatican City, 1963-71.

499. Formwork for pouring ceiling of private audience room (BSTA, 024-2766).

500. Unfinished intrados of private audience room ceiling floor (BSTA, 024-3291).

501. Intrados of private audience room ceiling floor (BSTA, 024-5113).

the design, the atrium of the *Aula per le Udienze* was intended to become a monumental room punctuated with columns with a cylindrical cross-section tapering off toward the bottom. These were not, however, to be in concrete. They were made out of drums of marble to be laid on top of one another with interposed steel disks that were clearly visible, in order to allow the expression of the device used to stabilise the system. The drums were perforated in the centre for the passage of the steel cables used for post-tensioning. The fact that Nervi envisaged having tests carried out at the ISMES indicated his desire to try out a new kind of structure: a contemporary reinforced stone. He emphasised the benefits of his system in portentous terms, almost foreshadowing a new era of construction following that of reinforced concrete. This was based on the use of marble drums reinforced to "increase" the "resilient capacity" of the columns and guarantee a "significant financial savings" compared to a conventional "monolithic column."[13] The principles he had pursued

[12] Pier Luigi Nervi, letter to Luigi Conti, 23 September 1970, Maxxi, private correspondence.
[13] Pier Luigi Nervi, letter to Enzo Lauletta, 26 September 1967, Maxxi, P83.

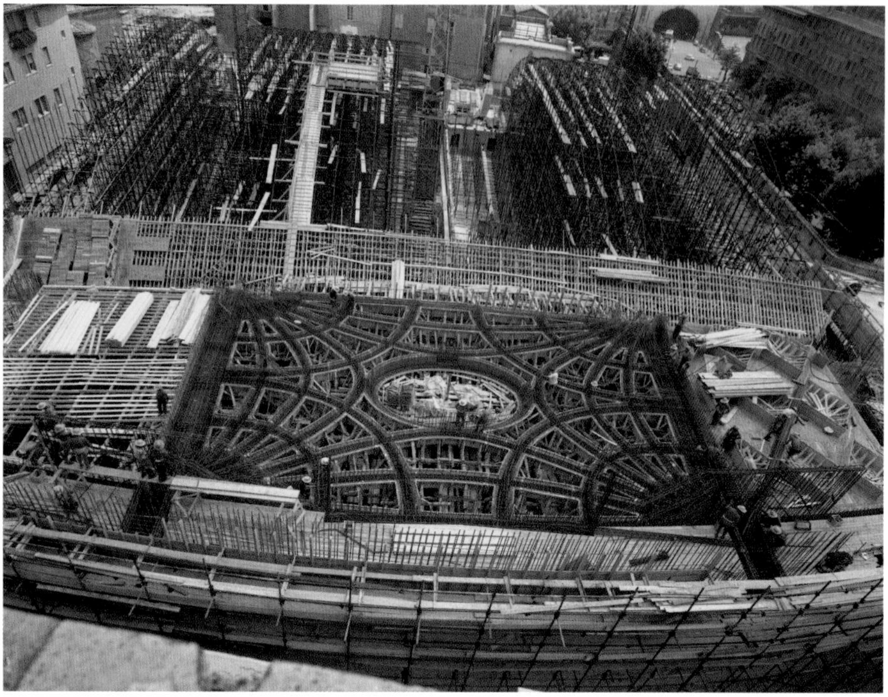

502-505. P. L. Nervi (with the cooperation of A. Nervi), *Aula per le Udienze*, Vatican City, 1963-71.

502. Formwork and reinforcement of papal tribune ceiling, photographed in July 1968 (BSTA, 024-3342).

503. Extrados of papal tribune ceiling (BSTA, 024-3407).

504. Intrados of papal tribune ceiling (BSTA, 024-4005).

505. Laying the waved ashlars of the vault, photographed in 1969 (BSTA, 024-4045).

506-510. P. L. Nervi (with the cooperation of A. Nervi), *Aula per le Udienze*, Vatican City, 1963-71.

506. View of vault wave cavity, photographed in 1969 (BSTA, 024-4424).

507. Precast reinforced concrete shading elements positioned on the vault extrados (BSTA, 024-4507).

508. Formwork for pouring the atrium ceiling (BSTA, 024-4107).

509. Unfinished intrados of the atrium ceiling floor (BSTA, 024-4271).

510. Intrados of the atrium ceiling floor (BSTA, 024-4749).

511-517. P. L. Nervi (with the cooperation of A. Nervi), *Aula per le Udienze*, Vatican City, 1963-71.

511. Laying the prefabricated baldachin *tavelloni* (BSTA, 024-4552).

512. Baldachin (BSTA, 024-5181).

513. Surface finishing of the floor intrados and entrance pillar (BSTA, 024-4394).

514. Surface finishing of variable cross-section pillars of the papal throne (BSTA, 024-4391).

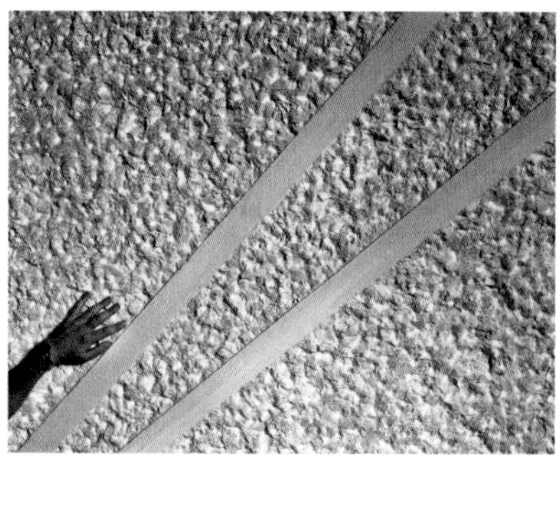

515. Papal tribune, photograph dated 2 July 1971 (BSTA, 024-4825).

516. View of the Hall from the entrance (BSTA, 024-4792-1).

517. View of the Hall from the papal tribune (BSTA, 024-4811).

during the period of autarky and the system designed for the *Monumento alla Bandiera* were re-emerging as part of this insight. The columns were built with a cylindrical stem in steel, thickened at the base and top in order to produce the image of a stem tapering toward the bottom, with a base and capital clad in strips of steel.

In the Papal *Aula per le Udienze* of the Vatican, Nervi's rhetoric, almost imperceptible in his great masterpieces of the 1930s and 1940s, ultimately assumed the connotations of a delirious baroque of coatings in travertine marble slabs or aluminium strips, gilded grilles, gorgeous white hammered surfaces, and with the Arcangeli floor transfigured into a bundle of false ceiling ribs, demonstrating the extent to which Nervi was now in thrall to figures that had become simulacra of one style of his technique. A statement made by Nervi about the Cathedral of Notre Dame in Paris might explain what had happened to the isostatic ribs in the false ceilings: "The most characteristic and expressive elements of the ceiling of the large central nave are the projecting ribs which clearly show the intersection of the vaults which form the cross vaulting. These ribs are not statically essential, but they do follow the lines of maximum concentration of the internal stresses. That the ribs were brought out and continued to the large supporting columns shows the clarity and depth of the imagination of those magnificent builders."[14]

Structures for skyscrapers

After his advice on the construction of the Pirelli building in Milan, Nervi also became famed as an engineer of reinforced concrete skyscrapers. Even Kahn sought his advice in developing the tetrahedral structure of his design for a skyscraper made out of precast reinforced concrete.[15] At the beginning of the 1960s, Nervi's name appeared alongside those of the architects famed for designing skyscrapers throughout the world. Nervi sought to reveal the lines of his structure as well as his materials, and even when forced to resort to cladding, he designed panels that still expressed the material of the load-bearing structure. In some cases, he experimented with state-of-the-art techniques in US reinforced concrete construction systems.

In a design drawn up between 1962 and 1970 for a skyscraper in the Schedel-Doekshaven complex in The Hague (developed in conjunction with the Jan A. Lucas and Henk E. Niemeijer firm),[16] his structure took the form of pilotis branched into four ribs, which then divided into pillars that remained continuous but gradually tapered off as they rose upward. The jack-block method that Nervi planned to use for this skyscraper made it possible to build the floors and pillars from the platform on the ground floor and then lift them as necessary with the aid of hydraulic jacks.

The structure became a colossal skeleton of pillars and Vierendeel girders located at the level of the four floors intended for the plant rooms, in a design for the Parcel 8 skyscraper in Boston, developed in 1964 with the Kelly & Gruzen firm of New York. Studio Nervi planned different versions of variable cross-section pilotis for the ground floor.

Nervi also acted as a consultant for the design and construction of three skyscrapers. One in Montreal, the Stock Exchange Tower, was built between

[14] Nervi, *Aesthetics and Technology in Building* cit., p. 16

[15] Gargiani, *op. cit.*, p. 77.

[16] Studio Nervi design No. 1135, n. inv. pra. 1125, n. id. 15057, coll. 119/3; Maxxi, P44, P52, P65, P96.

518. P. L. Nervi, A. Nervi and Architectenbureau Lucas & Niemijer (with the cooperation of F. M. Imbert), design for the Plan Schedel-Doekshaven complex, The Hague, 1961-69. Perspective drawing (BSTA, 024-3380).

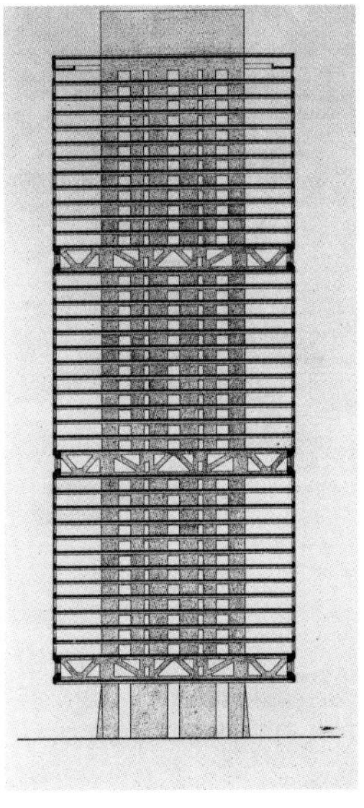

519. P. L. Nervi and Kelly & Gruzen Architects Engineers, competition design for Parcel 8 skyscraper, Boston, 1964. Cross-section through the structure (BSTA, 024-4967).

1962 and 1966, designed by Luigi Moretti; and two in Sydney, were designed by Harry Seidler: the Australia Square Building, built between 1961 and 1967, and the MLC Centre Building, built between 1972 and 1978.

For the Moretti skyscraper, Nervi invented a braced reinforced concrete structure divided into a central core and four corner pillars connected by means of truss girders arranged on the three plant floors – intermediate pillars were used to support the floors. It is significant to Nervi's method of conceiving structure that he explained the mechanical operation using the image of a skier's body that finds its own balance in movement through the use of sticks. When seen in these terms, the structure resembles the one he designed for the floating hotel project, with stabilisers to counter the fluctuations of wave motion. When constructing the four corner pillars, on which Nervi wished to leave the concrete exposed, it was decided to apply a cladding made out of precast reinforced concrete slabs spaced away from the pillars to ensure the circulation of warm air. The aim of this was to prevent different degrees of thermal expansion arising between the pillars and the core during the winter.

Nervi clad Seidler's design for the Australia Square Building with some of his trademark technical and structural solutions. For the pillars and beams, he used prefabricated slabs in reinforced concrete, with white cement as permanent forms, while he characterised the basement floor with triangular and diamond shaped ferrocement *tavelloni*, always in white. He used the same system of permanent forms in white precast ferrocement panels to build the structure of the MLC Centre Building skyscraper, because this enabled him to mould the beams to include a cavity with rounded ends.

Nervi also used beams similar to those of the MLC Centre Building, but post-tensioned with decorative studs for the Trade Group Offices (or Edmund Barton Building), in Canberra, which was again designed by Seidler and built between 1970 and 1974. One of the last versions designed by Studio Nervi for skyscraper structures was developed for the Sofitel Hotel in Cairo constructed between 1975 and 1978. The designs included cross-shaped prefabricated elements based on structures developed by the firm of Skidmore, Owings & Merrill.

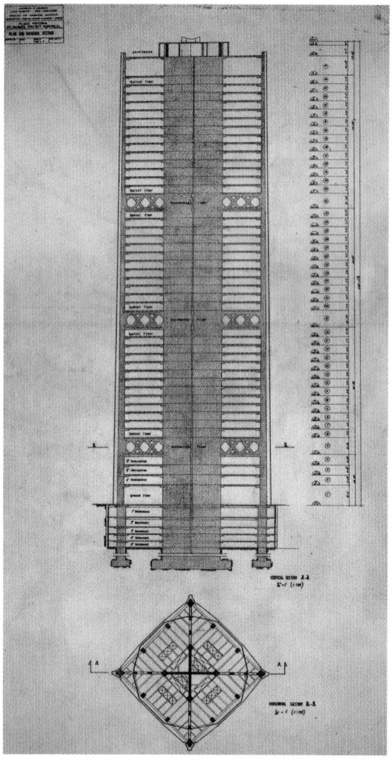

520-521. P. L. Nervi, structure of Stock Exchange Tower, design by L. Moretti with Greenspoon, Freedlander and Dunne, D'Allemagne & Barbacki, J. M. Morin, Letendre, Monti & Associates; Place Victoria, Montreal, 1961-66.

520. "Plan and diagonal section" (BSTA, 024-2116).

521. Construction stage of corner pillars (ACS – LM, design No. 211-1961, photo 4).

522. P. L. Nervi, consultancy for structure of Australia Square Building, designed by H. Seidler, George Street, Bond Street, Curtin Place, Sydney, 1961-67 (Maxxi - PLN, F07863).

523. Studio Nervi (with the cooperation of A. Arcangeli), design for Cultural Centre, Tripoli. Elevation, drawing dated 30 November 1967 (BSTA, 024-3375).

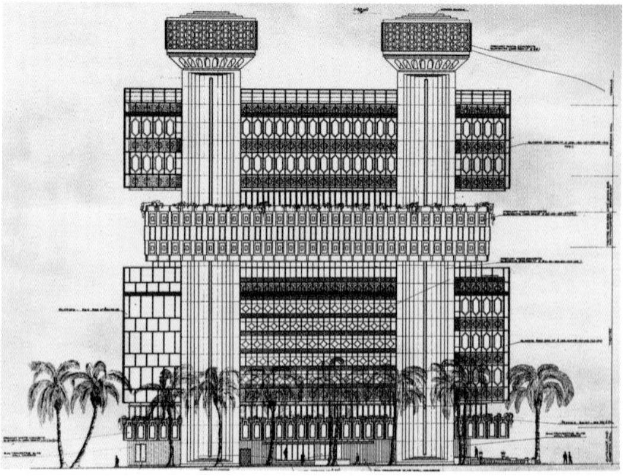

524-525. Studio Nervi, consultancy for structure of MLC Centre Building, designed by H. Seidler, Martin Place, Sydney, 1972-78.

524. View (BSTA, 024-7513).

525. Laying entrance floor *tavelloni* (BSTA, 024-7537).

526 Studio Nervi, consultancy for structure of Trade Group Offices Centre, designed by H. Seidler, Macquarie Street, Canberra, 1970-74. View of the building site (BSTA, 024-6627).

527. Studio Nervi with Cabinet Olympio Architecte, headquarters of Banque Africaine de Developpement, Avenue Lamblin, Abidjan, 1971-80. View of the building site (BSTA, 024-6578).

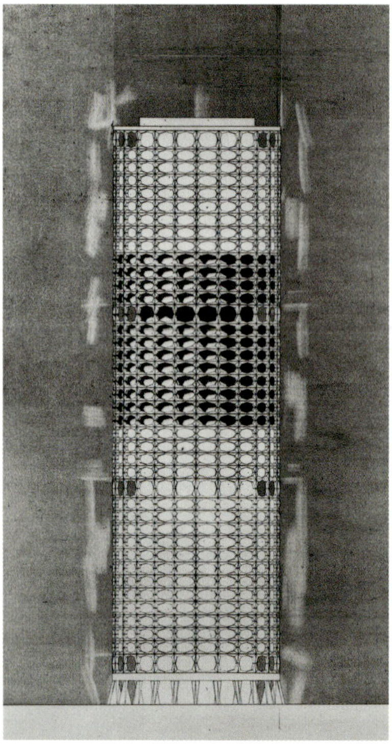

528-531. Studio Nervi, Sofitel Hotel, 3 Street Maglas Kiadat El Thawrah, El Gezirah Island, Cairo, 1968-78.

528. "Elevation," drawing dated 31 July 1968 (BSTA, 024-4885).

529. Model of prefabricated facade elements, about 1968 (BSTA, 024-3386).

530. Photograph of model (Unesco).

531. Preparatory model of ground floor pillars (BSTA, 024-6066).

Variable cross-section pilotis and decorative ribs

During the 1970s, Studio Nervi continued to be involved in designs for large structures – from skyscrapers to churches – in Italy, Ivory Coast, Saudi Arabia, Iraq, Nigeria, Libya, or South Africa. These were executed in cast-in-place reinforced concrete and left exposed, with some parts in precast concrete, sometimes also using steel. Sometimes the structural architecture was based on the presence of certain significant elements and trademark geometries of a Nervi system that was increasingly reduced to style.

An epilogue occurred in the story of the evolution of variable cross-section pilotis between the end of the 1960s and the beginning of the 1970s. This led to the design of pillars for a viaduct on Boulevard Atatürk in Ankara, built between 1969 and 1972; colossal hollow pillars in the form of a hyperbolic paraboloid for a competition design in 1969 for the suspension bridge over the Strait of Messina (the geometrical lines would have been visible due to the shape of the cladding slabs); or the pilotis of the Australian Embassy in Paris, built between 1975 and 1977, designed by Seidler and Peter Hirst with the advice of Breuer and Mario Jossa. A repertoire of figures typical of Nervi's structural architecture gained prominence in 1968-69 for the design of a Sports Center in Kuwait (variable cross-section inclined pillars, a grid of ribs, or a hyperbolic paraboloid). Arcangeli's isostatic lines also proved to be another source of inspiration for variants of decorative ribs, as in the 1968 design for the Gymnasium of the same Sports Center, in Kuwait; in the design of the Golden Acre complex in Cape Town, drawn up in 1975 with Louis Karol; or in the headquarters of the Banque Africaine de Developpement, in Abidjan, designed in conjunction with the Cabinet Olympio Architecte and built between 1971 and 1980.

Confirmation of the quest for symbolic structure that had imbued Nervi's work since the *Stadio Berta* included the Church of San Gaspare del Bufalo of the Congregazione dei Missionari del Preziosissimo Sangue, built between 1974 and 1981; and the funeral chapel for President Félix Houphouet Boigny, built between 1975 and 1978 in Yamoussoukro, Ivory Coast. The installation of prefabricated elements in the church of San Gaspare, as usual, was in accordance with the economic approach that had been evident in Nervi's building sites. However, the work ended up looking like an irrational stacking of mighty beams against a curved structure, conjuring up an image of an archaic tent.

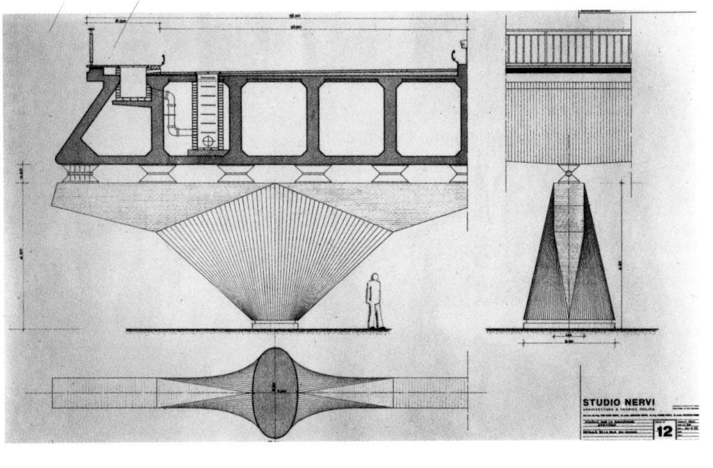

532. Studio Nervi (with the cooperation of A. Arcangeli), Viaduc de l'Opera, Boulevard Atatürk, Ankara, 1968-73. "Details de la pile du viaduc (Detail of the viaduct pylon)," drawing dated 30 June 1969 (BSTA, 024-4875).

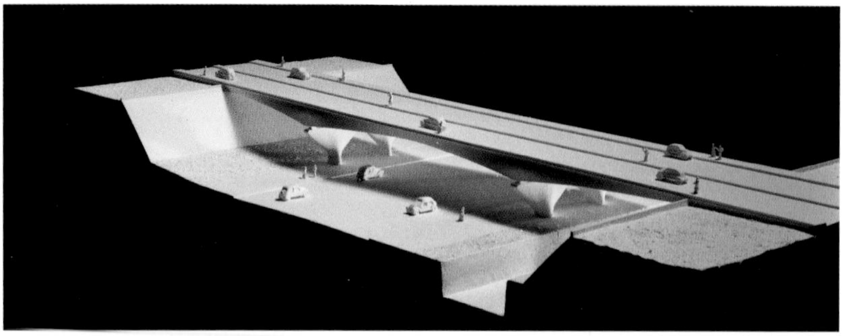

533. Studio Nervi (with the cooperation of A. Arcangeli), Viaduc de l'Opera, Boulevard Atatürk, Ankara, 1968-73. Model (BSTA, 024-4029).

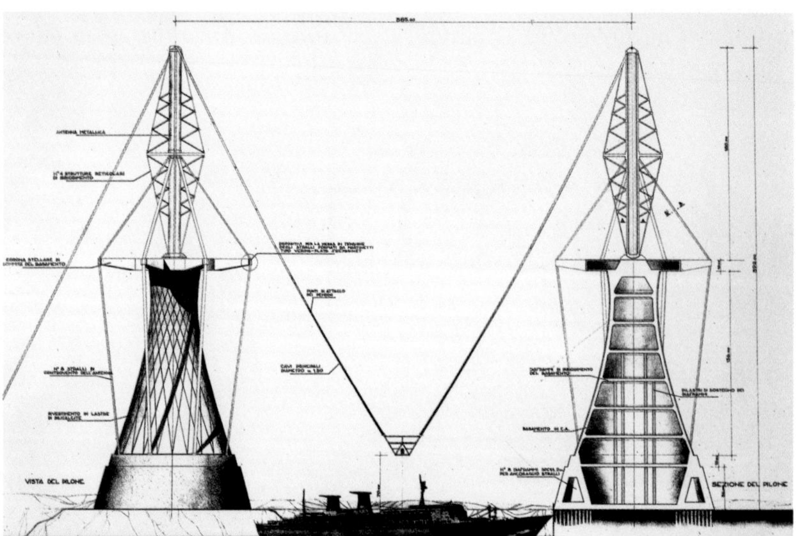

534-536. P. L. Nervi and A. Nervi (in cooperation with G. Tamaro), competition design for bridge over the straits of Messina, 1969.

534. Photograph of model (BSTA, 024-4170).

535. Cross-section through pylons and suspended deck (BSTA, 024-4341).

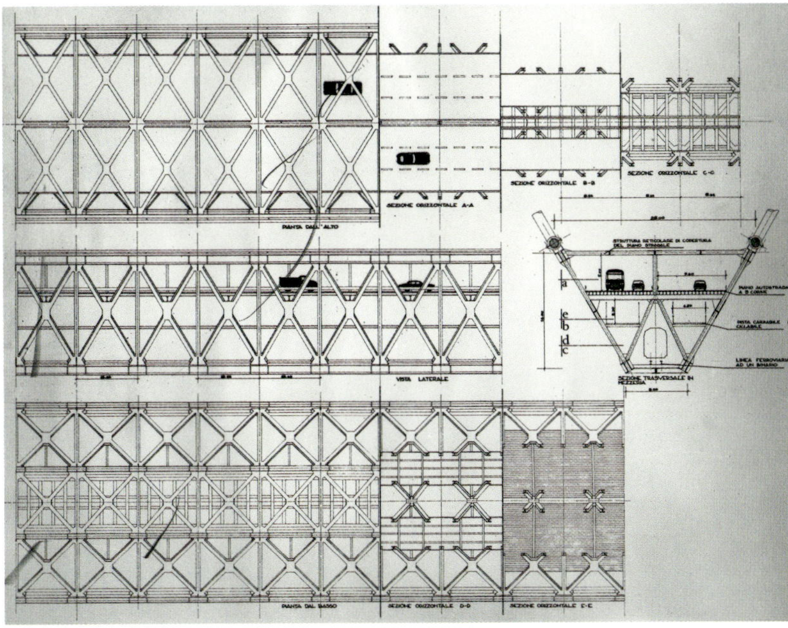

536. Plan, elevation and cross-sections through suspended deck (BSTA, 024-4339).

537-538. Studio Nervi, design for Kuwait Sports Centre, Kuwait, 1968.

537. Photograph of model (Unesco).

538. Model of pillars and abutment system of covered arena dome (BSTA, 024-3933).

539-541. P. L. Nervi and A. Nervi, consultancy for structure of Australian Embassy, designed by H. Seidler, P. Hirst, M. Breuer and M. Jossa, Rue J. Rey 4, Paris, 1975-77.

539. Study model for entrance pillar (BSTA, 024-4519).

540. Facade on Rue J. Rey (BSTA, 024-6731-3).

541. Detail of entrance pillar (BSTA, 024-7506).

542. P. L. Nervi, structure of Rio de Janeiro music centre, designed by O. Niemeyer, 1972 (BSTA 024-5306 (2)).

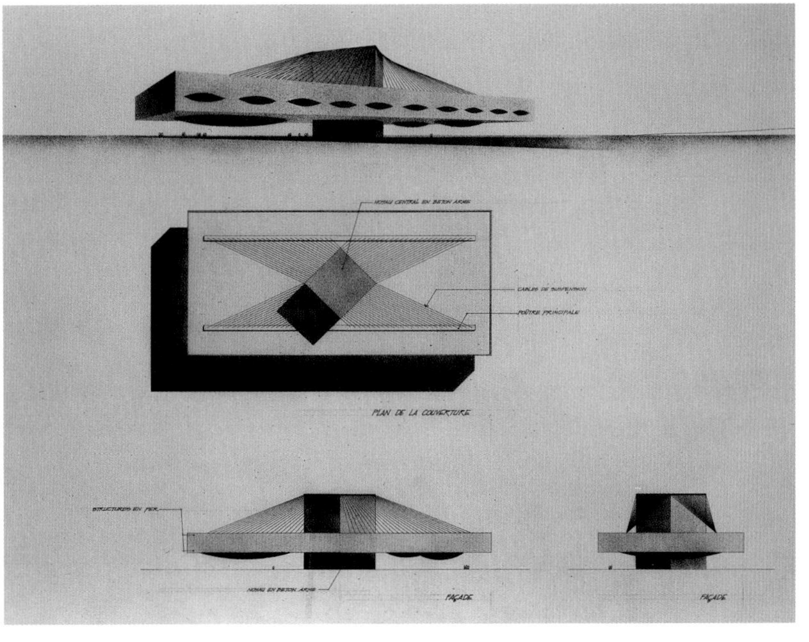

543-544. Studio Nervi with S. Rossi (in cooperation with C. Tropea and R. Moore), Cosenza railway station, Via Vaglio Lise, Cosenza, 1971-75.

543. View of atrium (BSTA, 0538-1).

544. Intrados of chapel floor (BSTA, 0538-2).

545-546. Studio Nervi, design for Development of Um-Al-Khanazeer Island, Baghdad, 1974.

545. Model (BSTA, 024-5440).

546. Model of walkways (BSTA, 024-3984).

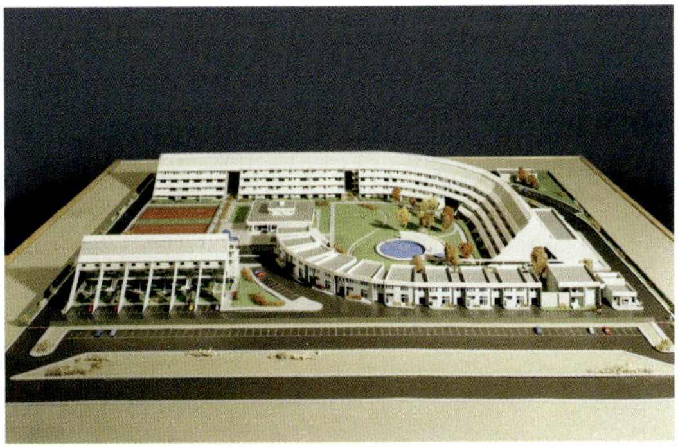

547-548. Studio Nervi (with the cooperation of A. Arcangeli), Arab Petroleum Investments Corporation residential complex, King Faisal Road, Al-Khobar, Dhahran (Saudi Arabia), 1977-81.

547. Model (BSTA, 024-6825).

548. View (BSTA, 024-7740-10).

549. P. L. Nervi and A. Nervi, Church of San Gaspare del Bufalo della Congregazione dei Missionari del Preziosissimo Sangue, Via Borgo Velino 1, Rome, 1974-81 (BSTA, 024-7746-10).

550. Studio Nervi, design for church of Notre Dame d'Afrique, Abidjan. View (BSTA, 024-7658).

551. Studio Nervi, Bus Garage Tor Sapienza for Azienda Tramvie e Autobus del Comune di Roma, Via Prenestina 607/609, Rome, 1967-71. View (BSTA, 024-5192).

The quadripod of the Italian Embassy in Brasilia

The headquarters of the Italian Embassy in Brasilia was one of the last grand works developed by Studio Nervi, designed between 1970 and 1973.[17] The offices and the Ambassador's residence are housed on a single storey raised off the ground to create a monumental arcade open to the landscape and intended for the contemplation of Nervi's exposed reinforced concrete structure – a proud display of Italian-engineered structural architecture standing against the backdrop of Neimeyer's joyful tropical structures. The high encircling walls, pillars, floors, or brise soleil transformed the monument into a local form of brutalist architecture. The concrete used for the Italian embassy nevertheless aspired to a level of finish different from that of the Brazilian works.

The pillar branching off into four arms is one of the last variants of the series summarised in the plate of drawings drawn up after the *Palazzo del Lavoro* – in the terminology used by Studio Nervi, it was defined as a "quadripod."[18] Despite the precedent of the pillars for the railway station in Naples and the skyscraper in The Hague, the quadripod for Brasilia is engaged in a dialogue with the branched pillars of Alcuin Library of St. John's University at Collegeville, by Breuer, and Durham Bridge, in the United Kingdom, by Ove Arup.

Work on the headquarters of the Italian Embassy, completed between 1974 and 1978, was carried out by Irfasa SA Construções Indústria e Comércio and managed by the engineer Andrea Toscano. Studio Nervi specified that the "formwork intended to receive the pourings will usually be constructed using wooden boards," while not ruling out the "use of formwork made out of different materials (iron, plastics, or other material) provided they are compatible with the appearance of the designed cement surface."[19] The construction of the pillars on the basement level provided an opportunity to experiment with the devices required to manufacture the formwork made out of wooden boards to ensure it was up to Nervi's quality standards (rubber strips were fitted in the joints to eliminate the wood essences that stained the concrete).[20]

552-553. P. L. Nervi and A. Nervi, Italian Embassy, Avenida das Nações Quadra 807, Lote 30, Brasilia, 1969-79.

552. "Progetto esecutivo. Sezione tipica (Working project. Typical cross-section)," drawing dated 1 December 1973 (BSTA, 024-6312).

553. Prototype of quadripod pillar (Maxxi - PLN, F13234).

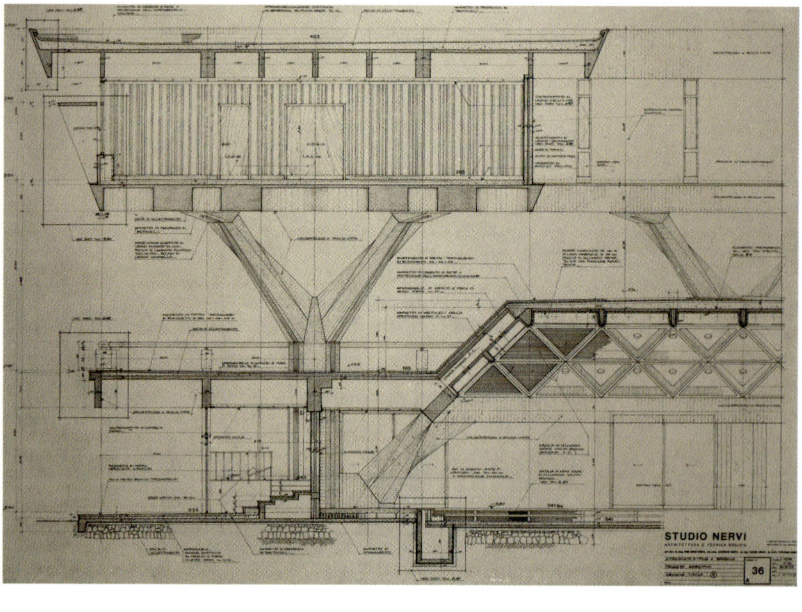

Before pouring the quadripod pillars, it was decided to construct a special prototype with only two branches. The end result revealed the workers' inexperience in processing the boards and joining them together properly (other defects concerned the pouring composition and vibration). Large panels of plywood fitted to obtain a regular design were used to produce the quadrangular fields of the ribbed floor.

Once the quadripods and floor had been stripped, the exposed concrete was still so imperfect that an order was issued to send a "technician specialising in mending work" to the building site.[21] Once the exposed concrete of the Italian Embassy had been mended and sandblasted, it became another expression of Nervi's controlled brutalist approach, which was unusual in Brazilian architecture.

Pier Luigi gradually took a back seat in the work of Studio Nervi from the beginning of the 1970s, overseeing designs produced by Antonio, while Mario was mainly concerned with structural calculation, together with Desideri, and Vittorio monitored the activities of Nervi & Bartoli. The death of Pier Luigi Nervi on 9 January 1979, and that of Antonio a few months later, put an end to the design activities of Studio Nervi. Works that were still under construction were brought to completion under the direction of Mario Nervi and his collaborators.

[17] See Tullia Iori, *L'Ambasciata d'Italia a Brasilia*, in, Tullia Iori, Sergio Poretti, *L'Ambasciata d'Italia a Brasilia*, Milano, Electa, 2008, pp. 53-81.

[18] Andrea Toscano, letter to Studio Nervi, 26 June 1975, Maxxi, P106.

[19] *Capitolato*, item 3.2, p. 4, *ibid*.

[20] Andrea Toscano, letter to Studio Nervi, 21 November 1974, *ibid*.

[21] Toscano, letter to Studio Nervi, 26 June 1975, cit.

554-556. P. L. Nervi and A. Nervi, Italian Embassy, Avenida das Nações Quadra 807, Lote 30, Brasilia, 1969-79.

554. View (BSTA, 024-6498-21).

555. Quadripods and ribbed floor (BSTA, 024-6504-35).

556. Spiral staircase (BSTA, 024-6504-27).

Acknowledgements

The authors wish to thank various people and institutions for their help in facilitating the archive research required for the critical appraisal of Nervi's work set out in this book.
All the archive research was conducted by Alberto Bologna, except for that relating to documents concerning the building of the Unesco headquarters. The writing of each chapter is the result of joint work of the two authors.
The authors thank the staff of the *Centro Studi e Archivio della Comunicazione* of the *Università degli Studi di Parma*, and in particular the Chairperson Francesca Zanella, and the archive managers, Paola Pagliari and Simona Riva; the staff of the *Biblioteca di Scienze Tecnologiche* of the *Università degli Studi di Firenze* Architecture Faculty and particularly the Director Maria Luisa Masetti and Gianna Frosali, Archive Manager; the staff of the *Museo Nazionale delle Arti del XXI Secolo* in Rome (Maxxi), and in particular the Director Margherita Guccione and Carla Zhara Buda of the *Centro Archivi di Architettura*; the Management of the Vatican City Governorate Technical Services, its Director Rafael Garcia de la Serrana Villalobos, and particularly Flaviano Antonio Vitali, Pietro Marco Secone, Enrico Businaro, members of the Technical Services, as well as Biagio Tamarrazzo for making his photographs available to the authors.
The authors would also like to thank: Stefano Pedullà, manager of the historical archive of the *Comitato Olimpico Nazionale Italiano* in Rome, for allowing us to study their documents; Pierpaolo Righero, archivist at the *Centro Storico Fiat* in Turin, for his willingness to help; and the staff of the *Archives de la Construction Moderne of the Ecole Polytechnique Fédérale de Lausanne* (EPFL). The authors thank the staff of the Unesco Archives and Jacques Duflos de St Amand, Head of the Architecture Unit, for allowing them to consult documents relating to the Unesco building site in Paris. Special thanks also go to: Giuseppe Cuscito, Manager of the Drawings Section of the *Archivio Storico del Comune di Firenze*; Roberto Fuda, Manager of the *Personalità dell'Architettura Contemporanea* sector of the *Archivio di Stato di Firenze*; Margherita Antonio Maria Martelli, Director of the study room, and Gabriella Sansonetti and Daniela Loyola of the *Archivio Centrale dello Stato* in Rome; Luca Devecchi, Administrative Secretary of the *Archivio Storico Civico Biblioteca Trivulziana* in Milan.
Our understanding of the way that Nervi used the various types of cement was partly aided by consulting the historical archive of Buzzi-Unicem in Casale Monferrato, made possible by Consolata Buzzi, and by explanations given to Alberto by Bruno Rossi, who took part

in meetings with Nervi and his colleagues during building work for the 1960 Rome Olympics. Further useful information on the methods employed by Nervi at his building sites was supplied to Alberto by Vittorio Nervi, son of Pier Luigi, and by Elpidio Marescotti, a Nervi & Bartoli employee. We would like to extend warm thanks to Irene Nervi for generously sharing information, memories, and anecdotes about the work of her grandfather and father.

Valuable contributions to our research were made by Adriana Arcangeli, Anna Daneri, Paolo Gherardi, Glauco Gresleri, Bruno Filippo Lapadula, Enrico Moncalvo, and Alberto Pecci who all allowed Alberto to consult their archives and agreed to the publication of their own photographs. The authors would like to thank one of the greatest experts on the work of Nervi, Claudio Greco, for giving them access to various materials in his possession. Gratitude goes out as well to Cristiana Chiorino, Michela Comba, Rita D'Attorre, Manuela Leoni, Roberta Martinis, and Gabriele Neri for having shared archive material from their research with Alberto over the years, and to Sergio Pace, Director of the *Biblioteca Centrale* of the *Politecnico di Torino* Architecture Faculty, for allowing Alberto to consult and make copies of rare books.

The authors thank Beatrice Lampariello for conducting research in the *Archivio Centrale dello Stato* in Rome on their behalf and remember with pleasure their discussions with Franz Graf and Aurelio Muttoni during educational activities and exhibitions on Nervi held at EPFL.

Production of the book, translation of the text, and its layout provided us the opportunity to work with *Presses polytechniques et universitaires romandes* of EPFL. We are sincerely grateful to the Director of the PPUR, Olivier Babel. Our thanks go out as well to the Deans of EPFL's School of Architecture, Civil and Environmental Engineering (ENAC), for generously supporting the book's publication.

The choice of the book's subject was prompted by research conducted by Roberto Gargiani under the auspices of the European Research Council on the topic of exposed concrete, which has for many years been part of studies conducted by his colleagues at EPFL's *Laboratoire de Théories et Histoire de l'Architecture 3* (LTH3), where he serves as department head.

This book was produced in conjunction with the *Biblioteca di Scienze Tecnologiche* of the University of Florence's School of Architecture. The authors and the publisher thank this institution for its support, and for granting permission to reproduce a large number of photographs.

Name Index

Figure credits and list of abbreviations

Archivio Aldo Arcangeli, Roma (AA)
Fig.#: 254, 434

Archive Marcel Breuer, Archives of American Art, Smithsonian Institution, Washington DC (AAA-SI)

Fonds Alberto Sartoris, Archives de la construction moderne, Ecole polytechnique fédérale de Lausanne (Acm-EPFL)
Fig.#: front cover, chapter two frontispiece, 27, 41, 42, 43, 45, 50, 51, chapter three frontispiece, 80, 81, chapter five frontispiece, 123, 124, 125, 126, 127, 129, 192, 206, 210, 213, 216, 217, 241, chapter eight frontispiece, 264, 272, 281, 282, 285, chapter ten frontispiece, 350, 351, 352, 366, 369, 373, chapter twelve frontispiece, 402, 403, 422, chapter thirteen frontispiece, 427, 429

Fondo Brevetti, Archivio Centrale dello Stato, Roma (ACS) authorization for the publication of the images No. 1298/2015
Fig.#: 5, 6, 63, 85, 87, 107, 114, 140, 161, 257, 258

Fondo Luigi Moretti, Archivio Centrale dello Stato, Roma (ACS – LM) authorization for the publication of the image No. 1298/2015
Fig.#: 521

Archivio Luigi Carlo Daneri, Genova (AD)
Fig.#: 232, 234

Archivio Fotografico Gherardi, Roma (AFG)
Fig.#: chapter eleven frontispiece, 377, 378, 379, 380, 383, 384, 385, 386, 387, 392, 393, 401, 417, 418, 419, 420, 421

Archivio Fotografico Ranfagni, Prato (AFR)
Fig.#: 8-9

Archivio Attilio Lapadula, Roma (AL)
Fig.#: 286

Archivio Pecci, Capalle, Prato (AP)

Archivio Parrocchiale della Chiesa di San Marcellino, Genova (APCSM)

Archivio Agenzia per il Turismo, Prato (APT)
Fig.#: 7

Archivio Riccardo Moncalvo, Torino (ARM)
Fig.#: 212, 214

Archivio Storico Buzzi Unicem, Casale Monferrato (ASBU)
Fig.#: 28

Archivio Storico Civico Biblioteca Trivulziana, Archivio Bardi, Milano (ASCBT) copyright@Comune di Milano
Fig.#: 266, 271

Archivio storico della Città di Firenze (ASCF)
Fig.#: 16, 17, 18, 20, 23, 25, 32, 38, 44

Archivio di Stato di Firenze, Fondo Enrico Bianchini (ASF) authorization for the publication of the images, Ministero dei Beni e delle Attività Culturali e del Turismo, No. 3452/762).
Fig.#: 3, 4

Archivio Studio Gresleri, Bologna (ASG)
Fig.#: 397, 398, 399

Archivio di Stato di Torino, Fondo Comitato Italia '61, Torino (AST)
Fig.#: 430

Archivio Azienda Trasporti Municipale, Torino (ATMT)
Fig.#: 293, 295

Archivio Ufficio Tecnico Cartiere Burgo, Mantova (AUTCB)
Fig.#: 439, 440, 442

Fondo Pier Luigi Nervi. Università degli Studi di Firenze, Biblioteca di Scienze Tecnologiche – Architettura, Firenze (BSTA)
Fig.#: chapter one frontispiece, 10, 11, 12, 14, 30, 52, 60, 61, 65, 73, 74, 75, chapter four frontispiece, 90, 92, 93, 95, 96, 97, 98, 99, 100, 101, 105, 109, 118, 119, 120, 131, 132, 133, 136, 174, 175, chapter seven frontispiece, 194, 195, 196, 198, 204, 205, 207, 208, 211, 218, 221, 222, 223, 225, 226, 229, 230, 236, 239, 240, 242, 243, 244, 245, 251, 261, 269, 270, 273, 297, 299, 306, 327, 328, 335, 336, 337, 359, 362, 367, 404, 405, 406, 407, 408, 412, 413, 414, 424, 426, 428, 431, 441, 445, 446, 447, 448, 449, 452, 453, 454, 455, 456, 458, 460, 461, 462, 463, 464, 465, 466, 467, 468, 469, 470, 471, 472, 473, 474, 475, 476, 477, 478, 481, 482, 483, 484, chapter fourteen frontispiece, 487, 488, 489, 490, 491, 492, 493, 494, 495, 496, 497, 498, 499, 500, 501, 502, 503, 504, 505, 506, 507, 508, 509, 510, 511, 512, 513, 514, 515, 516, 517, 518, 519, 520, 523, 524, 525, 526, 527, 528, 529, 531, 532, 533, 534, 535, 536, 538, 539, 540, 541, 542, 543, 544, 545, 546, 547, 548, 549, 550, 551, 552, 554, 555, 556

Archivio storico del Comitato Olimpico Nazionale Italiano, Roma (CONI)
Fig.#: 361, 363, 374, 375, 382, 388, 389, 390, 391

Archivio Pier Luigi Nervi, Centro Studi e Archivio della Comunicazione dell'Università degli Studi di Parma (CSAC)
Fig.#: 24, 34, 35, 37, 46, 47, 66, 67, 68, 72, 84, 86, 91, 94, 102, 106, 115, 116, 117, 130, 134, 135, 137, 138, 139, 143, 144, 154, 166, 167, 168, 169, 170, 173, 176, 177, 179, 180, 181, 182, 183, 190, 191, 193, 200, 219, 220, 224, 227, 228, 231, 237, 238, 260, 265, 267, 268, 274, 275, 276, 277, 283, 287, 291, 296, 298, 307, 332, 333, 339, 340, 345, 360, 395, 396, 400, 409, 410, 411, 416, 423, 425, 435, 436, 437, 450, 451, 459

Centro Storico Fiat, Torino (CSF)
Fig.#: 203

Fondazione Giovanni Michelucci, Fiesole, Firenze (FM)

Collezione MAXXI Architettura Archivio Enrico Del Debbio, MAXXI Museo nazionale delle arti del XXI secolo, Roma (Maxxi – EDD)
Fig.#: 33

Collezione MAXXI Architettura Archivio Pier Luigi Nervi, MAXXI Museo nazionale delle arti del XXI secolo, Roma (Maxxi - PLN)
Fig.#: 13, 15, 31, 53, 56, 57, 58, 59, 69, 76, 77, 78, 79, 82, 83, 88, 89, 104, 108, 110, 111, 112, 113, 121, 122, 128, 141, 142, chapter six frontispiece, 148, 150, 151, 155, 156, 157, 158, 159, 160, 171, 172, 178, 187, 188, 197, 201, 202, 209, 233, 246, 247, 248, 249, 250, 252, 253, 255, 256, 262, 263, 278, 279, 280, 284, 292, 294, 300, 301, 302, 303, 304, 305, 346, 347, 348, 349, 364, 365, 438, 443, 444, 457, 479, 480, 522, 553

Marcel Breuer Papers, Special Collections Research Center, Syracuse University Libraries, Syracuse NY (SCRC)

Direzione dei Servizi Tecnici del Governatorato Stato Città del Vaticano (SCV) © Governatorato SCV
Fig.#: 485, 486

United Nations Educational, Scientific and Cultural Organization, Paris (Unesco)
Fig.#: chapter nine frontispiece, 309, 310, 311, 312, 313, 314, 315, 316, 317, 318, 319, 320, 321, 322, 323, 324, 326, 329, 330, 368, 370, 371, 530, 537